Learning to Teach Art & Design in the Secondary School

Learning to Teach Art & Design in the Secondary School advocates art, craft and design as fundamental to a plural society; practices that are useful, pleasurable, critical and transforming. It provides support for mentors and tutors and assists students in the transition from studio to classroom practice. It suggests strategies to motivate and engage pupils in making, discussing and evaluating visual and material culture.

With reference to current debates *Learning to Teach Art & Design in the Secondary School* explores a range of approaches to teaching and learning; it raises issues, questions orthodoxies and identifies new directions. The chapters examine:

- ways of learning
- planning and resourcing
- attitudes to making
- historical and contextual studies
- values and critical pedagogy

Contributors are drawn from a range of practising teachers and artists who encourage readers to consider the role of Art & Design in the context of past, developing and emerging traditions.

Nicholas Addison is a lecturer in Art & Design Education at the Institute of Education, University of London. **Lesley Burgess** is a Lecturer in Art & Design Education and PGCE Art & Design Course Leader at the Institute of Education, University of London.

Related titles

Learning to Teach Subjects in the Secondary School Series

Series Editors
Susan Capel, Canterbury Christ Church College; Marilyn Leask, De Montfort University, Bedford; and Tony Turner, Institute of Education, University of London.

Designed for all students learning to teach in secondary schools, and particularly those on school-based initial teacher training courses, the books in this series complement *Learning to Teach in the Secondary School* and its companion, *Starting to Teach in the Secondary School*. Each book in the series applies underpinning theory and addresses practical issues to support students in school and in the training institution in learning how to teach a particular subject.

Learning to Teach English in the Secondary School
Jon Davison and Jane Dowson

Learning to Teach Modern Foreign Languages in the Secondary School
Norbert Pachler and Kit Field

Learning to Teach History in the Secondary School
Terry Haydn, James Arthur and Martin Hunt

Learning to Teach Physical Education in the Secondary School
Susan Capel

Learning to Teach Science in the Secondary School
Tony Turner and Wendy DiMarco

Learning to Teach Mathematics in the Secondary School
Edited by Sue Johnston-Wilder, Peter Johnston-Wilder, David Pimm and John Westwell

Learning to Teach Using ICT in the Secondary School
Edited by Marilyn Leask and Norbert Pachler

Learning to Teach Geography in the Secondary School
David Lambert and David Balderstone

Learning to Teach Art & Design in the Secondary School

A companion to school experience

**Edited by Nicholas Addison
and Lesley Burgess**

ROUTLEDGE / FALMER
Taylor & Francis Group

London and New York

First published 2000
by RoutledgeFalmer
11 New Fetter Lane, London EC4P 4EE

Simultaneously published in the USA and Canada
by RoutledgeFalmer
29 West 35th Street, New York, NY 10001

RoutledgeFalmer is an imprint of the Taylor & Francis Group

Typeset in Bembo by RefineCatch Limited, Bungay, Suffolk
Printed and bound in Great Britain by
TJ International Ltd, Padstow, Cornwall

British Library Cataloguing in Publication Data
A catalogue record for this book is available from the British Library

Library of Congress Cataloging in Publication Data
Addison, Nicholas, 1954–
    Learning to teach art and design in the secondary school /
    Nicholas Addison and Lesley Burgess.
            p.   cm. – (Learning to teach subjects in the secondary
    school)
    Includes bibliographical references and index.
    1. Art – Study and teaching (Secondary)   I. Burgess, Lesley, 1952–   .
    II. Title.   III. Series: Learning to teach subjects in the
    secondary school series.
    N363.A3   2000
    707'.1'2 – dc21   99–41054

ISBN 0–415–16881–3

Contents

Introduction to the series

Learning to Teach Art & Design in the Secondary School is one of a series of books entitled Learning to Teach Subjects in the Secondary School covering most subjects in the secondary school curriculum. The books in this series support and complement *Learning to Teach in the Secondary School: A Companion to School Experience* (Capel, Leask and Turner, 1995, 2nd ed. 1999), which addresses issues relevant to all secondary teachers. These books are designed for student teachers learning to teach on different types of initial teacher education courses and in different places. However, it is hoped that they will be equally useful to tutors and mentors in their work with student teachers. In 1997, a complementary book was published entitled *Starting to Teach in the Secondary School: A Companion for the Newly Qualified Teacher* (Capel, Leask and Turner, 1997). That second book was designed to support newly qualified teachers in their first post and covered aspects of teaching which are likely to be of concern in the first year of teaching.

The information in the subject books does not repeat that in *Learning to Teach*; rather, the content of that book is adapted and extended to address the needs of student teachers learning to teach a specific subject. In each of the subject books, therefore, reference is made to *Learning to Teach*, where appropriate. It is recommended that you have both books so that you can cross-reference when needed.

The positive feedback on *Learning to Teach*, particularly the way it has supported the learning of student teachers in their development into effective, reflective teachers, has encouraged us to retain the main features of that book in the subject series. Thus, the subject books are designed so that elements of appropriate theory introduce each behaviour or issue. Recent research into teaching and learning is incorporated into this. This material is interwoven with tasks designed to help you identify key features of the behaviour or issue and apply them to your own practice.

Although the basic content of each subject book is shared, each book is designed to address the unique nature of each subject. In this book, for example, some of the reasons for the special status often afforded to art and design within the curriculum becomes clearer. Many different groups have an interest in art and design education and so seek to have an influence on it. Among many other things, the book examines the consequences of this including some of the controversies that can sometimes develop. Art and design education has also acted as a significant stimulus for the development of new theories of learning and new teaching approaches and so the book also explores many of these developments.

We, as series editors, have found this project to be exciting. We hope that, whatever the type of initial teacher education course you are following and wherever you may be following that course, you find that this book is useful and supportive of your development into an effective, reflective art and design teacher.

Susan Capel, Marilyn Leask and Tony Turner
January 1999

Illustrations

FIGURES

TABLES

PLATES

All plates appear in the colour plate section between pp. 168–9

TASKS

Contributors

Nicholas Addison is a lecturer in Art & Design Education and teaches on both the PGCE and MA courses at the Institute of Education, University of London. For sixteen years he taught Art & Design and Art History in London in a comprehensive school and a sixth-form college. He has lectured in the School of History and Theory of Visual Culture at Middlesex University and is Chair of the Association of Art Historians Schools Group. His educational research centres around the integration of critical and contextual studies within studio-based learning.

Andy Ash is a part-time lecturer in Art & Design Education at the Institute of Education, University of London and also contributes to the PGCE Art & Design at Reading University and Homerton College, Cambridge. Other teaching commitments have included lecturing on Foundation, GAD, GNVQ and A Level courses at The Henley College, Henley-on-Thames. Prior to this he taught Art & Design, CDT and Photography in Berkshire secondary schools. He is presently completing his PhD whilst collaborating in a research project with the Tate Gallery, London.

Lesley Burgess is a lecturer in Art & Design Education at the Institute of Education, University of London. She is Course Leader for the PGCE course in Art & Design and Co-Director of the Artists in Schools Training Programme: in addition she teaches on the MA courses. Before moving to the Institute in 1990 she taught for fifteen years in London comprehensive schools. Her main research interests are curriculum development and resource-based learning. The focus of her current research is contemporary art and artists in education. She is a member of the Teacher Education Board for NSEAD, is on the Executive Committee of LAADE and is a trustee of Camden Arts Centre.

Paul Dash is a lecturer in Art & Design Education at Goldsmiths College, University of London. He contributes to the Goldsmiths' modular MA programme and the joint MA course run by Goldsmiths and the London Borough of Croydon. He taught in

London schools for twenty-two years before joining the staff at the Institute of Education in 1991. He moved to Goldsmiths in 1994. His main research interest is Art & Design education and Caribbean children. He is author of *Traditions from the Caribbean*. He is on the management committee of the 198 Gallery in Herne Hill and initiated the Elders exhibition which was shown at the South London Gallery and the Wolsey Gallery in Ipswich. He is a practising painter and has shown in the Summer Exhibition at the Royal Gallery, the Open Exhibition at the Whitechapel Gallery, the Commonwealth Institute and various commercial galleries in London.

David Gee is Head of the Art & Design Department at North Westminster Community School where he leads one of the largest visual arts departments in the country. His teaching career spans nearly thirty years in London comprehensive schools and adult education. He studied ceramics at Camberwell going on to Goldsmiths to complete his teacher training. He continues to practise as a potter building a collection of large earthenware pieces for a prestigious annual charity exhibition at the Man Galleries. He is interested in the development of contemporary thought and especially in theories of creativity and how these can be applied to the classroom. His long experience of teaching children and adults from a wide range of backgrounds, skills and interests has made him keenly aware of equal opportunity issues and the development of SEN practice through effective differentiation.

James Hall is Principal Lecturer in Art Education at Roehampton Institute, London. He coordinates the Art Teaching Studies Area in the Faculty of Education and teaches undergraduate and postgraduate programmes. He was previously lecturer in Art & Design Education at the University of Reading and Co-director of the PGCE (secondary) programme. He has also held posts as Inspector for the Arts in the London Borough of Hammersmith and Fulham and in secondary schools in Berkshire, Cambridgeshire, Durham, Cleveland and the USA. He has also been an external examiner for BA, PGCE and MA programmes. His research interests are in the education, training and professional development of teachers of Art & Design and in artists' prints. He is currently researching the development of the professional knowledge and expertise of teachers of Art & Design in secondary schools. As a practising printmaker, he is a regular contributor to artists' bookworks.

Richard Hickman is Director of Studies for Art at Homerton College, Cambridge and course leader for the art component of the University of Cambridge PGCE. His teaching experience includes ten years as a classroom teacher and as a lecturer in Art Education since 1985 (at the University of Reading and at Nanyang Technological University, Singapore). He gained his PhD in 1994, which built upon his earlier MA, focusing upon language and concepts in art education. His publications have appeared in a number of books and journals on art education and he has also presented papers at several international conferences. As an art practitioner, he has had five solo exhibitions. For his (many and various) sins he is also registered as an OFSTED inspector for Art & Design and has participated in secondary school inspections throughout the UK.

Pam Meecham is a senior lecturer in Art & Design Education and Course Leader for the PGCE in Art & Design at Liverpool John Moores University. In addition she lectures on the BA and MA Tourism and Leisure and Art History courses. She is a frequent guest lecturer at the Tate Gallery Liverpool and teaches at summer schools for the Open University course, 'Modern Art: Practices and Debates'. Her publications include *Working with Modern Art* (1996) for the Open University and *Working with Modern British Art* (1998) for the Tate Gallery both guides for teachers and gallery educators. *Modern Art: A Critical Introduction*, a Routledge book co-written with Dr Julie Sheldon, will be published in January 2000. Research interests include American art in the 1950s and photography and monument in the twentieth century.

Rose Montgomery-Whicher joined the Art & Design academic group at the Institute of Education, University of London in January 1998, after completing her PhD at the University of Alberta, Canada. She has taught in a range of community and museum settings, including Kettle's Yard in Cambridge and the National Gallery of Canada. Her research and publications focus on the experimental and pedagogical dimensions of drawing and the use of phenomenological and interpretative methodologies in Art & Design education research. She is currently directing the Institute-based research dimension of 'Visual Paths: Teaching Literacies in the Gallery', a project in partnership with the Tate Gallery.

Roy Prentice is a senior lecturer and Head of the Art & Design Education academic group at the Institute of Education, University of London. Formerly he was the art adviser for East Sussex LEA and led an Art & Design department in a comprehensive school in London. He was a member of the Art Advisory Group which reviewed the National Curriculum Order for Art for the School Curriculum and Assessment Authority. His main research interest is Art & Design and teacher education; in particular the role of studio-based work. He edited *Teaching Art & Design: Addressing Issues and Identifying Directions* published by Cassell in 1995. He is a practising painter.

Kate Schofield is a part-time lecturer in Art & Design Education at the Institute of Education, University of London. She is a PGCE tutor and a module tutor on the MA Museums and Galleries in Education course. (Between 1993 and 1997 she was Course Leader for the MA Museums and Galleries in Education.) After training as a textile designer, she gained extensive teaching experience in further education in Nottinghamshire before studying for an MA degree at London University. In her capacity as Chief Examiner, Senior Examiner and Reviser for each of three examination groups, she has gained additional experience of the examination system in Art & Design: she has also acted as a consultant for SCAA.

Amanda Starkey completed her degree in Graphic Design at Brighton Polytechnic in 1988. She went on to work as a freelance illustrator and taught Illustration and contextual studies part-time on the Foundation Studies course at Northbrooke College of Design and Technology. She worked as an illustrators' agent in London for a year before completing her PGCE in Art & Design at the Institute of Education in 1992. She taught at a co-educational grammar school in Bexley before moving to St Charles Sixth Form College where she was also the college GNVQ coordinator. She has recently moved to the new post of Courses Organiser at the Victoria and Albert Museum working with adults in formal education.

Foreword

In good times a well-founded publication can underpin optimism and imagination. In difficult times it reminds us of why we are here and where we are going. Thirty years ago such a book emerged from London University's Institute of Education. Entitled *Change in Art Education* by Dick Field it has remained a comfort and support to me since my days as a student teacher.

This book comes from the same institution at a time of massive changes in art, cultural practices and schooling. As such, I hope it strikes the same chord with many prospective teachers of Art & Design that Field's book achieved for me. The title of the book suggests that it is aimed at those often – but not exclusively – young adults about to embark on a career in our schools and I hope that they will grasp it eagerly and find that it informs a lifetime in arts education. Nonetheless I will be sad if that is the extent of its audience for I would recommend it to all those who are involved in the teaching of Art & Design.

This is because the book is a very clear, practical guide to ways of planning, delivering and assessing Art & Design in secondary schools but *more importantly* because it will stimulate and extend thinking about their work. We can organise teachers as much as we like but unless we produce thinking teachers, we can have no optimism for the future and so this publication is to be commended for it demands that we teachers 'question orthodoxy' *and* it helps us to achieve that.

This is not a surprise. The Art & Design department of the Institute of Education has a long and honourable tradition of work in the pedagogical field. The editors who inherit that tradition are committed, critical and clear-thinking and they have provided us with a timely and stimulating publication. I hope that in thirty years time we will be able to look back and celebrate its influence on a bright, new phase in Art & Design education.

Dr Dave Allen
Course Leader
Entertainment Technology
University of Portsmouth

Acknowledgements

We would like to thank the student teachers and teachers in partnership schools with whom we have worked over recent years: their continuing collaboration has made it possible to experiment, research and reflect critically on teaching and learning in Art & Design. Our partners in London galleries and museums merit special mention for providing us with insights into ways of making accessible to young audiences the expanding field of visual and material culture. Contributing authors represent some of the major centres for initial teacher education across England and our thanks go to them for their varying perspectives which complement the traditions and innovations of the Institute of Education (IoE). Colleagues and friends have provided invaluable insights, support and encouragement: they are too numerous to name individually. However, special mention must be given to Josephine Borradaile, secretary to the Art and Design Education academic group at the Institute of Education whose ingenuity, perspicacity, diplomacy and sheer dedication have made this book possible. The expertise of Peter Thomas and the graphics team at the Institute has also enabled its smooth organisation and we are indebted to their work.

We are particularly grateful to the following, artists, craftspeople and designers for their generous contributions, both in the form of statements and reproductions of their work: Michael Brennand-Wood, *Walking in Space*; Caroline Broadhead, *Steppenwolf*; Lucy Casson, *Playing with Fire*; Meera Chauda, *Untitled*; Thomas Heatherwick, *Gazebo*; Natasha Hobson, *Spice Boxes*; Cathy de Monchaux, *'Evidently Not'*; Keith Piper, *Robot Bodies*, commissioned by FACT, the Foundation for Art Creative Technology who have kindly granted permission for the reproduction of this image; Helen Storey, *Heart Development Hat*. We also wish to thank all PGCE students from the Institute who have given permission to reproduce their course work.

Finally, we would like thank Helen Fairlie and Jude Bowen from Routledge for their advice and encouragement.

Nicholas Addison and Lesley Burgess

Abbreviations

AAH	Association of Art Historians
AAIAD	Association of Advisers and Inspectors in Art & Design
AAVAA	African and Asian Visual Arts Archive
ACE	Arts Council of England
ACER	African Caribbean Education Resources
AEMS	Arts Education in a Multicultural Society
AS Level	Advanced Subsidiary Level
AT	Attainment Target
BECTa	British Educational Communications and Technology agency
CAD	Computer Assisted Design
CDT	Craft, Design and Technology
CEP	Career Entry Profile
CPS	Common Pay Spine
CV	Curriculum Vitae
DES	Department of Education and Science
DFE	Department of Education
DfEE	Department for Education and Employment
D&T	Design and Technology
DWEMs	Dead White European Males
EAL	English as an Additional Language
Engage	National Association for Gallery Education
EO	Equal Opportunities
ERA	Education Reform Act
FPD	Further Professional Development
GCE	General Certificate of Education
GCSE	General Certificate of Secondary Education
GNVQ	General National Vocational Qualification
HE	Higher Education

HEI	Higher Education Institute
HMI	Her Majesty's Inspector
HMSO	Her Majesty's Stationery Office
ICT	Information Communications Technology
IEP	Individual Education Plan
ILEA	Inner London Education Authority
inIVA	Institute of International Visual Arts
INSET	In-Service Training
ITE	Initial Teacher Education
IWM	Imperial War Museum
JADE	*Journal of Art and Design Education*
KS	Key Stage
LEA	Local Education Authority
LSA	Learning Support Assistants
NC	National Curriculum
NCET	National Council for Educational Technology
NSEAD	National Society for Education in Art & Design
NVQ	National Vocational Qualification
OFSTED	Office for Standards in Education
PGCE	Postgraduate Certificate in Education
PoS	Programmes of Study
PTE	Practical Teaching Experience
QCA	Qualifications and Curriculum Authority
QTS	Qualified Teacher Status
RAB	Regional Arts Board
RBL	Resource-Based Learning
RSA	Royal Society of Arts
SCAA	Schools Curriculum and Assessment Authority
SCT	Subject Co-Tutor
SEC	Secondary Examinations Council
SEN	Special Educational Needs
SENCO	Special Educational Needs Coordinator
SoW	Scheme of Work
TE	Teaching Experience
TES	*Times Educational Supplement*
TTA	Teacher Training Agency
WAL	Women's Arts Library
yBas	Young British Artists

1 Introduction

Nicholas Addison and Lesley Burgess

What is the role of art, craft and design in education?
What is the philosophical basis for the inclusion of Art & Design?
Why is Art & Design only a foundation subject at KS3?
How can the subject be developed to acknowledge changes in contemporary practice and the function of art, craft and design in the twenty-first century?

WHAT DOES ART & DESIGN DO?

It is important that as student teachers you advocate art, craft and design as fundamental to society, a useful, pleasurable, critical, and possibly transforming phenomenon. We ask you to question the orthodoxy, still prevalent in many schools, promoting art as an autonomous activity, something entirely distinct from other forms of social and cultural production. In acknowledging the diversity of cultural forms and practices Art & Design must accommodate the expanded and expanding field of material and visual culture, anything from functional footwear to live performance and virtual galleries. The significance of such forms lies in the uses to which they are put: utilitarian and symbolic, affective and discursive, physical and spiritual. These forms change depending on when and where they are produced, and how and by whom they are being used. Such change occasions shifts in function so that meaning becomes contingent on usage and intentions and definitions become difficult to fix. It is therefore not surprising that material and visual culture has generated discourses, from the purposeful construction of hierarchies and histories to their no less exacting deconstruction. Art & Design education provides opportunities for pupils to make and investigate art by exploring and discussing uses and by developing an understanding of values. It is only possible to develop this understanding if practice is placed critically in context so that pupils come to know how art can be more than self-expression, that it functions on different levels to support, critique and, significantly, produce the culture in which they live.

It is only during the first three years of secondary education that pupils are given the opportunity to engage with material and visual culture in its own right. Increasingly, Art & Design in primary education is seen as a vehicle for supporting the core curriculum, an adjunct to literacy, numeracy, science and ICT, learning through art. This instrumental role, the ability of the subject to contribute to the whole curriculum, can be beneficial, but you must not allow it, as a service subject, to deny the place of Art & Design as a different and fundamental part of knowledge and understanding. All pupils need to develop visual and aesthetic literacy, learning in art, craft and design.

The expanding field of art, craft and design offers challenges and possibilities beyond the technical and formalist limitations of orthodoxies in secondary Art & Design education. Makers and critics are questioning traditional boundaries and your own practice may already undermine historical distinctions you perceive as no longer valid. However, it is vital that you do not divorce your practice from the codes and conventions of the developing educational tradition. The past and the future should be in dialogue so that present practice can be negotiated rather than constituting a site where opposing ideologies battle out their differences. Such dialogue and reflection is certain to problematise practice because it introduces an element of doubt, raising the spectre of relative values and some uncertainty as to how a reconceptualised curriculum might work. It is you, student teachers, who need to be the agents of change. Living with ambiguity whilst managing the practicalities of the classroom is a risky business. It has been argued that to introduce an element of risk into the curriculum is counterproductive (Cunliffe 1999: 115–121): what pupils need is certainty and direction. This prescriptive approach, although eminently manageable, can provide a false impression of the nature of social and cultural production. If you believe that art, craft and design has the potential to develop creativity, then risk taking has to be an essential ingredient.

Education is not just a matter of what you teach; crucially it is concerned with how you teach. Critical pedagogy requires that you continue to reflect on both these questions but in addition ask the question why? How does teaching and learning in Art & Design address ethical questions? What are the cultural, ecological, moral, political, social and spiritual dimensions of the curriculum?

Contemporary practice in art, craft and design has already blurred the boundaries between art and life: the art object is no longer exclusively to be found in the gallery, the practitioner is no longer bound to their studio. Just as the sites of practice can be anywhere, from the natural environment to cyberspace, its methods can be interdisciplinary, from the anthropological to the psychoanalytical. As a student teacher taking on board an interdisciplinary approach you are invited to consider the methodological resources of other subjects in the school curriculum: Media Studies and Semiotics, Geography and Ecology, Religious Studies and Interculturalism. It is important that you are aware and open to such possibilities and recognise the potential reciprocity between Art & Design and other areas of the curriculum?

Any course in initial teacher education is a complex and multi-faceted process. It is essential that you take the opportunities it affords to develop your understanding and use of all available resources: the expertise of fellow students, tutors, teachers and pupils, the wide range of local, national and international facilities, libraries and new technologies, the natural and built environment, galleries and museums, studios, classrooms; learn to become resourceful.

The book aims to enable you to:

- acknowledge the diverse nature of art, craft and design and its implications for teaching and learning;
- reinforce and develop your subject knowledge;
- translate your practice in art, craft and design into good pedagogy;
- recognise that theory and practice are interdependent;
- understand and employ methods and strategies for effective learning;
- develop as a reflective practitioner: question, evaluate and modify your teaching;
- challenge existing orthodoxies and identify new directions;
- consider the function of Art & Design education in a plural society;
- formulate a personal philosophy for Art & Design education in the context of developing traditions.

WORKING DEFINITIONS

Art & Design indicates the subject in the school curriculum which includes art, craft and design.

On occasions the words art and artist are used generically to indicate all practice and practitioners in the production of visual and material culture: context should make this apparent.

'Art' is used to suggest the idea of a European, hierarchical tradition in which the fine arts: painting, sculpture, architecture are theorised as intellectual processes with possibly transcendental outcomes, and the applied arts, or 'crafts', are theorised as mechanical processes, producing utilitarian objects.

School art has come to suggest an insular tradition in which pupils' work is predicated on formalist and/or expressive modes perpetuated through exemplars. To the uninitiated it represents skill, verisimilitude or recreation and therapy. Its references are largely male and Western with tokenistic gestures to the art of others.

A BRIEF HISTORY: A DEVELOPING TRADITION

The idea that art, craft and design can significantly contribute to the education of all is a surprisingly recent phenomenon. The visual arts were excluded from the classical liberal arts, a hierarchy of disciplines divided into two domains, the 'quadrivium' and 'trivium', which identified the necessary components of an academic education, respectively: arithmetic, geometry, astronomy and music; grammar, rhetoric, and logic. Not until the modern period were the sciences introduced as a separate and expanded domain. The visual arts are notably absent from academe until the Italian Renaissance when they surfaced as a parallel but isolated programme. It was the industrial revolution that forced modern governments to consider the role of design in educating a workforce capable of competing in world markets. This process began in Britain from the beginning of the Victorian period (Thistlewood 1992; Selwood *et al.* 1994).

Three Victorians in particular were responsible for conditioning the way Art & Design education was to develop in Britain. Henry Cole conceived it as having an instrumental role in the development of the industrial modern nation state, John Ruskin proposed the arts as an aesthetic and moral entity, William Morris as a social and political activity. In the twentieth century, such theorists and educators as the philosopher John Dewey and the critic Herbert Read, the educators Franck Cizeck and Marion Richardson, focused on the experiential and expressive potential of art, its role in the education of the whole child, a means towards self-actualisation. More recently, educators have been keen to establish art as a cognitive activity, a distinct and unique way of coming to know the world. There are many permeations, and deviations from these philosophies and they have affected Art & Design in schools to varying degrees. But it is important to understand that what happens in schools is in part a reflection of, or response to, broader practices, usually at one or two paces behind.

The promotion of critical approaches to the art curriculum first advocated in schools during the 1970s (Field 1970; Eisner 1972) marked a notable shift in practice. Since then there have been many initiatives promoting curriculum development. The report, *The Arts in Schools* (Robinson 1982) provided a rationale for the central place of art in the curriculum. It advocates: 'The arts have an essential place in the balanced education of our children and young people . . . [it provides] for a broad-based curriculum rather than one that is too occupied with academic learning' (pp. 3–4). There was also an attempt to accommodate the principles of equal opportunities through strategies of visibility, both multicultural and gendered, and, more significantly, antiracist, anti-sexist initiatives (1980s). Before this could be consolidated the requirement to address the critical and contextual components of the National Curriculum (NC) (DES 1992) diverted attention to the canonic in relation to 'our' cultural heritage. However, the NC (DFE 1995) directive for pupils to respond to the 'work of others' necessitated a liaison between schools and gallery/museum education departments which, depending on the venue, included multicultural examples. Some of this may have been pragmatic or only tokenistic, visibility and celebrationist tactics taking the place of a critical engagement with difference, but teachers were forever having to reconcile their personal ideals with external constraints.

Elsewhere (Hickman forthcoming) we have categorised current approaches to the Art & Design curriculum (11–18). The categories are not exclusive, they indicate:

> approaches that can be combined, for example genre-based outcomes can be managed through formalist, perceptualist or expressive means. However, in some schools one approach dominates. There are two which are no longer evident: one, basic design, a cogent programme in the 60s and 70s, is lost because its workshop-based strategies, if not its holistic philosophy, have been subsumed by technology (Baynes 1982; Green 1982); second, anti-racist art education, promoted in the 70s and 80s, but now diffused and

dissipated, is no longer perceived as urgent in a post-apartheid era. However, post the Macpherson Report (1999) inquiring into the death of Stephen Lawrence, this can be seen to have been premature . . .

1	Perceptualist	mimetic procedures, a search for the 'absolute copy' reduction to appearances (Clement 1993);
2	Formalist	a reduction to the visual elements, exercise driven, representational and/or abstract (Palmer 1989);
3	Expressive	intuitive making through affective and/or material exploration: privileging the essential and individual (Witkin 1974);
4	Genre-based	preconceived types perpetuated by teacher expertise and the imitation of exemplars, the successful work of past pupils' e.g. still-life, life-drawing, landscape, CD covers, ceramic figures;
5	Pastiche	the imitation of canonic exemplars, occasionally taking on board the postmodern practice of parody (SCAA 1996);
6	Technical	the development of a succession of discrete technical skills: drawing followed by print-making, followed by batik, etc.;
7	Object-based	a response to common, sometimes themed, often spectacular artefacts in the form of a big still life/installation e.g. natural and made forms, a multicultural potpourri (Taylor and Taylor 1990);
8	Critical and contextual	an investigation of art as a means of social and cultural production privileging cognitive and analytical procedures (Field 1970; Dyson 1989; Taylor 1989);
9	Issue-based	an integration of the personal with the social, political and moral through responses to current and contentious issues (Kennedy 1995a);
10	Postmodern	promoting plural perspectives and approaches and embracing the new technologies (Efland *et al.* 1996; Swift and Steers 1999).
		(Addison and Burgess in Hickman forthcoming)

PGCE COURSES IN ART & DESIGN

PGCE courses, extended over one academic year of full-time study, are student-centred and based on an organic structure through which student teachers gain insight into the principles, processes and practice of Art & Design education. They are founded on a model of partnership in which local schools and Higher Education (HE) collaborate to provide a context in which theory and practice can be investigated and implemented.

Teaching experience

TE takes place in partnership secondary schools where Art & Design teachers are responsible for supporting and monitoring your progress. You are introduced gradually to teaching by means of observation and team teaching. It is likely that opportunities also exist for you to draw upon the expertise of staff in education departments in galleries and museums.

Tutor groups

The tutor group, to which you are assigned at the beginning of the course, is your main support group. It is important that you view it as a vital resource for collaboration and actively participate in activities and discussions, sharing experiences, knowledge and skills.

You should receive a copy of the Standards for the Award of Qualified Teacher

---

OBJECTIVES

PGCE courses aim to extend your knowledge and understanding of:

- your own creative processes to provide a basis for teaching;
- theoretical issues which inform practice in Art & Design education and help you to develop teaching skills;
- the contributions to be made by art, craft, design and technology to the education of all children and young people in a plural society;
- recent curriculum projects and initiatives in Art & Design and how current thinking in the field relates to wider educational debates;
- the relationship between art, craft, design and technology and other areas of the curriculum;
- the use of tools and materials to ensure safe practice;
- the importance of communication skills in teaching and learning;
- rationales for teaching and the ability to articulate a personal educational philosophy;
- what it means to function as committed professionals, particularly in the matter of the relationship between the PGCE year, Career Entry Profiles (CEP), the induction period and Further Professional Development (FPD);
- the Standards for the Award of Qualified Teacher Status (DfEE 4/98) and their specific application in Art & Design education.

Status (DfEE 4/98) as part of your Teacher Training Agency (TTA) Career Entry Profile (CEP).

HOW TO USE THE BOOK

The structure of the book is likely to correspond to the sequence of your course. It begins with the transition from artist to teacher, moves on to the nature of learning, explores planning, resourcing, management and assessment, before considering issues such as attitudes to making and values in Art & Design education. Throughout the book, especially through tasks, we ask you to reflect on your experience, question orthodoxies and critically evaluate your own and others' practice.

We ask you to relate your subject knowledge to the 'expanding field'. The book helps you to identify areas for development and suggests strategies to extend your knowledge.

Although we do not make continuous reference to the Standards (DfEE 4/98), to which all courses of initial teacher education subscribe, they are implicit. However, the Standards are only a base line, it is our intention that you should reach beyond them.

A series of tasks is integral to each chapter. They function in various ways: collecting data through observation, asking questions and discussing issues with other students, your tutors, and colleagues in schools, devising and resourcing schemes of work and trialling practical and critical activities. It is important that you question the culture of selfish isolation fostered by those practitioners of art, craft and design who privilege originality and genius above cooperation and partnership. Often we ask you to work collaboratively, for we believe that effective teaching is practised as a collective and reciprocal activity.

2 Making Connections between Subject Knowledge and Pedagogy: the role of workshops

Roy Prentice

In teaching we do not merely pass on a free-standing package of knowledge of the different periods, cultures and traditions in art, say, or the skills involved in working with different materials. What we do is rather to offer, however indirectly, a sense of the personal meaning which our curriculum has for us – its value, its relevance, its implications for us as particular human beings. In teaching . . . we represent not only our subject, more importantly, the stance we take towards it!

(Salmon 1995: 24)

INTRODUCTION

The purpose of this chapter is to support your growing understanding of the factors that help to shape the complex relationship between your knowledge of art, craft and design and your developing skills as a teacher. A PGCE course provides a structure within which connections between personal knowledge of subject content and pedagogy can be explored in increasing depth and be strengthened and made explicit.

To assist you in arriving at a considered response, located in a theoretical framework, some key issues are identified and discussed within the following units:

Unit 2.1   the relationship between previous and present experience;
Unit 2.2   the nature of teaching and learning;
Unit 2.3   the role of reflective practice;
Unit 2.4   workshops: environments for enquiry.

OBJECTIVES

By the end of this chapter you should be able to:

• consider how to articulate the ways in which experience of professional practice informs the teaching of Art & Design;
• understand the significance of the wider, ongoing political and professional debates about educational standards and teacher effectiveness;
• bring more sharply into focus the nature and importance of the relationship between subject knowledge and subject application, particularly in the context of courses of initial teacher education;
• answer the fundamental question: how does what you know about art, craft and design, and the way you have come to know it, provide the basis for your professional development as a teacher and your evolving philosophy of Art & Design education?

UNIT 2.1 THE RELATIONSHIP BETWEEN PREVIOUS AND PRESENT EXPERIENCE

Whilst it is recognised that teachers, whatever their subject specialism, have in common a need to make genuinely creative and sustainable connections between their previous and present experience, for teachers of Art & Design there is sometimes insufficient acknowledgement of the impact subject-specific influences have on this process. It is useful to reflect upon the educational route which the majority of Art & Design teachers have followed in order to understand more clearly the nature of the issues involved. The pattern of national provision for undergraduate courses in art, craft and design is broad and complex. Degree courses differ enormously in their declared aims, content, structure and ethos. The majority of courses are highly specialised within a given area of activity, e.g. graphic design, painting, three-dimensional design. However, modular courses continue to develop and these offer students a wider range of experiences at the expense of working in a chosen area in greater depth. In addition an increasing number of opportunities are available to students who wish to pursue alternative routes that are more vocationally oriented. Overall, they represent a continuum of professional practice in art, craft and design that constitutes the subject field. Such courses are predominantly practical, their nature being studio-workshop rather than lecture and text-based. They aim to develop students' creative capacities and technical skills as artists, craftspeople and designers – thus the emphasis is on making.

For an artist or designer the decision to train as a teacher raises fundamental and complex questions about professional integrity, creative energy, belief systems and self-image. Attitudes that influence responses to such questions include those which support a strong personal commitment to creative work. This is even more apparent when an increasing number of PGCE students have gained invaluable and often substantial experience of employment in a field of art, craft and design practice. It is significant that the majority of PGCE Art & Design students are motivated by a particularly strong subject allegiance and a well developed sense of personal identity.

As well as helping you to understand the complex connections between your previous and present experience it is essential at the outset that all those with whom you work, university tutors and school-based mentors, are equally aware of the unique combination of factors that have an impact on the transitional experience of intending teachers of Art & Design. Given the range of provision at undergraduate level, different courses provide students with different experiences of the 'same' subject. This, in turn, has a powerful influence on the stance adopted by secondary specialist teachers towards their subject:

> For young people in school, art and design is inseparable from the art and design teacher: the tasks, materials, goals of the lesson come infused with the teacher's personal identity . . . To teach is to reveal, both intentionally and unwittingly, what the curriculum really means to the teacher.
>
> (Salmon 1995: 24)

However, the formal curricula for Art & Design at school, undergraduate and postgraduate levels cannot be held solely responsible for the way the subject is perceived, valued, understood and ultimately taught. Teachers' constructions and reconstructions of 'what the curriculum really means' to them, are shaped by an amalgam of experiences encountered and revisited in a variety of formal and informal settings and relationships over an extended period of time.

Task 2.1.1 Identifying your personal position

In order to deepen your understanding of your personal position identify and reflect upon the idiosyncratic baggage of beliefs, feelings, ideas, attitudes and skills you carry with you to the PGCE course. They underpin your present orientation towards art, craft and design and provide the growth points for your future development as an Art & Design educator.

UNIT 2.2 THE NATURE OF TEACHING AND LEARNING

It is apparent from the previous remarks that a model of Art & Design teaching based on a simplistic concept of transmission, e.g. of an acquired body of knowledge or set of practical skills, is flawed. Successful teachers demonstrate an ability to transform their knowledge of subject content into teaching material. They use appropriate representations of content to match the particular needs, interests and abilities of learners. Effective teaching of this kind relies on what Shulman (1986) refers to as professional understanding, a complex combination of knowledge of subject matter and knowledge of pedagogy.

From different perspectives an increasing number of contributors to the debates about teaching and teachers are focusing their attention on subject knowledge. Indeed it is significant that the first criterion (from a list of eight criteria) that OFSTED inspectors are required to address when making judgements about the quality of teaching in a school relates to subject knowledge. Inspectors are concerned about the extent to which teachers have a secure knowledge and understanding of the subjects they teach (OFSTED 1995a). No longer is it assumed that a teacher's competence is determined by pedagogical skills alone. The complex nature of teachers' knowledge of subject content has been investigated by Shulman and his colleagues at Stanford University (Wilson *et al.* 1987; Grossman *et al.* 1989). The ways in which teachers' personal belief systems about learning influence their approaches to teaching have been explored by Brophy (1991) and a further contribution to the discussion about what teachers should know is made by Aubrey (1994). Questions about the content of lessons, the issues they raise and the nature of teachers' responses are addressed by Shulman. In an attempt to demonstrate how teachers reveal different levels of subject knowledge through their classroom behaviour, Aubrey (1994) says, 'Where subject knowledge is richer, deeper and better integrated it is more likely that the teacher will be confident and more open to childrens' ideas, contributions, questions and comments' (p. 5).

While all teachers use representations to teach particular topics, through which subject content is transformed into teaching material, it is possible to identify qualitative differences in the representations chosen by teachers who function from a strong subject knowledge base and those who do not. Predictably the former reveal deeper understanding rooted in concepts, principles and underlying themes. As a result they are able to adopt a more flexible approach to their teaching, make more imaginative connections between different topics and lessons and avoid the perpetuation of stereotyped and preconceived ideas about a given subject. This ability to transform subject matter, through carefully selected representations, using metaphor, analogy, illustration and demonstration displays the depth of understanding and capacity to generate alternatives that knowledgeable teachers share.

In order to provide a theoretical framework within which the inter-relationships between three components of subject knowledge in a given discipline can be identified, the work of Grossman *et al.* (1989) is helpful.

1   Content knowledge includes:

- factual information;
- central concepts;
- organising principles and ideas.

2   Substantive knowledge includes:

- explanatory models or paradigms;
- conceptual tools used to guide enquiry and make sense of data.

3   Syntactic knowledge includes:

- relevant forms of methodology;
- ways of introducing new knowledge – justification and evaluation.

This model usefully informs a theoretical framework for a professional knowledge base for teaching. It embraces both subject content and pedagogy and includes:

1   General pedagogical knowledge

- knowledge of theories and principles of teaching and learning;
- knowledge of learners;
- knowledge of principles and techniques of classroom behaviour and management.

2   Subject content knowledge

- ideas, facts, concepts of the field;
- relationships between ideas, facts and concepts;
- knowledge of ways new knowledge is created and evaluated.

3   Pedagogical content knowledge

- understanding of what it means to teach a given topic;
- understanding of principles and techniques to teach a given topic.

For teachers of Art & Design their understanding of pedagogical content knowledge determines how the Art & Design curriculum is approached, resourced, taught and evaluated. It also determines how learners are engaged and learn; with reference to subject-specific difficulties, attitudes, skills, requirements and misconceptions. Such pedagogical content knowledge is informed by knowledge of Art & Design and general pedagogical knowledge. Thus throughout your PGCE year it is important that you make creative connections between your developing knowledge of art, craft and design, your experience of Art & Design teaching and your wider awareness of the nature of teaching and learning.

Task 2.2.1 Identifying your knowledge base

With reference to the theoretical framework suggested by Grossman *et al.* (1989) identify your knowledge base in relation to the categories: content, substantive, syntactic.
In pairs examine which areas require extensive development. How can you ensure that this need is addressed during your PGCE year?

UNIT 2.3 THE ROLE OF REFLECTIVE PRACTICE

Through reflective practice it is possible to make explicit those aspects of professional practice in Art & Design and teaching that would otherwise remain implicit. The concept of the reflective practitioner proposed by Schon (1987) relies on a dynamic interrelationship between three phases of reflection:

- reflection-in-action
- reflection-on-action
- reflection *on* reflection-on-action.

The point is made by Schon that:

> . . . reflection-in-action is a process we can deliver without being able to say what we are doing. Skilful improvisers often become tongue-tied or give obviously inadequate accounts when asked to say what they do. Clearly, it is one thing to be able to reflect-in-action and quite another to be able to reflect *on* our reflection-in-action so as to produce a good verbal description of it; and it is still another thing to be able to reflect on the resulting description.
>
> (p. 31)

An example of reflection-in-action is a smoothly integrated performance by a group of accomplished jazz musicians. This is described by Schon as:

> Listening to one another, listening to themselves, they *feel* where the music is going and adjust their playing accordingly . . . Improvisation consists in varying, combining, and recombining a set of figures within a schema that gives coherence to the whole piece. As the musicians feel the direction in which the music is developing, they make new sense of it. They reflect-in-action on the music they are collectively making – though not of course in the medium of words.
>
> (*ibid.*: 30)

The development of a reciprocal relationship between an artist and the work-in-progress may be likened to the evolution of a fruitful conversation. Ideas and feelings are presented and articulated in ways that were unknown at the outset. Thus, there emerges through the interdependence of content and form 'a discovery in union' (Reid 1969: 279). Ben Shahn, a painter who has experienced this condition from the inside says:

> From the moment at which a painter begins to strike figures of colour upon a surface he [sic] must become acutely sensitive to the feel, the texture, the light, the relationships which arise before him. At one point he will mould the material according to an intention – perhaps his whole concept to emerging forms, to new implications within the painted surface. Idea itself, many ideas move back and forth across his mind as a constant traffic . . . This idea rises to the surface, grows, changes as a painting grows and develops.
>
> (Shahn 1967: 49)

Beyond reflection-in-action, the basis on which their professional practice exists, neither the jazz musicians nor the painter are required to articulate ideas about the nature of their creative process. As a teacher of Art & Design it is precisely this level of reflection on professional practice, in art, craft and design and in education, that you are required to develop. You are encouraged to think about each session you teach as a complex and subtle performance that is determined by your knowledge and understanding, skills and attitudes. As a reflective teacher you come to recognise the problematic nature of teaching Art & Design and systematically reflect upon your practice in order to improve it. In so doing you simultaneously become engaged in teaching and learning: a relationship that echoes the quality of creative activity in art, craft and design.

Such a view of teaching acknowledges the range of personal experience that teachers – as well as pupils – bring to the educational enterprise in which they are involved and as Elliot (1991) points out:

> Learning to be a reflective practitioner is learning to reflect about one's experience of complex human situations holistically. It is always a form of experiential learning. The outcome of such learning is not knowledge stored in memory in propositional form, but *holistic understandings* of particular situations which are stored in memory as case repertoires.
>
> (p. 313)

Central to a creative model of professionalism, to which Elliot subscribes, is the view that the personal growth and professional development of teachers is inextricably entwined. A reflective teacher is valued for being a resourceful and developing individual rather than someone who functions routinely in a predetermined role that merely reinforces the image of the teacher as an infallible expert.

However, the notion of the reflective practitioner in education has sometimes been discredited by critics who claim too much is attached to subjective judgements made in a vacuum and that a preoccupation with process and means is at the expense of concern for curriculum content and outcomes. Effective reflective behaviour requires teachers to subject their own practice to scrutiny within a conceptual framework that allows alternative theoretical interpretations and practical approaches to coexist. Traditionally the main focus for reflection has been on process, but, increasingly, the importance of reflecting on aspects of content is recognised in order to challenge assumptions about subject knowledge. Above all it is necessary to recognise and value insights into experience that can be gained through modes of human behaviour other than those that rely on patterns of processing that are linear, logical and rational.

UNIT 2.4 WORKSHOPS: ENVIRONMENTS FOR ENQUIRY

Through your direct involvement in PGCE workshop studies you can discover opportunities to deepen your understanding of the issues raised in this chapter. The section which follows draws your attention to different dimensions of workshop activity. Throughout your course you are expected to maximise the potential of the different kinds of workshops in which you engage by making creative connections between your previous and present experience, theory and practice, art, craft and design and education. In order to address ideas relating to subject knowledge and subject application simultaneously and 'in action', a strong case is made for the centrality of workshop studies in PGCE courses. It is to a detailed discussion of workshops that this unit is devoted.

Within the organisational structure of each PGCE Art & Design course workshop studies are approached in a variety of ways. Workshops may take place in the universities, schools, museums and galleries or in combinations of these locations. They may involve groups of students, students and teachers from partnership schools, groups of pupils and other artists or designers working in education through residency schemes. It is likely that you will also be encouraged to engage in workshop activity as an 'individual', a necessary component of your private study or directed study time. You are invited to consider the implications for teaching and learning that such varied organisational frameworks may have, along with the value of workshops in which you are able to share your experience as a member of a group as opposed to those in which you work alone.

The stated intentions of workshop studies are also likely to differ from course to course and the emphasis placed on the acquisition and development of skills and an exploration of ideas through issues-based work, for example, may shift within the same course. However, fundamental concerns and values are shared and these serve to underpin the importance attached to workshops and their role in the initial and continuing education of all specialist teachers of Art & Design.

Whatever the particular requirements of the workshops in which you participate, it is important that you exploit their rich potential as environments for enquiry. Through your engagement in different kinds of workshop activities you *become* able to enquire into the making of and response to art, craft and design and ways of teaching and learning in Art & Design. Workshops reaffirm the importance of active learning (learning through doing, practical knowledge) they provide 'a transaction with a situation in which knowing and doing are inseparable' (Schon 1987: 78). They foster learning modes that are experiential to ensure 'knowledge is not divorced from knowers' (Salmon 1995: 24). As Salmon says:

> Learning from experience calls for educational modes that are as far removed as they could be from traditional classroom teaching. Instead of didactic transmission of information such modes need to engage learners actively and purposefully in their own learning. In place of top-down knowledge pupils must construct things for themselves. And what is learned must go beyond merely doing things; the learner must come to reflect on that practical experience, to articulate something of what it means.
>
> (*ibid.*: 22)

It is through your developing capacity to reflect on practical experience that you gain greater insight into your own creative functioning as an artist, craftsperson or designer. This in turn informs your teaching and your growing capacity to make finely tuned decisions about how best to structure learning experiences for others, in order to meet the individual needs of a wide range of pupils. This reflective dimension of workshop studies enables you to make and sustain vital and creative connections between your previous and present experience as a learner, on the basis of which your present and future approach as a teacher is shaped. Those aspects of your creative processes that have remained implicit in the past, become focal points of attention; they are made explicit. Their functions are articulated, shared, critically analysed, interpreted and made available for modification. Throughout your deepening awareness of your creative processes and preferred working methodologies it is important to recognise alternative ways of thinking and working in art, craft and design along with their implications for pedagogy. As a result it is possible to demonstrate how workshops in the field of Art & Design education can play a powerful role in helping you simultaneously to address subject knowledge and subject application.

Whatever the nature of the workshops in which you participate as a PGCE student, they provide you with opportunities to engage in practical art, craft and design activity and at the same time develop your critical awareness of the factors that influence the quality of learning. As a teacher of Art & Design you are required to create conditions that maximise the potential for pupils' experiential learning in the lessons, projects and courses for which you are responsible. To achieve this it is necessary to make numerous decisions about content and organisational structure. In order to provide an informed basis for such decision-making related to your teaching, you are asked to reflect upon and critically consider the content and structure of those PGCE workshops in which you participate as a learner. For this purpose it is recommended that you keep a detailed log or diary to provide an ongoing record of your workshop experience, your reflections-on-action. By reflecting on your creative behaviour in this way you are able to identify the most significant factors that determine your preferred patterns of working. You are able, as a result, to articulate more clearly your understanding of the ways in which ideas are generated, shaped, supported, developed and refined. Strategies through which connections are made between concepts and skills, process and product, visual, written and verbal modes of communication can be explored. The insights gained through such explorations inform your growing awareness of the interdependence of skills, knowledge and understanding in the National Curriculum for Art. Additionally, when planning projects for teaching practice, you can investigate the influences on creative activity of such constraints as the nature of the brief, time, space and materials along with ways of dealing with variables of this kind.

In order to provide a framework for your reflections-on-action your attention is drawn to the following key factors on which you are invited to focus:

TIME

Creative activity requires time so that it can be paced and structured in such a way that ideas have opportunities to evolve, to take shape through impact on material and, above all, time to explore alternatives through investigative research. Time is needed in order to become familiar with materials, tools, equipment and technical processes to acquire and develop practical skills and to apply them in new situations. Above all, workshops require sufficient time for art, craft and design activities to be engaged in, sustained and for personal satisfaction to be experienced through struggle. As Schon (1987) reaffirms:

. . . nothing is so indicative of progress in the acquisition of artistry as the student's discovery of the *time* it takes – time to live through the initial shocks of confusion and mystery, unlearn initial expectations, and begin to master the practice of the practicum; time to live through the learning cycles involved in any design like task; and time to shift repeatedly back and forth between reflection on and in action.

(p. 31)

Task 2.4.1 Addressing time

In tutor groups:
   Consider the basis on which time is allocated for Art & Design workshops and how it is organised and managed. You are constantly reminded that time is a precious commodity and not to be wasted. There is never enough of it for you to achieve what you know you are capable of achieving: if only you had more time. Compare what you, as a specialist, are able to achieve in a PGCE workshop and what a pupil might achieve within the constraints of their timetable. What are the implications for planning?
   Widely available and advanced information and communication technology allows instant, simultaneous and continuous access to a vast and rapidly increasing bank of information. The everyday expectation of an increasing number of individuals is that gratification is instant and that results are rapidly and easily won in their interactions with the material world and in their interpersonal relationships.
   How can you enable pupils to question this expectation?

STRUCTURE

Task 2.4.2 Relating content to structure

In tutor groups:
   Consider the relationship between the organisational structure and the content of the workshops in which you are involved. Analyse their overall structure. Consider the under-lying rationales on which they are based along with the appropriateness of their design to effectively achieve their declared intentions. Critically examine the main similarities and differences between your responses in these workshops and your experience of studio-based work during your degree course.

The organisational possibilities for workshop studies are infinite and it is important that you draw upon the range of workshops you have experienced to help you make informed decisions about how best to manage multiple variables in the lessons and projects that you plan for teaching practice.

   Some of the most challenging and far-reaching decisions that have to be made by both teachers and learners in Art & Design relate to *ways of getting started*. As a teacher it is necessary to address the advantages and disadvantages of different 'ways in' to each project you plan to teach. You may decide to introduce a whole class to a project through a common brief; alternatively it may be considered more appropriate for individuals to invent their own briefs, thus generating a range of starting points within an agreed framework. Ideas may be generated through the direct manipulation of media and processes, discussion, writing, brainstorming exercises, visits, the critical interrogation of artefacts or research material collected prior to the introductory session. Each approach requires of learners a different structure and demands different kinds of skills. It is also important to be aware of the relationships between pupils' past and present experience. Different ways of introducing and developing projects help to challenge orthodoxies and disrupt routine ways of working and avoid uncritical and stereotyped responses. As work in Art & Design evolves it is necessary to make ongoing decisions about the management of variables in relation to time, content, media and working methods. In order to sustain a successful workshop project you need to structure it in such a way that an appropriate balance is achieved between freedom and constraint, skills and issues, individual and group endeavour, making and critical analysis.

SPACE

Task 2.4.3 The environment as determinant

In tutor groups:
  Consider the nature and organisation of the physical environment in which you work. How does it influence the kind of work you do, the way you approach it and the way you feel about it?

The physical setting in and through which a teacher's knowledge of and attitudes towards Art & Design, pupils, teaching and learning, are made explicit, has an unavoidable effect on the quality and mode of interaction between a teacher and pupils, and between pupils and their art work. The degree of ease or difficulty with which connections can be made and explored between ideas, feelings, perceptions, media and processes is significantly influenced by the setting within which these elements coexist. A prescribed working area should operate as a coherent resource. The layout and visual, tactile and spatial richness of a room, for example, encourages and supports certain behaviours whilst it discourages and is unsupportive of others. The way in which space is allocated and arranged in relation to specific activities, the nature and amount of mobility it allows and interaction with others it facilitates all operate as controls over creative responses. So too does the availability of materials and equipment. The ways in which we perceive, think about and are motivated to use the resources in our immediate surroundings are influenced by the manner in which they are presented to us. Familiar things placed in unfamiliar contexts, and unfamiliar things placed in familiar contexts, provide powerful triggers for the release of creative energy.

RELATIONSHIPS

Table 2.4.4 Reflection-on-action
In tutor groups: Consider how other members of your workshop group contribute to your subject knowledge and growing understanding about teaching and learning. Through your log or diary record the reflective dimensions of the workshop activities in which you have engaged.

This reflection-on-action can be likened to a conversation with yourself: an activity that helps you gain deeper insight into the conditions that favour creative thought and action. As a member of a group you are exposed to alternative views about art, craft and design rooted in different belief systems. Different ways of working coexist in the same workshop. To the extent that your subject knowledge base can accommodate different, sometimes conflicting, philosophies of art, you are able to draw upon richer and deeper levels of understanding of subject content which allows increasing openness to accept pupils' ideas.

Through the opportunities that workshops offer to share ideas with others, attitudes are modified. By shaping ideas to clearly communicate them, by making the implicit explicit they enter the public domain, become available for scrutiny. Above all, such interaction provides you with opportunities to reflect on reflection-on-action and reaffirms the value of exposure to alternative ways of seeing, thinking and doing.

CONCLUSION

The majority of teachers declare, verbally, that they encourage pupils to be creative in Art & Design lessons. Unfortunately, in practice, there is often an absence of those conditions which maximise opportunities for genuinely creative behaviour; for various reasons they are unacceptable in operational terms. A teacher's verbal encouragement of risk taking, the flexible use of ideas, materials, time and space must be accompanied by personal behaviour which embodies these qualities and a climate which allows such behaviour to flourish.

Anyone who has not 'lived through' experiences in Art & Design of the same order as those in which pupils engage and struggle to make sense of the process through reflection, 'is not likely to ignite others with an intrinsic excitement of the subject' (Bruner 1960: 90). To achieve such ignition workshops have a central role to play in courses of initial teacher education.

FURTHER READING

Salmon, P. (1995) 'Experiential learning', in R. Prentice, (ed.) *Teaching Art & Design: Addressing Issues and Identifying Directions*, London: Cassell.

Schon, D. (1987) *Educating the Reflective Practitioner*, San Francisco: Jossey Bass.

3   Learning in Art & Design

Nicholas Addison and
Lesley Burgess

This chapter explores the nature of learning and teaching in art, craft and design. You are introduced to a range of learning theories, some of which are general while others are specific to the visual and material arts. Issues relating to special educational needs (SEN) are addressed within each unit.

The units in this chapter examine:

Unit 3.1   definitions of learning and teaching (the didactic/heuristic continuum);
Unit 3.2   learning in relation to developmental psychology and theories of intelligence;
Unit 3.3   active and experiential learning;
Unit 3.4   aesthetic literacy;
Unit 3.5   the tacit and the intuitive: complementary ways of learning;
Unit 3.6   learning: language, strategies and motivation;
Unit 3.7   enabling learning: transforming subject knowledge into pedagogical practice.

3.1 DEFINITIONS OF LEARNING AND TEACHING
(THE DIDACTIC/HEURISTIC CONTINUUM)

Most theorists agree that learning is primarily defined as a process of acquisition, assimilation and application of knowledge (Capel *et al.* 1995: Unit 5.1). Cognition is the word that best describes this process, a process that includes acts of perception, intuition and reason. Acts of perception are the basis of experience. Art, craft and design provide ways to represent and embody experience. Acts of intuition and reason are the basis of imagination. They are the processes through which perceptions are reworked to form concepts and precepts, in other words, are transformed into ideas

---

OBJECTIVES

By the end of this unit you should be able to:

● understand the function and potential of different pedagogic methods and their effect on learning;
● use conceptual frameworks to focus your classroom observation and inform your lesson planning;
● consider how conceptual and sensuous responses to the world affect the way pupils learn.

---

and values. The processes of art, craft and design are transformative acts which produce and develop ideas and values in material and multi-sensory forms. These acts are, in themselves, evidence of learning. The result is a physical or virtual outcome that not only represents but also embodies knowledge and understanding:

> The odd notion that an artist does not think and a scientific inquirer does nothing else is the result of converting a difference in tempo and emphasis into a difference in kind. The thinker has his [sic] esthetic [sic] moment when his ideas cease to be mere ideas and become the corporate meaning of objects. The artist has his problems and thinks as he works. But his thought is more immediately embodied in the object . . . The artist does his thinking in the very qualitative media he works in, and the terms are so close to the object that he is producing that they directly merge with it.
>
> (Dewey 1934: 15–16)

The process of applying and embodying knowledge in a work of art suggests that the maker has understood that knowledge; you might call this understanding made concrete. It is therefore important to realise that process and product are interrelated concepts and that it is counterproductive to develop and assess one without reference to the other. It is also important to recognise that knowledge is value laden and dependent on cultural and social factors.

In Art & Design the kinds of knowledge presented, investigated and applied have been divided into three domains:

a   A conceptual domain that is concerned with the formation and development of ideas and concepts.

b   A productive domain that is concerned with the abilities to select, control and use the formal and technical aspects of Art & Design in the realisation of ideas, feelings and intentions.

c   A contextual and critical domain that is concerned with those aspects of Art & Design which enable candidates to express ideas and insights which reflect a developing awareness of their own work and that of others.

(SEC 1986: 6–7)

The National Curriculum for Art (NC) (DFE 1995 and QCA 1999) and the Art Core at GCSE and A/S and A Level (1998) identify how these domains relate to classroom practice. Many earlier models defined Art & Design education in terms of autonomous and discrete activities, whereas these later developments implicitly acknowledge art, craft and design as interrelated forms of social production. The NC (DFE 1995) states: 'Pupils should be taught to relate art, craft and design to its social, historical and cultural context' (p. 7).

The SEC model should not be read as promoting three discrete areas that need to be taught separately. It represents three constituent parts of a complex and inter-dependent field. In order to integrate these domains you need to employ different and changing teaching methods, from the didactic to the heuristic. A varied and system-atic approach will help you to break down the oppositions between child- and adult-centred, or progressive and traditional means of learning and allow you to satisfy the specific requirements of different tasks and the needs of individuals: (for an Art & Design specific advocacy of the Progressive movement see Lowenfeld and Brittain 1987; for a critique see Abbs 1987: 32–46). Table 3.1.1 identifies the characteristics of the didactic/heuristic continuum. Extreme positions can be held regarding the effi-cacy of each method. We contend that no one method is entirely sufficient for all pupils or for every task. Therefore, it is important to maintain a balance between the methods in this and the other continuums presented and referenced below. Classroom observations during your course provide you with the opportunity to define the type of learning and teaching taking place.

The implications of this continuum for teaching and learning are explored in Unit 3.7 where examples of practice demonstrate how these theoretical positions are applied to the classroom. It is useful to refer to variations on this continuum such as Mosston and Ashworth's 'Pupil participation and teaching styles' and 'Continuum of teaching styles' (Capel *et al.* 1999: Unit 5.3).

These types of teaching and learning have implications for SEN. Warnock (1978) suggests that it is pupils' needs, not their disabilities, that should be identified when differentiating strategies for learning. She proposes 'common goals' and 'a common purpose' in education (p. 5). Common elements in the context of differing needs means that the routes to achieve understanding have to be different. For example, pupils with literacy needs are likely to find experiential forms of learning more sympathetic in developing understanding of art, craft and design than in annotating drawings. In activities that demand written analysis, including self-assessment, these pupils require exactly the same literacy support that they receive elsewhere in the curriculum. In this instance, if specialist support is not available, it is necessary to build in opportunities for oral assessment so that they can try to articulate the understand-ing evidenced in their practice during group evaluations. However, if a pupil is unable to contribute to evaluation in either written or oral form it does not necessarily signify a lack of understanding. In the introduction to this unit we referred to the notion of embodied knowledge as understanding made concrete: it is important that you identify and record this, and acknowledge it in class discussions. Exceptionally able pupils require a different type of support, needing continuous challenges, tasks that stretch their abilities. In many curriculum subjects this group is identified with the most literate, logocentric pupils: it is often the same group who excel in Art &

Table 3.1.1 The didactic/heuristic continuum

centred	LEARNING	TEACHING	CHARACTERISTIC FORMS	pupil	teacher	JUSTIFICATION	DRAWBACKS
adult teacher	PASSIVE	DIDACTIC	• instruction • information • lecture • demonstration • closed procedures and structures	dependent memoriser imitator	expert provider (differentiation by testing)	• outcomes are certain (appropriate to factual syllabi) • introduces techniques, establishes a sense of shared beliefs and values • conditions pupils to become receptive, to observe, listen and record • encourages memory skills • confirms the expert status of the teacher	• can be authoritarian single perspective • can alienate because it fails to acknowledge difference (abilities/backgrounds) • pupils may become dependent • results in conformity and normative outcomes • knowledge may be lost unless reinforced using other methods
	RESPONSIVE AND ACTIVITY BASED	DIRECTED	• rehearsing and imitating activities • responses to given stimuli, e.g. still life, work of others, design brief • working to exemplars • investigation (probable findings already known by teacher) • conditioned/determined structures	responder	trainer director resourcer (differentiation by outcome and taste)	• provides common experience NC Art (e.g. core skills) **enables:** • continuity and progression • identification of pupils not on task • ease of assessment • efficient transfer of skills • activity: individual/pair/group	• knowledge is given/fixed, determined by teacher's experience: often privileging making **neglects:** • pupils' prior knowledge • individual needs

ACTIVE AND EXPERIENTIAL NEGOTIATED	• discussion/debate • collaborative work • purposeful investigation • critical evaluation • multi-faceted and flexible structures • interaction • reflexivity	contributor interactor	facilitator motivator guide negotiator supporter (differentiation by individual learning routes)	**provides:** • intelligent making • critical thinking • learning as social activity; art as social practice • mutual respect/trust • learner as teacher/teacher as learner **acknowledges pupils':** • prior knowledge • individual needs **enables pupils to:** • communicate ideas • evaluate their own and others' work • negotiate their own learning • crosses boundaries	• time consuming • difficult to coordinate and resource • difficult to monitor and assess • teacher requires breadth of knowledge • teacher needs to be ready to relinquish a degree of control
HEURISTIC EMPOWERING AND VALIDATING	• meeting needs • answering hypotheses • experimentation • unknown findings • discovery • problem solving • investigation	researcher self-motivator inventor discoverer	coordinator reciprocator (differentiation by role contribution)	**encourages:** • the application of knowledge to practical contexts • pupils as planners • divergent thinking • risk-taking	• pupils need to be ready to take on initiatives • difficult to resource • only works with pupil self-motivation • teacher may feel insecure • teacher needs to acknowledge self as learner
OPEN DEPENDENT REDUNDANT	• self-determined structures, motivated by pupil interest • exploration	agent director	attendant technician	**appropriate for:** • highly motivated highly resourceful learners	• can be chaotic, unfocused • lacks boundaries • can invite stereotypical responses and/or a rejection of learning
child pupil					

Design. However, there are exceptional cases where pupils with literacy and numeracy needs display highly developed mimetic skills, or advanced motor skills with which they can transform materials. Because they have been supported by a sympathetic curriculum in a particular department, they have managed to thrive.

THE VISUAL/HAPTIC CONTINUUM
(Lowenfeld and Brittain 1987: 356–368)

There is evidence to suggest that the way people learn about the world can be categorised into two extreme modes; the visual 'the observer, who usually learns about things from their appearance' and the haptic who 'utilises muscular sensations, kinesthetic [sic] experiences, impressions of touch, taste, smells, weights, temperatures, and all the experiences of the self' (Lowenfeld and Brittain 1987: 357). Lowenfeld believes that mental growth is only possible if pupils are allowed to interact with their environment on a sensuous level. He states that most people fall somewhere between these two extremes, but warns that Art & Design teaching can privilege the analytical and visual at the expense of the emotional and haptic.

Table 3.1.2 The visual/haptic continuum, characteristics of the two extremes

Visual	Haptic
spectator	participant
analytical	intuitive
detached	emotional
objective	subjective
abstract	concrete
mimetic	affective
optical	tactile and kinaesthetic
perceptual	synaesthetic

You must remember that although nobody is exclusively visual or haptic, particular Schemes of Work (SoW), and the way they are assessed, may isolate and promote activities which prioritise one or more of these characteristics. If the curriculum is geared exclusively to one extreme or the other, for example the visual, those pupils who learn effectively through the haptic mode will have SEN in predominantly visual SoW. To accommodate all pupils you should:

- build in changes of activity that address the two modes;
- provide open routes so that pupils can choose the most appropriate mode;
- identify and target pupils who require additional support.

Haptic modes of learning are often associated with the way SEN pupils come to understand the world. Handling collections at museums can provide one forum in which the combination of direct contact with artefacts and discussion about their function and contexts can facilitate learning. For example the Imperial War Museum (IWM) in London provides the following information on special needs in their education service leaflet (1999):

> Handling Sessions
> On Mondays handling sessions using artefacts can be arranged for pupils with learning and/or physical difficulties. The sessions are on either the First or the Second World War and can be adapted to the specific needs of each group.
>
> (IWM 1999: 10)

Task 3.1.1 The visual/haptic continuum

When planning SoW for your teaching placements.
   Look at the PoS for year 7 in your placement schools and identify the extent to which SoW and their learning objectives relate to the categories in Table 3.1.2.
   Can you add to the table?
   Repeat the exercise for year 10.
Can you identify any differences between the two year groups?
   How can you ensure that your lessons are not weighted exclusively to any one side of this continuum? Discuss these issues with your school mentor.

UNIT 3.2 LEARNING IN RELATION TO DEVELOPMENTAL PSYCHOLOGY AND THEORIES OF INTELLIGENCE

OBJECTIVES

By the end of this unit you should be able to:

- evaluate the teaching in Art & Design that is informed by theories of learning based on developmental psychology;
- consider the place of syncretistic vision in a curriculum which promotes analytical procedures;
- consider how theories of multiple intelligence produce different types of learning and learners.

Many educational theories subscribe to the notion of developmental stages which determine the ability of an individual to assimilate forms of knowledge at different moments in their psychological development: for example, it would seem inappropriate to ask a year 7 pupil to produce a detailed design for a suspension bridge and provide all the technical data required for its construction. Clearly the degree of abstract and mathematical know-how combined with knowledge of specific materials and their behaviour would be outside the understanding of most 11 year olds. Developmental psychologists refer to the idea of 'readiness' as a necessary condition for assimilation and integration (Capel *et al.* 1995: Table 5.1.1). For a learner to be 'ready' their educational experiences must match their level of understanding so that, with help, they can come to recognise and process information in such a way that it can be retrieved and applied to different contexts. Piaget (1962) formulated a theory that divides the maturing child's ability to reason into different, clearly defined stages. The first, the 'sensory-motor period', lasts into the second year and denotes a time when children, unable to conceptualise their experience, can only react to the world in response to sensations. From the age of 2 they begin to understand the world through interaction, by exploring their immediate environment actively, using all their senses, what Piaget calls 'concrete operations'. Only with time, and only once the child is ready, do they develop more abstract ways of thinking, 'formal operations', constructing and testing hypotheses, imagining the possible effects of actions not yet experienced. It follows from Piaget's theories that learning is not the same as knowing. Learning involves a process of understanding whereby knowledge is developed through interaction with an environment. His sequential stages signal different ways of understanding and each of these conditions the way people see. He coined the term 'syncretistic' to differentiate a child's vision from that of the adult's. For Piaget the syncretistic is a primitive way of seeing that is superseded by adult analytical processes. The syncretistic allows a person to see holistically and recognise objects through cues without the need to match part to part cumulatively as in analytical vision. Ehrenzweig (1967) is less dismissive than Piaget suggesting that this way of seeing is dependent on a process of 'unconscious scanning':

> . . . unconscious scanning makes use of undifferentiated modes of vision that to normal awareness would seem chaotic. Hence comes the impression that the primary process merely produces chaotic phantasy material that has to be ordered and shaped by the ego's secondary process. On the contrary, the primary process is a precision instrument for creative scanning that is far superior to discursive reason and logic.
>
> (Ehrenzweig 1967: 5)

The child's syncretistic vision utilises these undifferentiated processes producing a view of the world which is unhindered by the analytical, conscious mechanisms which, Ehrenzweig believes, can inhibit creative thinking. In the artist or other creative practitioner, unconscious scanning is supplemented by more conscious processes so the that those 'happy accidents' which appear to arrive from nowhere can be tested by empirical and analytical procedures at some later date. He considers in detail the implications of these ways of seeing for art education:

> The child's more primitive syncretistic vision does not, as the adult's does, differentiate abstract details . . . This gives the younger artist the freedom to

> distort colour and shapes in the most imaginative, and, to us, unrealistic manner.
>
> (*ibid.*: 6)

From about the age of 8 children become more aware of adult expectations and begin to analyse their own work against:

> . . . the art of the adult which he [sic] finds in magazines, books and pictures. He usually finds his own work deficient. His work becomes duller in colour, more anxious in draughtsmanship. Much of the earlier vigour is lost. Art education seems helpless to stop this rot.
>
> (*ibid.*: 6)

Unlike Piaget, Ehrenzweig does not believe that in passing from one stage to the next the former stage is lost, merely that it is suppressed. Gardner supports the belief that early forms of knowing are not eradicated or transformed: 'they simply travel underground; like repressed memories of early childhood, they reassert themselves in settings where they seem appropriate' (Gardner 1993: 29). Ehrenzweig does not suggest that child art should be seen as a paradigm for art education as this would only lead to an aesthetics of regression. He suggests that anyone can retrieve their syncretistic facility but usually in situations where analytical processes are bypassed, as in the case of humour. He cites the example of caricature where perceived features can be acutely distorted, yet the sense of likeness enhanced.

Ehrenzweig argues that the undifferentiated processes of syncretistic vision are essential to creative thinking of all types (Ehrenzweig 1967: 32–46) and provides evidence in the work of scientists and modernist artists. From the late 1960s he influenced a tradition of process-led, conceptual practice, especially in the USA. Robert Morris, acknowledging Ehrenzweig's concern with the whole visual field as opposed to figure/ground differentiation shifted from using minimal geometric forms, with their strong Gestalt, to formless more heterogeneous materials like thread waste, a by-product of the textile industry used in packaging. Morris saw boundary-crossing as essential to dedifferentiation and with Robert Smithson began to explore the possibilities of ephemeral materials such as steam and time-based projects recorded photographically. One effect of this shift was to move practice outside the studio and gallery to any site or environment (Taylor 1995). This expanded field has transformed contemporary practice, but has had little impact in secondary schools.

It is important for you to consider the implications of the change from syncretistic to analytical perception for your planning and teaching, in general, but also in teaching SEN pupils. When pupils first enter secondary school the 'rot' that Ehrenzweig identifies may have already set in: how often have you heard the plea, 'but Miss, I can't draw'. What the child is articulating here is their inability to draw in a particular way, usually the analytical manner of observational tonal drawing which is so ubiquitous in schools. Pupils will undoubtedly be aware of older pupils' work and may in comparison feel intimidated by their lack of technical and imitative skill. A limited sense of what is good or right can be reinforced by your approach to practice, both as artist and as teacher. Such rigid boundaries are quickly communicated to pupils:

Those teachers who were unable to tolerate their own spontaneity and the loosening up of their rigid planning could not tolerate the spontaneous and wilful reaction of their young pupils during their teaching practice either . . . But what has perhaps not been sufficiently realised is the close correlation between the two kinds of ego rigidity, the trainee's intolerance of the independent life of his [sic] own work of art and his intolerance of his pupils' independent contributions to his teaching programme. The unconscious fear of loosing control underlies both.

(Ehrenzweig 1967: 101–102)

The Art & Design curriculum frequently privileges analytical modes of production suggesting they are of a 'higher order' than affective and emotional modes. Pupils with learning needs often display a syncretistic approach to representation and find it difficult to conform to the prevailing norms of school art. This inability can manifest itself in a negative, 'primitive', self-image. It is therefore important that you discuss the potential of both syncretistic and analytical modes: provide reproductions of artists who work in affective and syncretistic ways and include tasks in SoW that have affective as well as analytical criteria. Ensure that you examine analytical modes of representation by using them yourself. This enables you to break down the activity into realisable steps and helps you to explain the process to pupils who have not developed the mimetic strategies or facility of those pupils who 'can do school art' even without your help.

Task 3.2.1 Recognising and valuing the syncretistic

With your Art & Design mentor:

- look at the reproductions of artists' work on display throughout the department and identify their syncretistic and analytical elements;
- go through the same process with examples of work by year 7 pupils;
- discuss ways to explain to pupils the value of the syncretistic in their own and others' work;
- devise a lesson which enables pupils to approach a task in either a syncretistic or an analytical mode;
- consider the implications of syncretistic expression for assessment?

THEORIES OF INTELLIGENCE

The emphasis in the secondary curriculum on factual knowledge and measurable outcomes privileges subjects and approaches that promote the ability to reiterate given knowledge. This is believed to provide a knowledge base from which pupils can construct arguments, solve problems and organise experience. The critical curriculum advocated by this book recognises the importance of cognitive processes for developing visual and aesthetic literacy but invites you to question whether logical and sequential processes and abstract reasoning hold all the answers to the creative curriculum. The dominant culture of accountability has driven teachers back to the certainties of traditional, tried and tested, pedagogic methods. Systems of assessment that measure abstractions through tests, like the IQ, are still applied in educational contexts but are recognised as biased and woefully inadequate for determining the potential of pupils to contribute to society (Capel *et al.* 1995: Unit 6.2, especially 'The Art & Design portfolio'). More inclusive theories of intelligence that acknowledge different ways of knowing and understanding do exist and they have influenced practice in schools. Gardner's theory of multiple intelligences (1983) proposes that other faculties besides reasoning can be defined as intelligences. He categorises intelligence into seven types:

> linguistic
> musical
> logical–mathematical
> spatial
> bodily–kinaesthetic
> interpersonal
> intrapersonal

This more inclusive taxonomy embraces 'ways of knowing the world' that recognise artistic processes, although, at first sight, only spatial intelligence specifically belongs to

---

Task 3.2.2 Investigating multiple intelligences

In your placement school:

- observe and record the relationship between Gardner's models and what happens in the classroom.

Discuss in tutor groups:

- how you might assess SEN pupils differently if you formulated criteria based on Gardner's intelligences;
- why traditionally the 'intuitive' learner has found a sympathetic home in the Art & Design department and what this infers about the status of the subject in schools;
- how your own learning in Art & Design relates to Gardner's models.

In what educational situations have you been encouraged to take risks?

Art & Design. However, when you read Gardner's definitions, 'bodily-kinesthetic' intelligence is described as 'the use of the body to solve problems or to make things' (Gardner 1993: 12) and clearly also belongs to the subject.

Gardner (1993: 6–7) points out that there are different types of learners: intuitive, traditional and disciplinary. He describes the intuitive learner as natural, naive and universal; the traditional student as scholastic, one who works comfortably within school systems; the disciplinary expert as one who can successfully apply their knowledge to new contexts. He feels it is essential that teachers and pupils should be willing to take risks, including the possibility that they might fail, rather than reiterate the safe formulae known to produce standard outcomes. He suggests that 'such a compromise is not a happy one, for genuine understandings cannot come about so long as one accepts ritualised, rote, or conventionalised performances' (*ibid.*: 150).

UNIT 3.3 ACTIVE AND EXPERIENTIAL LEARNING

OBJECTIVES

By the end of this unit you should be able to:

- define active and experiential learning;
- develop strategies to facilitate active and experiential learning.

It is important to recognise that making in Art & Design is always active in the sense that pupils are engaged in doing something, whether self-generated or in response to direction. However, doing is not everything. It is wrong to assume that because pupils are busy they are learning something – they may only be reinforcing existing knowledge: activity is not active learning. Passive learning, that achieved through listening, has traditionally been suspect: I hear, I forget; I see, I remember; I do, I understand (anon, Chinese proverb). Lowenfeld suggests that activities such as drawing enable pupils to transform passive into active knowledge:

> The child draws only what is actively in his [sic] mind . . . A child knows a great deal more in a passive way than is included in the drawing. Part of a teacher's responsibility is to make this passive knowledge more active.
> (Lowenfeld and Brittain 1987: 36)

The implication here is that you should not spoon-feed pupils with ready-made formulae as this can repress their personal interests and inhibit their motivations. Instead, you must find ways to encourage them to take on increasing responsibility for their learning.

Similarly experiential learning is commonly defined as learning by doing. This definition is partly true, but from the point of view of construct theory (Kelly 1955; Rogers 1969) experiential learning is recognised as a reflexive activity, where action and reflection are coexistent, both interdependent and interactive. Only where pupils are 'engaged actively and purposively in their own learning' (Salmon in Prentice 1995: 22) is the term 'experiential' appropriate. Usher and Edwards (1994) suggest:

> Experience is most often accorded importance as the 'authentic' representation and voice of the individual. Experiential learning has been constructed as a progressive and emancipatory movement in education, a shift away from the learning of canons of knowledge which, it is argued, marginalises the majority of learners by not giving value to their voices and thereby demotivating them.
>
> (p. 187)

Developing strategies to enable pupils to learn experientially requires effort. Pupils do not assimilate knowledge as presented, as though it were something separate and 'out there', information to be recorded and retrieved at will. They interpret it in relation to their existing conceptual models. Experiential strategies acknowledge:

a   prior knowledge and learning, both in and outside school;
b   social and cultural backgrounds.

It follows that knowledge is actively and differently constructed by each individual. Therefore it is important for you to recognise these differences by valuing pupils' ideas and contributions and by providing opportunities for them to:

- take responsibility for their own learning;
- devise and give presentations;
- reflect, discuss and evaluate;
- negotiate meanings.

To achieve this the teacher needs to rethink their traditional role as knowledge provider by making the relationship between teacher and learner more reciprocal (Freire 1985; Giroux 1992), although you must remember this relationship is always 'asymmetrical' (Reid 1986). However, it is essential that you help pupils to understand dominant signifying systems so that they can negotiate their position within them. The NC Art 2000 requires pupils to identify 'codes and conventions and how these are used to represent ideas, beliefs and values' (p. 7): the representation of space is an accessible example. To a year 7 group you might be tempted to explain this didactic-ally, as a body of given knowledge, by comparing oblique and axonometric projection in Japanese Ukiyo-e prints to one-point perspective in painting from the early Italian Renaissance (Willats 1997: 37–69). The mathematical concepts implicit in this explanation are unlikely to be remembered if delivered in this way. To develop a more experiential approach you need to devise questions inviting pupils to compare and contrast the spatial organisation in these different traditions. For example:

> How can you tell that a figure or object is in the background or the foreground?
> Where do the lines leading you back from the foreground finish?

Pupils can then formulate rules governing these differences in spatial organisation and apply them to their own representations. Not until this point should you provide a mathematical explanation against which pupils can test their findings. This could be extended by asking pupils to investigate the use of these systems in contemporary forms, e.g. computer games and Manga comics (Schodt 1993). In addition pupils can explore alternative spatial systems, for example aerial and oblique perspective (Willats 1997; Cole 1992).

As a teacher you need to consider your own position as far from neutral (hooks 1994a: Chapter 6). You bring with you your own preconceptions which are embedded in everything you say and do. Your beliefs about the appropriate role of teachers and pupils will affect the interactions that take place. A statement such as: 'This class is never responsive' says as much about the way you have devised your lesson as it does about the pupils themselves. It may be that pupils are not usually expected to be responsive, in which case it will take you time to gain their trust and convince them that responsiveness is beneficial to their learning.

Setting up situations and environments which encourage pupils to take ownership of ideas, and thus their learning, is equally demanding. Pupils are more likely to be motivated if the learning environment is stimulating and well resourced. This is not to be confused with an environment of pure spectacle manufactured in the hope that an overwhelmed audience will be stunned into reverential compliance, e.g. one consisting solely of the virtuoso displays of artefacts selected by teachers or work by star pupils from the past decade. Pupils are more likely to be motivated if they contribute to the environment themselves, one they have helped to construct. Such contributory practice is particularly useful in building self-esteem. In this context learning is not the transmission of knowledge from expert to novice but an active and productive partnership where meanings are constructed, questioned and negotiated (see Unit 3.6; see also Capel *et al.* 1999: Unit 3.2). This is not to suggest that teachers should surrender their responsibility to provide new and different forms of knowledge. A complete reliance on pupils' prior knowledge and experience would clearly limit the Art & Design curriculum in ways which would leave it open to accusations of subjectivity and introspection.

> **Task 3.3.1 Differentiating between activity-based and active/experiential learning**
>
> Look carefully at the SoW you have observed or taught in your first teaching practice.
>
> Identify those activities that encourage responsive, active and experiential learning. Ensure that you differentiate between activity and active/experiential learning when revising existing SoW and planning new ones.

FURTHER READING

hooks, b. (1994a) *Teaching to Transgress*, New York and London: Routledge.

Giroux, H. (1992) *Border Crossings*, New York and London: Routledge.

Salmon, P. (1989) *Psychology for Teachers: An Alternative View*, London: Hutchinson.

3.4 AESTHETIC LITERACY

OBJECTIVES

By the end of this unit you should be able to:

- define the aesthetic;
- identify its significance in Art & Design.

Aesthetic experience is often regarded as beyond words. It is a very complex phenomenon; any generalisations are as likely to obscure as to illuminate. However, it has been been associated with the following terms: auratic, sublime, intense, illuminating, intuitive, emotionally gratifying, uplifting, enervating, sensual, significant, spiritual, liminal, disturbing.

It is defined by the Pocket Oxford Dictionary (1992) as:

aesthetic *-adj.* **1** of or sensitive to beauty. **2** artistic; tasteful.
-n. (in *pl.*) philosophy of beauty, esp. in art.

> **Task 3.4.1 Aesthetics and your education**
>
> What role did aesthetics play in your own art education?
> Was the term referred to during your degree course; if so in what context?
> Record your answers and compare them with other members of your tutor group.
> Try and identify any common experiences.
> Did the following have any influence – degree specialism, year of degree?

Parsons (1987) and Taylor (1992) are noted for promoting the aesthetic dimension of Art & Design education. Parsons proposed a cognitive developmental account of aesthetic experience which brings together Piaget's stages of scientific thought with Kolhberg's stages of moral judgement (pp. xii–xiii). Parson's stages are summarised under the following headings:

- favouritism
- beauty and realism
- expressiveness
- style and form
- autonomy

Like Piaget and Kolhberg, Parsons sees development as progressive or incremental: a sequence of steps, each one a new insight towards a mature understanding of art, in his terms, a 'more adequate' understanding. Taylor (1992) promotes Abbs's concept of a dynamic aesthetic field in which responding, evaluating, making and presenting, form a 'highly complex web of energy linking the artist to the audience' (p. 3). He devises a framework for pupils to engage with artworks: content, form, process, mood. Taylor asserts that this 'provides an invaluable means of empowering young people so as to enable them to enter effectively into the aesthetic field' (*ibid.:* 69). Taylor was influential in the 1980s for promoting a resource-based approach to Art & Design education. However, in line with the NC Art Order he continued to promote an approach which, beginning with Fry's 'significant form' culminated in Greenberg's formalism (Frascina and Harrison 1982). This is based on the assumption that responses to high art and some natural forms are universal, that art's intrinsic properties can profoundly affect the viewer on an emotional and even transforming level. Taylor's more recent work with SCAA on universal themes can be seen as a continuation of this somewhat limited ideology. He also builds on Hargreaves's 'traumatic theory of aesthetic learning' to promote the related notion of 'illuminating experience' and 'aversive experience' (Taylor 1986: 18–34).

D. H. Hargreaves (1983) makes the distinction between the incremental and the traumatic theories of learning in art. The traumatic accounts for those aspects of aesthetic learning which cannot be explained by the incremental. The viewer finds their response to the artwork intense, even disturbing, and this has a powerful impact on their learning and long-term memory. Although it may be possible to learn from such an experience it is not one that can be taught or in any way be accurately predicted. The 'traumatic experience' occurs only occasionally, and many people never respond in this way. It is a reaction reserved for the work of others, reception, not one experienced through engagement with materials and processes, production.

Abbs reiterates the arguments outlined in the influential Calouste Gulbenkian Report (Robinson 1982) reinforcing the notion of the aesthetic as a distinct way of knowing. He agrees with Fuller (1993) that aesthetic experience begins with the sense responses of individuals and then radiates out to become a 'shared symbolic order'. Critics of this view assert that aesthetic values can be construed as elitist and should be replaced with a more critical, sociological analysis of art (see Unit 14.1). Abbs (1987) defines the term:

> Aesthetic derives from the Greek word *aesthetika* meaning *things perceptible through the senses*, with the verb stem *aisthe* meaning: to feel, to apprehend through the senses. Here in this small cluster of words: perception, sensing, apprehending, feeling, we begin to discern the nature of the aesthetic mode.
> (p. 53)

This cluster makes it clear that for Abbs the aesthetic is a fully cognitive mode, parallel but distinct from logical and discursive modes.

In differentiating between aesthetics and logic you have to consider how these two ways of knowing differ and relate. A percept is the mental product of perceiving and is peculiar to a specific experience. Aesthetic response is the means to differentiate between percepts that are, or are not, pleasing: it is a process of synthesis that appears to be immediate, total. A concept is the mental product of conceiving, an abstraction, a system of classification which assists in forming patterns of predictability. Logic is analytical and cumulative; it organises concepts to solve problems. However, it is not the only means, for example, you have already encountered the process of unconscious scanning (Unit 3.2). Aesthetics defines the taste for things; logic defines the application of concepts. Both ways of knowing are liable to inform decision making in the production and reception of the arts. We call this 'aesthetic literacy' (see Unit 14.1).

FURTHER READING

Abbs, P. (1987) *Living Powers*, London: Falmer Press.

Parsons, M. (1987) *How We Understand Art*, Cambridge: Cambridge University Press.

Taylor, R. (1992) *Visual Arts in Education*, London: Falmer Press.

3.5 THE TACIT AND INTUITIVE: COMPLEMENTARY WAYS OF LEARNING

OBJECTIVES

By the end of this unit you should be able to:

- define tacit learning and intuition;
- identify their significance in Art & Design;
- consider the extent to which tacit learning and intuition have informed your practice in art, craft and design.

TACIT LEARNING

Polanyi's theory of cognition (1964) successfully overcomes the traditional dichotomy between making and thinking. His explanation of tacit knowledge, the kind of knowledge that we cannot fully articulate, is commonly used by artists to account for their difficulty in explaining how they have worked their materials and the skills they have deployed in making. Cognition, according to Polanyi, constitutes a continuum between tacit and, its opposite, explicit knowledge.

In formal education, explicit knowledge is highly valued. Explicit knowledge can be articulated conceptually and in most areas of the curriculum it is thought to be the only kind of knowledge there is. In contrast, tacit knowledge arises in and through what Polanyi calls 'indwelling', a kind of empathetic participation: learning through example, learning by trial and error. Learning a tacit skill involves going through a series of 'integrative acts' by which a student grasps the full meaning of a process. This requires imitation, practice, repetition and complete immersion: it takes time. Heidegger (1954) talks about the relationship between making and thinking: 'Every motion of the hand in every one of its works carries itself through the element of thinking, every bearing of the hand is rooted in thinking' (p. 16).

How then can this be taught? The timetable at KS3 militates against prolonged engagement with making. Tacit learning recalls the apprenticeship model which is still practised in vocational education at FE and may increasingly return in the form of GNVQs in schools. Unless you can provide some continuity in practice, learning with materials is liable to be superficial and quickly forgotten. There are voices in Art & Design education arguing that it may be more productive to limit the range of materials and processes so that pupils come to know a few, and know them well (Mason 1995).

Task 3.5.1 Tacit learning

Look at SoW used in your placement school, identify:

- those practices that can be associated with tacit learning;
- the extent to which continuity is considered both in terms of prior knowledge and available time;
- in what ways tacit learning can enable progression.

It is essential that you recognise and build on the tacit knowledge pupils have developed in other curriculum areas and outside the school. For example, many pupils have particular skills in ICT which can provide a foundation from which to develop Art & Design specific exploration. We predict that not until approximately 2015, once the present generation of primary-school children begin to enter the teaching profession, will the majority of teachers' ICT skills match those of their pupils: here is a particular case for reciprocity. There is a middle way between tacit and explicit learning. The terms 'intelligent making' and 'practical thinking' (Burgess and Schofield 1998) provide this bridge.

INTUITION

Art & Design is one of the few subjects in the school curriculum that has recognised the significance of preconscious cognitive processes for the development of the imagination. The American philosopher John Dewey (1934) emphasised the central position of intuition for creativity; although he would concur with the notion that creative acts are 90 per cent perspiration and only 10 per cent inspiration:

> 'Intuition' is that meeting of the old and the new in which the readjustment involved in every form of consciousness is effected suddenly by means of a quick and unexpected harmony which in its bright abruptness is like a flash of revelation; although in fact it is prepared for by long and slow incubation.
>
> (p. 266)

Bastick (1982) explores intuition by identifying a series of associated properties which he also relates to the term insight:

1   Quick, immediate, sudden appearance
2   Emotional involvement
3   Preconscious process
4   Contrast with abstract reasoning, logic or analytic thought
5   Influenced by experience
6   Understanding of feeling
7   Associations with creativity
8   Associations with egocentricity
9   Intuition need not be correct
10   Subjective certainty of correctness
11   Recentring
12   Empathy, kinaesthetic or other
13   Innate, instinctive knowledge or ability
14   Preverbal concept
15   Global knowledge
16   Incomplete knowledge
17   Hypnogogic reverie
18   Sense of relations
19   Dependence on environment
20   Transfer and transposition.

(p. 25)

Task 3.5.2 Intuition

Which of Bastick's properties do you relate to your practice as an artist?
To what extent do you think intuition is antithetical to critical practice?
Identify in your teaching, actions which you would ascribe to intuition.

Bruner (1960) contrasts analytical and intuitive thinking suggesting that they are complementary in nature:

. . . intuitive thinking characteristically does not advance in careful, well defined steps. Indeed, it tends to involve manoeuvres [sic] based seemingly on an implicit perception of the total problem. The thinker arrives at an answer which may be right or wrong with little if any awareness of the process by which he [sic] reached it. He rarely can provide an adequate account of how he obtained his answer, and he may be unaware of just what aspects of the problem situation he is responding to. Usually intuitive thinking rests on familiarity with the domain of knowledge involved and with its structure, which makes it possible for the thinker to leap about, skipping steps and employing short cuts in a manner that requires a later rechecking of conclusions by more analytical means whether deductive or inductive.

(p. 58)

Many educators at the turn of the century find the notion of intuition unhelpful because it is both vague and all-embracing. The properties that Bastick identifies can easily be confused with such concepts as the imagination and creativity, and Bruner's explanation bears a close resemblance to Polanyi's definition of tacit learning.

The ability of the mind to process information in ways other than cognitive reasoning is frequently overlooked or marginalised in the school curriculum. 'Intuition' 'is abused by ordinary people who want to avoid thinking, by philosophers who, without much examination, dismiss it as mere subjective hunch opposed to reason' (Reid 1986: 28). It is often used as a catch-all phrase to describe pupils whose learning does not easily correspond to the logocentric curriculum: pupils who may find themselves being diagnosed as having learning difficulties. We wish to propose a definition of intuition as a synthesis between the preconscious and unconscious processes identified by Bruner, Ehrenzweig and Polyani. Ehrenzweig advocated that Art & Design educators recognise the importance of unconscious processes for creativity: Bruner's reference to 'the implicit perception of the total problem' can be related to Ehrenzweig's 'syncretism' which he describes as pre-analytical; Bruner's reference to 'familiarity with the domain of knowledge involved' can be related to Polanyi's 'indwelling'; Bruner's reference to 'skipping steps and employing short cuts' equates with Ehrenzweig's notion of unconscious scanning.

3.6 LEARNING: LANGUAGE, STRATEGIES AND MOTIVATION

OBJECTIVES

By the end of this unit you should be able to:

- identify different methods and strategies to motivate learning in Art and Design;
- consider the role of language for learning in and through Art & Design.

It is evident from the diverse approaches and methods outlined in the previous units in this chapter that learning is a multi-faceted process. Within the limitations of the school timetable how is it possible to ensure a range of methods to differentiate for pupils' learning needs?

MOTIVATION

Motivation, or the will to do, is central to education: without it any learning is liable to be short term or superficial. Until the 1960s theories of motivation were dominated by two schools, the psychoanalytical and the behaviourist. For Freud motivation was conditioned by the tension between the primary, innate drives of survival e.g. hunger, sex, communication, and the secondary, rational processes of the ego which modify instinct in relation to social and cultural conventions. The emphasis here is on internal drives constrained by external factors, the social possibilities of satisfying the pleasure principle. Behaviourists suggest that motivation is essentially a reaction to external stimuli and that these provoke certain patterns of behaviour. Thus, habits of learning can be formed through such processes as repetition and imitation and are reinforced positively or negatively according to past experience. Since the 1960s these deterministic models have been layered and questioned by theories that promote more interactive processes. Social psychology proposes that an individual's personality interacts with variables such as class, race, geography, culture, so that motivation is conditioned by social environment. Cognitive psychology also recognises the social environment as a constructive element but emphasises the formation of knowledge through a dynamic interaction between it and an individual's innate cognitive processes. The educational implications of these general theories can be found in Capel *et al.* (1995: Unit 3.2 Table 1).

There are many ways in which you can motivate pupils to learn: the following list indicates some of the most effective:

- differentiate individual needs to enable self-actualisation;
- create a safe environment in which risk can take place;
- infect through your enthusiasm;
- communicate your high expectations to pupils;
- enable pupils to share and take ownership of ideas;
- develop critical reflection so that pupils can effect change;
- develop pupils' competence in subject-specific and transferable skills;
- provide pupils with constructive feedback;
- acknowledge successful learning through praise and positive reinforcement;
- catch pupils doing things right;
- recognise the significance of self and peer-group esteem;
- allow for the possibility of pleasure.

LANGUAGE FOR LEARNING IN ART & DESIGN

Learning in Art & Design mostly requires working with objects that communicate through non-verbal means. It is something of a paradox therefore, that you often have recourse to words in order to communicate clearly and efficiently. The mediation of language is central to a critical approach to learning and your command of verbal skills will not only benefit your teaching but your advocacy of the subject within your school and the wider educational community.

For Vygotsky (1986) the potential for learning is revealed and realised in inter-actions with 'more knowledgeable others'. One of his main contributions to the understanding of learning is the concept of the 'zone of proximal development' which refers to the gap between what an individual can do alone and what can be achieved with the help of 'more knowledgeable others'. For Vygotsky, the foundation of successful learning is cooperation and the basis of that success is communication and language. He believes that children solve practical tasks with the help of their speech as well as their eyes and hands. This includes inner speech, thinking things through for yourself, as well as explaining to others.

The recognition of the important role of language for learning in art is well established. Field (1970) insists that art is more than making; it is also about appreciat-ing. He notes that art educators find it extremely difficult to grasp the 'inwardness of the aesthetic experience mode' without experiencing it for themselves, without 'making' art. Field goes on to insist experience is not enough. Art & Design education should not merely enable pupils to 'think' in the subject, to approach it from within, but also help them to see it in a wider context. To do this a pupil needs to learn to become articulate 'to be able to discuss the nature of art experience, the criteria for art, the purpose of art' (p. 111). Field believes this must go beyond private conversa-tions between pupils and their teachers to the formation of a common language, a language which pupils must learn to apply in formal discussion.

Right from the start, you need to consider the language you use in the classroom and find a means to communicate clearly without diluting the specificities of the subject. The NC Order 2000 provides a framework. However, what at first sight may seem clearly defined is open to interpretation.

Task 3.6.1 Interpreting the language of the National Curriculum

In small groups, define NC terms or phrases.

In your tutor group, compare the explanations. Try to come to some consensus.

Write down your agreed terms and consider whether they are accessible to a KS3 pupil. If not, 'translate' them. Test their efficacy in the classroom: record pupils' responses and usage.

Try to be consistent in the way you use and explain NC terms. You can determine whether the 'official' language has been assimilated by pupils when you analyse the language they use to evaluate their own and others' work. You must begin from this critical point of view and expect your pupils to do likewise. Encourage them to question at all times. If they have not understood your explanation do not dismiss their concerns but work with them to construct definitions.

DEVELOPING LANGUAGE

SCAA (1997a) *Art and the Use of Language* recommends that art teachers consider some of the ways that work in art, craft and design develops speaking, listening, writing and reading skills, alongside ways in which language can enhance an understanding of the subject. The emphasis here is on transferability, a one way process. Eisner (1998) has pointed out the dangers of justifying Art & Design by the way it serves other subjects. He warns teachers against the temptation to defend the subject in ways that indicate its subservience to core skills. You need to consider the reciprocal relationship between word and image, number and artefact. Literacy and numeracy can serve art just as effectively as art can serve them.

Osbourn (1991) has suggested that learning to respond to visual stimuli is 'a two way process in which language is used to communicate perceptions and understandings while perceptions and understanding serve to stimulate language' (p. 33). She challenges a traditional view of art appreciation, one that involves a 'silent viewer, cut off from reality and wrapped in his or her own thoughts' (*ibid.*). Instead she insists that responses to works of art should be discussed and debated and that by encouraging pupils to do this you develop their ability to discriminate, analyse, scrutinise, interpret and communicate understanding.

Forming questions

You have already introduced pupils to a subject-specific language and explained and or modified it depending on their responses. Use it consistently in your introductions, demonstrations, class discussions and reviews. You need to reinforce these terms by highlighting key words and phrases in your planning and class reference materials and by using them in oral and written communication, including assessment. If you use unfamiliar jargon or a laissez-faire approach to questioning, pupils can feel threatened. It is therefore necessary to prepare questions carefully, aligning them to the learning objectives you have identified for your pupils (Capel *et al.* 1999: Unit 3.1).
Questions can be defined as:

- open: open to experience and interpretation e.g. 'How do you think women are represented in advertising?'
- closed: there is only one answer, often factual e.g. 'What material is this sculpture made from?'
- pseudo: there is only one answer and it is known to the questioner e.g. 'What is my name?'
- framed: open to interpretation within a given framework e.g. 'Among the contemporary crafts-work we have investigated which artefact bridges the fine/applied art divide most effectively?'

When instigating discussion and debate or when questioning pupils about their own and others' work teachers often use questions such as: 'What do you think?' 'What can you tell me about this?' or 'How does it make you feel?' (Meecham 1996:

72). The pupils' response is usually an unqualified value judgement: 'I like it', or, 'It's rubbish'. The further question, 'why?' can yield embarrassed inarticulacy or evasion.

> Given particular subject matter or a particular concept, it is easy to ask trivial questions or to lead the child to ask trivial questions. It is also easy to ask impossibly difficult questions. The trick is to find the medium questions that can be answered and that take you somewhere.
>
> (Bruner 1960: 40)

Initially, most pupils respond more readily if questions are concise, focused and related to their own experience and prior learning. These do not have to be questions of known 'fact' (pseudo) but should encourage pupils' descriptive, analytical, deductive or speculative skills. Open questions allow for individual, subjective answers. These have their function, but only once pupils feel confident that their opinions will be taken seriously and used to extend the debate. However, on some occasions it may be necessary for you to challenge pupils' preconceptions. You should avoid being confrontational. Refer to Chapter 10 for examples of the types of question you can use to investigate works of art; refer to Chapter 7 for examples relating to the evaluation of pupils' work.

Enabling discussion

To promote a critical curriculum in Art & Design it is essential that you develop the classroom as a site for dialogue, a place where talk is productive and meanings negotiable. Small-group discussion is likely to facilitate exploratory talk, particularly when you are introducing new concepts. In contrast, whole-class discussion may sometimes be intimidating and inhibits pupils from contributing. You need to acknowledge that making meaning through talking takes time, just as when writing, several drafts are required. Initial discussion is likely to be searching, and can include half or incomplete sentences, awkward syntax, even muddled or contradictory ideas: this is particularly noticeable when pupils are asked to provide personal responses or information. It is important as a teacher that you recognise and value such hesitant, tentative, half-formed ideas. The teacher does not always have to know the solution. In Art & Design there does not always have to be a right answer. You need to provide pupils with the opportunity to discuss and test out new concepts in the safety of small peer or friendship groups before they feel confident enough to present ideas in a wider forum.

Task 3.6.2 Concepts and key words

Read the activities extracted from a year 9 SoW (below) devised by student teachers.
Identify and record the underlying concepts that inform each activity.
Identify and record key words which help to explain the concepts.

Scheme of work: beyond weaving

Aims

- pupils consider and define craft practice;
- pupils experiment with and develop weaving techniques in relation to three-dimensional portraiture;
- pupils respond to given materials;
- pupils use assessment and evaluation to develop work.

Activities

Research and investigation

- brainstorm craft practice;
- discussion around selected representational images;
- introduce weaving and construction materials and techniques;
- in groups investigate initial responses and experiment with materials by categorising, comparing, contrasting and considering their representational possibilities;
- introduce sketchbooks as a research resource;
- homework: select found materials to incorporate in weaving.

Developing ideas through making and evaluation

- experiment further to identify materials suited to a representational function;
- construct a framework for three-dimensional weaving;
- begin weaving;
- pupils review their work by assessing it with a partner and evaluating it in their sketchbooks;
- the teacher introduces further resources and contextual information each session, inviting pupils to question boundaries between fine art, craft and design;
- as the SoW progresses pupils present their work to larger groups: final critique in the form of a whole class evaluation.

(PGCE students 1998)

Task 3.6.3 Forming questions

Having completed Task 3.6.2 construct questions to:

- ascertain pupils' prior knowledge in relation to the SoW;
- encourage pupils to discuss the concepts;
- review pupils' understanding as the SoW develops;
- evaluate their work.

BRAINSTORMING AND CONCEPT MAPPING

Brainstorming is usefully deployed as a whole-class or group activity in order to explore ideas for practical and critical investigation. It involves choosing a key word to represent a theme or issue. A key word is presented to pupils as a trigger to invite related ideas. It should be a quick activity, almost a word association game, so that ideas come thick and fast. The key concept is written in the centre of a flip chart and associated words recorded around it: this is often referred to as a spider diagram (Figure 3.6.1). Where one association derives from a word other than the key, it is connected to its originating term. You may find that initial responses take the form of clichés and stereotypes: these should be acknowledged but you can encourage pupils to extend their frames of reference by repeating the activity without duplicating words from the first version.

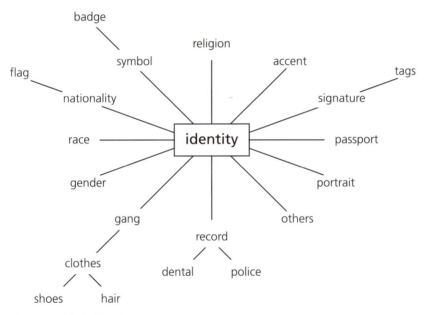

Figure 3.6.1 Spider diagram

Buzan (1982) provides a useful practical guide to these and similar processes although it is not Art & Design specific. Creber (1990) discusses Buzan's methods: '[brainstorming] offers what Witkin might call "a holding form" without the interference caused by a premature inner demand for explicitness and order in what should be the crucial, thinking, stage of the operation' (p. 49).

Concept mapping takes this process a stage further by grouping words into categories and defining them in terms of priority and possible developments. Figure 3.6.2 groups the associations made in Figure 3.6.1 into four categories. This grouping helps pupils identify an area for investigation. They might consider why some terms fall into more than one category and why others are closely identified with just one.

Group	Personal	Records	Signs
others	gender	passport	flags
nationality	shoes	dental	badge
religion	tag	police	tags
gang	signature		signature
race	clothes		symbol
gender	hair		shoes
shoes	passport		clothes
badge			hair
passport			
clothes			
hair			

Figure 3.6.2 Identities

Task 3.6.4 Brainstorming and concept mapping

In pairs, brainstorm the theme 'sense of place'.
   Exchange your diagram with a neighbouring pair and translate it into a concept map.

Task 3.6.5 The role of language in Art & Design

Consider the following quotations from artists, critics and educators concerning the complementary and conflicting role of language in Art & Design education.
   Which one most succinctly represents your view?
   Which one do you find most problematic?
   Compare your thoughts with other students.

The initial appraisal of images is therefore a complex cognitive process which integrates the visual and the verbal questioning in such a way as to contribute extensively to the development of cognitive skills. The cognitive value of instruction in the arts is seriously underestimated. By promoting talk and thereby making understandings and responses more public, this view can be effectively changed.

(Osbourn 1991: 54)

The published statements of artists are often found to be elliptical, contradictory, evasive and rhetorical . . . yet it is rare to find an inarticulate artist . . . there is a way in which the artist too is an onlooker, a beholder of his or her work, and suffers, no less than any other member of the audience, from the problem of defining just what has been thought, made and achieved. A mixture of fantasy and conjecture, anecdote and metaphor are the likely companions of a working process which is as much concerned with concrete things as it is with words. It is often the case, too, that the interpretation and the theoretical position of the critic, the interlocutor, is at some variance with what the artist believes; they stand at different sides of the artwork. The work itself – seeming all too often in danger of disappearing under a superfluity of words – is where the artist and audience meet. This is where the discussion begins.

(Searle 1993: 1)

Language and visual communication are very different. Yet they are part of the same culture and it is ultimately the culture that limits what can and cannot be meant. So often they can express the same meanings but in different ways . . . different periods, different cultures and different social contexts within the same culture and period, can assign different roles to the verbal and the visual, giving them different jobs to do, and they can also value them differently, assigning them a different importance in the semiotic scheme of things.

(Kress and Leeuwen 1996: 102–103)

I think of words as an invisible material; manipulating punctuation is like focussing the attention on certain points in the painted surface of a pot.

(Britton 1991: 4)

Turning art experience into language also carries its dangers. Words are prisons as well as searchlights and pigeonholes, for what we see and comprehend. So a vocabulary which is provided by the teacher rather than invented by the pupil may constrain and regiment our seeing and interpreting. The rate at which new and specialist vocabulary is added to pupils' existing personal vocabulary needs sensitive and teacherly monitoring, so that the elastic between the new to the familiar is never stretched to breaking.

(Stibbs 1998: 203)

3.7 ENABLING LEARNING: TRANSFORMING SUBJECT KNOWLEDGE INTO PEDAGOGICAL PRACTICE

This unit explores ways in which you can can reinforce, extend and develop your subject knowledge. It is important that you reflect upon your own learning in art, craft and design before attempting to transfer, wholesale, practice that may be inappropriate for learning in a classroom context. However, it is possible that, although your practice is seen as falling outside the orthodoxies of 'school art', it corresponds to the NC and examination learning objectives. You should consider the range of practice which the NC and the examination syllabuses endorse and how your own practice can inform and develop it.

> **OBJECTIVES**
>
> By the end of this unit you should be able to:
>
> - recognise the origins of existing orthodoxies their uses and limitations;
> - consider ways to transfer knowledge and understanding into good pedagogic practice;
> - investigate methods and strategies and consider ways to use them in your teaching.

The following list, although not exhaustive, highlights productive and receptive practices which have become marginalised or neglected in Art & Design:

Productive:

- audience and site-specific problem solving
- visualisation and memory training – mind and word picturing
- metaphor and metonym
- metaphor and material exploration
- synaesthesia
- associative, automatic and chance processes

Receptive:

- comparison
- semiotics

You are asked to question the current validity of these practices and consider their application to the curriculum.

AUDIENCE AND SITE-SPECIFIC PROBLEM SOLVING

When you direct pupils' attention to the needs of others you help them to understand the social functions of art, craft and design. This encourages pupils to consider the relationship between their own needs, tastes and preferences and those of potential audiences. For example, setting the task of designing packaging for household goods intended for the visually impaired ensures that pupils investigate the haptic, tactile and spatial properties of materials.

For a site-specific task you can ask pupils to design a playground for pre-school children. To do this pupils have to take into account: materials, safety, ergonomics, economics, aesthetics, etc., survey local playground provision and identify the needs of users, both children and adults. Following this research, they feedback findings before brainstorming possible solutions. You can differentiate tasks and allocate them to individuals and small teams. More ambitious projects of this kind might involve pupils working with designers, including work placements, or with community artists on funded public art initiatives. A project of this type, to develop a crazy-golf course, attracted design submissions by such well known yBas as Gavin Turk and Damien Hirst (Millar 1998: 8). Pupils could be asked to develop their own design solutions and then compare and contrast their efforts with professional schemes.

Task 3.7.1 Site-specific SoW

Devise a SoW in which pupils design and construct a public monument (or model) for their school grounds. The monument should commemorate a significant event in the history of the school or local community.

VISUALISATION AND MEMORY TRAINING

Working from observation is one of the major orthodoxies in British secondary schools. This has its roots in a number of traditions: the academic, with its emphasis on emulating canonic exemplars; the perceptual, in particular Ruskin's advocacy of working directly from nature, and vocational education with its reliance on copying to facilitate technical competence. However, there is an alternative tradition that promotes the practice of working from memory. Horace Lecoq de Boisbaudran (1862) wrote a treatise propounding a method that was influential for Whistler's painting. His friend, T. R. Way, recalled the significance of this approach:

> I shall never forget a lesson which he gave me one evening. We had left the studio when it was quite dusk, and we were walking along the road by the gardens of Chelsea Hospital, when he suddenly stopped, and pointing to a group of buildings in the distance, an old public house at the corner of the road, with windows and shops showing golden lights through the gathering mist of twilight, said, 'Look!' As he did not seem to have anything to sketch or make notes on I offered him my notebook, 'No, no, be quiet,' was his answer; and after a long pause he turned and walked a few yards; then, with his back to the scene at which I was looking, he said, 'Now see if I have learned it', and repeated a full description of the scene, even as one might repeat a poem one had learned by heart. Then we went on, and soon there came another picture that appealed to me more than the former. I tried to call his attention to it, but he would not look at it, saying, 'No, no, one thing at a time'. In a few days I was at the studio again, and there on the easel was the realization of the picture.
>
> (Spencer 1989: 106)

Dependence on working directly from the object, graphically or by using a camera, can blunt the power of the memory to recall sights and filter out incidental details. As you devise programmes for image making you are advised to consider the difference between recording from observation, with its scientific and documentary credentials, and recording from memory, which is more selective. Your teaching, should include opportunities for both to be developed.

In the sphere of children's education Marion Richardson explored methods of visualisation that depend, to some extent, on memory training. She was particularly keen that pupils develop imaginative responses to different forms of stimuli:

> Mind-Picturing involved the learner closing the eyes and allowing images of any type – figurative, non-figurative, ornamental, etc. – to appear in the 'mind's eye' whereas Word-Picturing consisted of carefully worded descriptions of actual events or paintings recalled by Richardson, or poems read by her, acting as stimulants for pictorial work. As in mind-picturing, the product was not predetermined, indeed the word-picture might or might not relate to the described image.
>
> (Swift 1992: 118)

The way in which mind-picturing is described here suggests a process similar to a sort of willed daydreaming in which memories, in the form of images, are allowed to flood the mind. Such images can either be recorded immediately, using spontaneous methods, or developed over time using premeditated processes. In word-picturing the learner has to recall or imagine forms associated with or equivalent to given text.

Task 3.7.2 Using memory and imagination

Recall and record; the front elevation of your home, the face of a 'significant other'.

Attempt to conjure images using mind-picturing techniques: what materials have you made available to record your sensations/memories as they occur?

Choose a brief text, song, piece of music or smell as a stimulus and apply word-picturing techniques.

METAPHOR AND METONYM

Metaphor and metonym are linguistic terms which denote a process of explaining one thing by relating it, respectively, to something similar (similarity in difference; her fortitude is explained with reference to an oak), or, something with which it is associated (a connection or attribute; throne stands in for royalty). Kress and Leeuwen (1996) suggest that all representation is a process of metaphoric transformation:

> The process of sign-making is the process of the constitution of metaphor in two steps; 'a car is (most like) wheels' and 'wheels are (most like) circles'. Signs thus result from a double metaphoric process in which analogy is the

constitutive principle. An analogy, in turn, is a process of classification: X is like Y (in criterial ways). Which metaphors carry the day and pass into the semiotic system as 'natural', neutral classifications, is then governed by social relations. Like adults, children are ceaselessly engaged in the construction of metaphors. Unlike adults they are, on the one hand, less constricted by culture, by already existing metaphors, but, on the other hand, usually in a position of less power so that their metaphors are less likely to carry the day.

(p. 7)

Visual and material metaphors and metonyms are analogous to linguistic ones in that they are representations: an image of a basket of fruit stands in for fertility; an abundance of hair, animality. Traces of the making process can also be recognised as metaphoric/metonymic: erratic mark-making can stand in for agitation; associations can be invited between applied and rendered surfaces and attractive or repellent experiences. Metaphoric processes, while layering and enriching an idea, 'can simultaneously grasp the familiar and make it strange' (Davison and Dowson 1998: 253). Hercules' lion-skin cloak reinforces his physical strength but the neutral garb of a Victorian angel obscures his biblical gender.

The following task suggests ways for you to explore the metaphoric potential of the properties of materials without any preconceived outcome in mind (see Plates: 1, 2, 3, 4).

Task 3.7.3 Exploring metaphor through material resources

A scrap bank has provided you with: rolls of telephone wire, damaged sponges, different sizes of cardboard tubes, skeins of wool, buttons, hooks, eyes and zips. These examples suggest types of material that might come your way.

In tutor groups collect a similar array of found materials and work through the following exercise:

Consider each material separately: what do they signify by their look? what do they signify by touch? what do you associate with them? Record your sensations and thoughts.

Place the materials in pairs and other multiples: follow the same process; see whether these combinations suggest different meanings.

Investigate each of the materials individually, testing their resiliance, durability, malleability, etc.

Explore the materials in different combinations so that they interact, e.g. penetrate, envelop, entwine, intervene by cutting, distressing, piercing, etc. Record your associations.

Use these experiments as a basis for designing and constructing a garment or body adornment which is intended to either attract or repel an identified audience.

Devise a similar task or adapt this one for use with GCSE pupils.

In the above task the focus is on recyclable materials collected from scrap schemes and resource banks. These are sites where industry offloads surplus materials. For an annual fee schools can have regular access to these materials as long as they can provide their own transport. The result of this can be that you have a selection of arbitrary materials that do not obviously relate to your existing SoW. The task should enable you to take advantage of this situation to help pupils learn.

SYNAESTHESIA

Synaesthetic approaches are concerned with the correspondence between different forms of sensory experience, the way that sights, sounds, surfaces, tastes and smells relate, or seem to relate, to one another. Synaesthetic responses are types of cross or multi-sensory metaphors in which one sense experience can suggest or even realise another: for example, a particular colour, deep purple, finds its metaphoric equivalent in a low chord on a cello, and the feel of thick velvet. The way young children learn about the world is mostly through sensory exploration and they begin to form associations between different sense experiences:

> Size, colour, texture, temperature, weight and plasticity are all aspects of the objects which a child can sensorily enjoy . . . This sensory awareness and discrimination in a young child is the basis of an intelligence which helps the child to bond to the earth and all the perplexing variety of sensations which life provides.

> (Gentle 1988: 37)

This notion of correspondence became very important to artists in the nineteenth century. In their respective fields Gautier, Baudelaire, Wagner and the Symbolists experimented with synaesthetic evocation and affect, Seurat with the relationship between angles and vectors and their related colours; Whistler gave his paintings musical titles; Debussy provided his Preludes for piano with pictorial ones. For many artists in the twentieth century, synaesthetic experiment was a key method for developing 'non-referential' modes. Kandinsky describes his synaesthetic theories amid the dubious metaphysics of his spiritual ruminations (Chipp 1968: 152–155) and more recent Abstract artists such as Gillian Ayres have continued such experiments. Synaesthetic correspondence can be one way to make sense of the ambiguous combinations of image and material in 'surrealist' art, with its deployment of automatic processes and strategies of defamiliarisation: from Ithell Colquhoun's Sea-Star 1 (1944) to Cathy de Monchaux's Evidently Not (1998) (Plate 21). Film, and more recently video, have been ideal vehicles for synaesthetic expression, although the emphasis on the spoken word in the commercial cinema has limited its development. However, pop video and multimedia CDs provide an ideal outlet for synaesthetic correspondence, for example, that between visual and musical rhythm or colour/graphic fields and ambient sounds. To what extent these correspondences can be seen as universal, culturally specific or personal and contingent is open to debate: for instance how might Helen Chadwick's Chocolate Fountain (1994) encourage synaesthetic responses?

1   Devise exercises for your pupils which explore the relationships beween sound, shape and colour, texture, smells and colour, or any other combination.

Pupils can extend these exercises by exploring the correspondences beween sense experiences and other affective states such as moods and emotions. If you or your pupils desire, representational elements can be introduced.

Compare and discuss the results with fellow students.

2   Devise a lesson introducing the concept of synaethesia to a KS3 class to help them respond to a text, a memory or a sense experience.

3   Consider how you might introduce synaesthetic elements into an animation project using multimedia.

ASSOCIATIVE, AUTOMATIC AND CHANCE PROCESSES

An immediate and accessible way to motivate pupils, particularly those who feel insecure in their technical abilities, is to introduce automatic and semi-automatic techniques as a starting point. We do not propose this strategy to subvert analytical procedures and aesthetic practice in a nihilistic way like some of the Dada artists, but more in the manner of the Surrealists for whom chance processes acted as a method for freeing up the imagination and loosening the hold of codes and conventions. As early as the fifteenth century Leonardo da Vinci in his Treatise on Painting suggested that artists look at chance configurations on stained walls to conjure images. Even as controlled an artist as Degas is said to have been frightened by fresh sheets of paper and he would stain them with coffee grounds before beginning a drawing.

Ernst often used rubbings, the process of 'frottage', to form chance or arbitrary configurations into which he would see or project images: his Histoire Naturelle (1925–26) is entirely dependent on this process. Wolheim (1987) discusses the concept of 'seeing in' theorising its central place in human perception:

> Seeing–in as I have described it, precedes representation: it is prior to it, logically and historically. Seeing–in is prior to representation logically in that I can see something in surfaces that neither are nor are believed by me to be representations . . . I can, for instance, see headless torsos in clouds ranged against the vault of the sky. And seeing–in is prior to representation historically in that surely our remotest ancestors engaged in these exercises long before they thought to decorate their caves with images of the animals they hunted.
>
> (pp. 47–48)

Whether seeing-in precedes representation or is a type of transformative mental activity that is itself a form of representation, is open to debate. What is significant is that seeing-in appears to be an innate faculty and is therefore an accessible method for all to use. Inviting pupils to apply this faculty to their immediate surroundings can convince them that even the most prosaic and banal of environments may hold the potential for imaginative transformation. The school environment can be used as a resource to stimulate imaginative responses beginning with the rubbing of surfaces such as wood-grain, stone and brick. The process of seeing-in can be applied to any circumstance. Ithell Colquhoun's view of herself lying in a bath (Chadwick 1991: 104), transformed from a nude into a rocky but nonetheless gendered seascape, is an amusing and troubling work, the process of which many pupils will both readily understand and be able to apply to their own work.

Another favoured method practised by Dominguez, Ernst, Leonor Fini and Ithell Colquhoun was decalcomania (Chadwick 1991). You can follow this process by spreading paint or other pigments over a non-porous surface such as glass or perspex. You then press either paper, card, canvas, etc. against this surface, pulling it off to reveal a tracery of interconnecting lines. The resulting forms are not unlike the veins on a leaf or certain fan corals. These forms can be used either as stimulus for 'seeing in', as with Ernst's *Europe after the Rain* (1940–42), or as discrete elements for use in a collage or decorative schemes.

A more fully automatic process is the common practice of doodling; the type of drawing you produce when your conscious mind is engaged in another activity such as phoning a friend. Masson, Dali, Fini, Matta, Bacon, are particularly noted for using this process, although, once again they use it as a stimulus for seeing-in. The Abstract Expressionists developed a doodling-like process as a means of bringing to the surface and realising sub- and unconscious forms and images, although, at its limits, this strategy was used by artists like Pollock to subvert the figural potential of painting, to bypass representation (Clark 1999: 299–369). In schools the wholesale imitation of the 'look' of such work, without the contexts and motivations of its making, is an absurd spectacle. Such strategies may however provide pupils with ways of working that they can use to make their own metaphors.

An engaging Surrealist practice, the 'exquisite corpse' is more familiarly known in the UK as the party game 'consequences'. Recall this game and relate it to the Surrealist activity. The sort of chance encounters that are the result of this process can be similarly activated by choosing images at random and juxtaposing them to form unlikely, amusing, contradictory or troubling combinations. This can be developed as a more conscious activity and used to construct meanings which require the viewer to make or seek connections which may not, at first sight, appear logical or coherent. Man Ray and Meret Oppenheim produced notable objects which subvert the utility of domestic objects, respectively Object to be Destroyed (1923) and the Fur-lined Object (1936). Ernst conjured unexpected, dreamlike images using photomontage. Hannah Hoch brutally satirised bourgeois conventions using popular sources to subvert its proprietous tastes. Heartfield juxtaposed contrasting and opposing images to undermine the credibility of Nazi ideology (Ades 1976). More recently artists have taken up the decentring and dislocation of Dada photomontage and the Surrealist object to construct phobic and abject installations and objects, e.g. Louise Bourgeois (Bernadec 1999), Robert Gober (Nesbitt 1993), Damien Hirst and Sarah Lucas (Rosenthal 1997).

Task 3.7.5 Chance and automatic techniques

Try out some of the chance and automatic techniques cited.
Go through the process of elaborating the outcomes by seeing-in.
Devise a SoW which incorporates one or more of the techniques as a starting point.

COMPARISON

Comparative methods can prove useful in engaging pupils' attention_and inviting them to investigate the work of others (Dyson 1989: 129–132). You may wish to set up a comparative task by presenting pairs (or more) of reproductions designed to focus on points of similarity and difference. For example, Dyson proposes six types of comparison:

- art objects/everyday objects;
- different art objects with the same subject matter;
- pupils' own work/appropriate art objects;
- artefacts of different periods;
- objects, texts, etc., of the same period;
- art objects of a particular school or period.

Task 3.7.6 Making comparisons

In your tutor group: think of additional combinations and their implicit questions.
Note these and decide in what context and for what reason (learning objective) you would use them in a lesson.

Chapter 10 develops these ideas and provides additional methods.

SEMIOTICS

Semiotics, or the science of signs, is increasingly recognised as an important field for Art & Design education. In recent years it has been associated more with theories of communication (Fiske 1982) and in schools, media studies (Buckingham and Sefton-Green 1994). However most critical practice in art, craft and design can be recognised as a form of semiotic inquiry because it involves processes of analysing and decoding visual and other sensory signs. Although increasingly there is good critical practice in Art & Design it is often dependent on individual teachers' interest and training and can lack a coherent theoretical basis. Kress and Leeuwen (1996) have formulated a method for analysing the objects of visual culture and claim that any two-dimensional (2D) image from the west can be analysed by applying their system. Although the authors emphasise the differences between visual and verbal modes of communication their method is analogous to linguistic models. However, it manages to avoid the hierarchical dangers of some art historical investigations because it treats all manifestations of visual culture in the same way; it does not suggest a different method for investigating a special 'expressive' practice called art. Some critics have suggested that while Kress's theory of social semiotics is convincing when applied to media images, advertising, illustrations and to a lesser extent painting, it neglects three-dimensional forms, craft and most contemporary art practice. You should always approach universalising methods with caution but this book is well worth investigating.

See Chapter 10 for a development of this issue.

4 Curriculum Planning

Lesley Burgess and
David Gee

INTRODUCTION

What is a curriculum?
What constitutes an effective Art & Design curriculum?
What do I need to do before discussing the curriculum with colleagues in my placement schools?

This chapter introduces you to ways of planning an Art & Design curriculum for pupils at key stage 3 (KS3). A brief examination of the development of the National Curriculum (NC) Art Order helps you to identify the implications for teaching and learning as prescribed in the NC Art Order 2000.

This chapter is designed to help you look critically at curriculum planning and consider a number of methods used by secondary-school art departments to decide the content and delivery of the curriculum. You are introduced to some key concepts to build your understanding. This enables you to contribute to existing schemes of work (SoW) and devise new ones.

You are encouraged to identify, understand and, where necessary, challenge existing orthodoxies by developing SoW and lesson plans to promote ways of working which give full consideration to pupils' prior learning, interests, and their social and cultural capital.

Without the support of an agreed curriculum, planning SoW can be a bewildering and isolating experience. Sharing curriculum planning with others, and knowing what and how you can contribute through your teaching, empowers you in the classroom and gives a clear message to your pupils about your purpose and professionalism.

OBJECTIVES

This chapter should help you to:

- develop a working definition of the curriculum;
- become conscious of the variety of aims and values informing the curriculum;
- understand how individual departments plan their curriculum;
- contribute to the curriculum when planning SoW, lessons and homework;
- contribute to the development of a pluralist and fair society through the curriculum.

4.1 WHAT IS THE CURRICULUM?

OBJECTIVES

By the end of this unit you should be able to:

- define the term 'curriculum';
- examine the development of the curriculum since the Education Reform Act (ERA) 1988 to understand the development of NC Art, programmes of study (PoS), SoW, and lesson plans;
- develop SoW and plan lessons in which the learning objectives are carefully formulated and differentiated to take into consideration the specific needs of the pupils involved.

Put simply, the curriculum is a plan or framework for that which is taught or learned. Pupils gain access to it through the ways they are taught and the conditions in which they learn.

The ERA (1988) introduced the NC to provide a 'common learning experience' for the majority of pupils who attend state schools. PoS describe the 'matters, skills and processes' which are required to be taught for each subject. Teachers use these to plan the content of SoW and individual lessons.

The curriculum in its broadest sense includes all the conscious and unconscious influences on pupils' learning. Learning does not take place in a vacuum. The ethos of the school, its rules, regulations, shared values, the individual beliefs and interests of teachers all form part of the pupils' curriculum. Schools plan some of these aspects through agreed aims and whole school policies; however, increasingly these are dictated by government agencies: the DfEE, TTA and QCA. As Steers (1998) points out, 'demands for greater accountability from the teaching profession lead inexorably to ever tighter control, if not specification, of the curriculum and its assessment and, through these mechanisms, to control of teachers' (p. 2).

Since the introduction of the NC teachers have found it increasingly necessary to discuss and plan their delivery of the curriculum and share responsibility for devising and resourcing SoW. Departments are required to record and review these to provide evidence that they are managing resources effectively and striving to raise pupils' achievement.

When taking on responsibility for classes it is useful to refer to previous SoW or curriculum frameworks in order to develop what has gone before and avoid needless repetition. It is dispiriting to hear, 'not portraits again', as a result of you not knowing what has been taught before. Consulting written records also helps you to identify the previous learning objectives and ensure some continuity and progression of learning.

WHERE CAN I FIND THE CURRICULUM?

The following list identifies some of the places where you can find evidence:

- NC;
- the Art & Design department's aims (spoken and written);
- SoW;
- lesson plans;
- archives of pupils' work;
- selections of resources/equipment/materials;
- Art & Design department and whole school policies, e.g. homework; teaching methods and styles;
- department/school ethos;
- collective memory of department members;
- range of colleagues' subject knowledge/specialisms;
- department discussion/meetings.

Task 4.1.1 Where can I find the curriculum?

Ask a number of Art & Design teachers what the 'curriculum' means and where it can be found. Check their replies against the list provided above. Add to the list where appropriate.

THE NATIONAL CURRICULUM ART ORDER

Introduction

The publishing of PoS in 1995, their subsequent revision in 2000 from two discrete yet interdependent attainment targets (ATs) to one, has obliged all state schools in

England to reconsider how they plan, monitor and assess both what is taught and learned. Most departments endeavour to provide a balanced programme of art, craft and design work with knowledge and understanding of artists, craftspeople and designers developed along with practical work. Schools differ in the way they plan and teach. This is often determined by the range of specialists within the department, different teaching styles, prior achievement and learning needs of pupils, locality of school and availability of resources. However, many educators including Meecham (1996), Cunliffe (1996), Hughes (1998a) and Steers and Swift (1999) suggest current practice based upon the NC Order rarely challenges the prevailing orthodoxy or 'school art' which prioritises a perceptual or formalist, modernist approach.

> As we move towards the Millennium, we are still delivering art curricula in our schools predicated largely upon procedures and practices which reach back to the nineteenth century – practices and procedures which cling to the comfortable and uncontentious view of art and its purposes. As a result, secondary Art & Design education in England and Wales is, in general, static, safe and predictable.
>
> (Hughes 1998a: 41)

It is your responsibility as a student teacher to ensure you understand the rationales underpinning existing practice so that you can both value and build upon its strengths, recognise its limitations and identify areas of potential development.

Task 4.1.2 Planning SoW

- Find out how SoW are planned and recorded in Art & Design departments in your placement schools.
- If the departments work to set schemes, find out when, how and who compiled them.
- Find out how and when existing schemes are reviewed and new ones introduced.

The PoS outlined in the NC Order covers the range of experiences and opportunities which schools should make available to pupils. It does not specifically set out the manner in which this is to be done, nor place them in any particular hierarchy. When planning SoW or lessons it is possible to use the PoS both to stimulate ideas for SoW and as a means of checking whether you are working within its framework. As you become more familiar with its terms, and can readily identify practical applications, it becomes less of a document and more of a conceptual and organisational framework on which you will build your practice.

A background to the NC Art Order

The proposals for the NC Art Order (DES 1991a) claim to provide a broad and flexible framework to enable teachers to develop their own SoW. It was developed by a working party representing teachers, academics, artists and industrialists who identified the main concerns and aims for Art & Design education. These were identified in the NC Art Working Group Interim Report (DES 1991b) as: 'visual, communication, aesthetic, sensibility, sensory perception, emotional and intellectual development, physical competence and critical judgement' (pp. 7, 3, 6).

The following aims were drawn up by the Art Working Group and formed the basis of the PoS:

We take the view that from age 5 to 16 art education should:

- enable pupils to become visually literate: to use and understand art as a form of visual and tactile communication; to have confidence and competence in reading and evaluating visual images and artefacts;
- develop particular intellectual and technical skills so that ideas can be realised and artefacts produced;
- develop pupils' aesthetic sensibilities and enable them to make informed aesthetic judgements in art and design;
- develop pupils' design capability;
- develop pupils' capacity for original thought and experimentation;
- increase pupils' capacity to enjoy and value the visual, tactile and other sensory dimensions of the natural and made environment;
- develop pupils' ability to articulate and communicate ideas, opinions and feelings about their own work and that of others;
- develop pupils' ability to respond thoughtfully and critically to ideas, images and objects of many kinds and from many cultures.

(ibid.: 13, 4.1)

Task 4.1.3 The National Curriculum aims for Art & Design education

In your tutor groups consider the following:

- are all the aims equally important?
- are some aims more relevant to pupils at KS1 and 2 than KS3?
- how do you prioritise the aims in relation to pupils at KS3?
- what are the implications for planning KS3 SoW?

Task 4.1.4 Reflecting on the aims of your own education

Speculate on and make a list of the aims which governed your Art & Design education. Compare your list with the Art Working Group's.

THE LEGACY OF RECORDING FROM OBSERVATION

High expectations were placed within schools by the Art Working Group (DES 1991b) on the role of 'observation and recording of visual images'. It was thought that this would enable pupils to express feelings and emotions, transform materials into images and objects, plan, visualise and design. Recording, and especially drawing from primary and secondary sources, was, and still is, a major feature of pupils' Art & Design education. It is promoted as the primary means for developing visual communication. Recording also has a part to play in the acquisition of formal, technical and conceptual languages, as well as the assessment of technical and critical skills. This view is reiterated in the NC Art Order (DFE 1995) and the NC Art Order 2000.

Task 4.1.5 Observing and recording visual images

Identify the extent to which observing and recording visual images plays a part in the SoW in which you have been involved or observed.

Discuss other ways of recording and communicating ideas and feelings and how these could be incorporated within a scheme of work. Refer to contemporary practice and a range of cultural/artistic traditions.

EMBEDDED VALUES WITHIN THE ART ORDERS

The earlier NC Art Order (DFE 1995) was written in a particular place by a particular group at a particular time and cannot help reflecting attitudes and values which may to some extent have shifted. The PoS and attainment targets depend, for their coherence, on shared assumptions about child development and the role of art, craft and design in education and beyond. These assumptions are culturally determined and may be different for other societies and indeed, within them, perceived differently by various groups.

The UK is often construed as a liberal democracy. Within this ideological framework childhood is seen as a time for self-discovery and development. The 'self' is seen as autonomous and generative and if provided with an appropriate environment 'naturally' is motivated to create, construct images, experiment with materials and seek out stimulus and challenges. Evidence for this is easy to find through observing very young children at play who become stimulated by drawing, painting and assembling things. Most activity of this kind is assumed to be performed for intrinsic reasons: adults only become involved when they show interest or intervene.

THE FORMATION OF THE ART & DESIGN CURRICULUM IN THE EARLY YEARS

From the first occasion when adults assist or direct activities or attempt to provide meanings or interpret childrens' intentions, two sets of conflicting motives come into play.

Children's motives: intrinsic/inherent

- to seek sensory gratification;
- to seek mental stimulation;
- to explore the environment;
- to begin to construct personal meaning.

Adults' motives: extrinsic/instrumental

- to aid social awareness/development;
- to develop skills;
- to set and solve problems;
- to provide means of production/exchange;
- to transfer or exchange meaning.

Mixed motives!

The role of the curriculum is seen by many as a means of directing pupils' interest towards experiences which they can communicate and share with others. The arbitrary nature of subjective and personal experience is made coherent through developing a common language, e.g. formal, technical and conceptual languages must be systematically learned if skills, knowledge and understanding are to develop in a coherent way. In this way intrinsic and extrinsic needs are addressed simultaneously.

> **Task 4.1.6 What motivates you?**
>
> Reflect on your own art, craft or design activity and compare your attitude and motivation when working on personal projects with those prescribed by others, e.g. commissions, examination criteria, OFSTED.
>
> How can you include these motivational factors in your planning? Discuss with other students.

BREADTH OR DEPTH? THE DILEMMA FOR ART & DESIGN DEPARTMENTS

Arriving at a consensus about what a 'broad and balanced' Art & Design curriculum might contain is not easy. The sometimes conflicting needs of teachers, students and society have to be resolved. Whatever is stressed gives rise to criticism of neglect in other areas: if the focus is on measurable outcomes, the individuality of pupils can be overlooked or highlighted in a negative way; if too finely differentiated on the specific needs of individuals, it can be impossible to make valid comparisons between pupils or establish common expectations or standards.

With the limited number of periods for teaching the subject in most schools, departments are obliged to use their time wisely and think carefully about what they can or cannot include in their SoW.

They are often pulled in one of the following directions:

- to limit the range of activities materials and resources and rerun and refine existing SoW. In this instance pupils are motivated by developing specific skills in response to tried and tested, high-quality resources supported by exemplars;
- to provide a broad range of activities, materials and resources to promote experimentation and innovation. In this instance pupils are motivated by *novelty* and the opportunity to take responsibility for the development of their ideas.

> **Task 4.1.7 Breadth or depth? Real or false choice?**
>
> - Discuss the above models with Art & Design educators.
> - Which of the above directions best describes their practice?
> - Are the models adopted consistent or are they age/key stage related?
> - Identify and discuss alternative models.

The Manifesto for Art (Swift and Steers 1999) proposes 'a postmodern solution for postmodern situation' recommending a shift from the existing school art orthodoxy towards an art curriculum that promotes 'difference, plurality and independence of mind' with an 'emphasis on negotiation of ideas which arise from asking pertinent questions, and testing provisional answers, rather than seeking predetermined ones' (p. 1).

Consider the recommendations from the 1999 Manifesto (available from *http://www.nsead.org.uk*). How do they compare with your findings and deliberations to date?

PLANNING BY DEPARTMENTS

Departments vary in their ability to discuss and plan the curriculum together. This is often made easier where there is informal day-to-day contact, a difficulty when schools are on split sites or employ part-time teachers. Approaches to curriculum planning vary. The following examples characterise some alternatives, but are by no means exclusive.

Example department A

Teachers devise separate SoW, all using the NC as a guide. Each attempts to provide a broad and balanced programme independently.

Advantages:

- teachers are free to develop SoW which build upon their interests and areas of expertise;
- teachers take responsibility for broadening their own subject knowledge and are personally accountable for pupils' achievement;
- a variety of teaching styles is almost guaranteed.

Limitations:

- possible duplication of resources;
- static teaching styles: self-review may not be sufficient to encourage variety or development;
- isolation: lack of common themes or purpose, the department becomes 'invisible', lacks identity/presence; strengths of individuals are not maximised for departmental good.

Example department B

The department meets before each term or year and decides on a theme, either for each or all year groups, from which SoW are planned.

Advantages:

- themes provide both content and context for whatever material and process is used;
- resources can be shared and built up collectively;
- common themes provide the department with a strong identity within the school.

Limitations:

- an orthodoxy of method, style, delivery can establish itself: shared practice can become self-justifying and closed to internal or external criticism;
- Only those involved in the original discussion can fully claim ownership of the theme: much of the content of the discussion is not recorded and is thus inaccessible to other teachers.

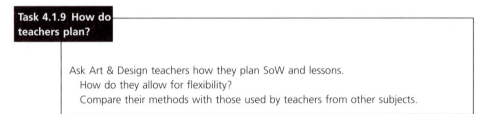

Task 4.1.8 Organising the curriculum

Study the examples provided above.
Identify the strategies your departments have adopted for developing their curriculum.
How do they relate to the given examples?
What alternative models can you identify (see Barrett 1982)?

A GUIDE TO PLANNING SoW

Introduction

The ability to plan, record and review schemes of work and individual lessons is increasingly regarded as a key requirement for effective teaching and learning.

Teachers vary in the way they plan: some record in detail the content, resources and methods to be used, others prepare simple check lists of inputs and activities. Obviously not everything that occurs during lessons can be planned for or predicted: indeed, pupils sometimes respond in ways very different from those you anticipate. Therefore, your planning should not be too rigid. Good planning 'builds in' flexibility and allows for serendipity. Planning provides both a structure and a set of objectives with which to guide events and also a means to measure your pupils' and your own performance.

Task 4.1.9 How do teachers plan?

Ask Art & Design teachers how they plan SoW and lessons.
How do they allow for flexibility?
Compare their methods with those used by teachers from other subjects.

Your first SoW

When planning SoW for the first time you need to consult with the subject teachers with whom you are working. Identify and discuss the area of the PoS you are expected to cover. It is useful to 'brainstorm' ideas with the subject teacher, school tutor/mentor or another student teacher. Conduct your research at a level suited to your classes. Remember, although you may be aiming for depth and some degree of complexity, you are planning for secondary-school pupils; you need to give careful consideration to both the 'pitch' and the 'pace' of your projects.

Decide exactly what your aims are; think about a 'way in'. Ask an experienced teacher whether your aims are suitable and if they align with the NC and/or examination syllabuses.

Divide your ideas into sections. Find: resources, books, posters, stimulus, display materials. Ideas for SoW can sometimes come about when browsing. Record these in an 'ideas book'. If using techniques and materials which are new to you, practise these yourself using the school materials and equipment so that you are aware of their qualities and limitations! Decide how you are going to set tasks and activities and plan a homework programme to provide a range of differentiated tasks.

Getting your aims right!

Establishing aims for SoW is not straightforward. Pupils' competence and maturity can vary enormously within the same group. How realistic, ambitious and comprehensive can you expect your aims to be? Can you expect all children to meet them or should you make exceptions? In general, the more clearly your aims are differentiated to the specific learning needs within the group the more chance they have of being realised. The way you word and use aims in your SoW can either assist or hinder you. If appropriate, they provide a helpful focus; if inappropriate, they become a distraction.

Aims about aims

Think clearly about the support you offer and the resources you need to realise your aims. Bearing in mind that lowering expectation becomes self-fulfilling decide whether you expect:

- all pupils to meet your aims to a comparable standard;
- all pupils to meet your aims at some level;
- the majority to meet your aims and make provision or adapt the aims for those who do not;
- only a few pupils to meet your aims.

To teach confidently you need to be able to justify your aims educationally and ethically. It is easy to corner yourself unintentionally by not fully thinking through your aims. In the same way that the NC requires pupils to review and modify their work as it progresses, you may find it necessary to change the way you plan your SoW or adjust them as they progress.

Teaching for diversity

As a general rule the more homogeneous the ability and experience of the group, the easier it is to narrow the focus of your aims. Within a wider range of pupils, wider terms of reference are often needed. Often teachers regard aims as reminders of what they should provide. If some choice or variety of outcome is built into a SoW, many of the problems associated with too narrow an aim or focus can be avoided. A tension always exists between consistency and flexibility. You can tell whether aims are appropriate or not by monitoring the response of your pupils.

Using pupils' responses to help differentiate needs and monitor the effectiveness of your aims

If you focus too much on your aims without having thought through the support needed to achieve them:

- some pupils make a half-hearted attempt to meet them, are not fully engaged and look for the easiest solutions;
- others feel daunted from the start; they might be unable or unwilling to make any effort and 'play up' accordingly.

If insufficient emphasis is given to your overall aims and the SoW is allowed to completely alter course:

- enterprising pupils make good use of the opportunity to adapt resources to their own purposes;
- others pretend to do this, but with the sole intention of exhausting your supply of materials;
- the more insecure feel that you have let them down in your duty to direct, stimulate, entertain them and will switch off and stop working.

CHECKLIST FOR PLANNING SoW

Consider the following before planning:

- the number of lessons available;
- what you want pupils to learn;
- the areas of the PoS you are covering;
- pupils' prior learning;
- the materials, processes, skills and techniques, you need to gather, organise and rehearse;
- the critical, contextual and historical resources needed to support, guide and reinforce learning;
- the way you balance and integrate the requirements of the attainment target: skills, knowledge and understanding;
- how to make each lesson memorable and significant;
- assessment and evaluation.

Look for opportunities to:

- build on pupils' existing knowledge, skills and interests;
- stimulate and maintain motivation through a variety of tasks, materials and teaching methods;
- develop technical skills through instruction;
- arrange pupils to work in pairs and groups as well as whole-class teaching;

- provide opportunities for group discussion and peer appraisal;
- involve pupils in discussion about the content and direction of their work; provide a degree of choice;
- broaden pupils' frames of reference;
- refer to a range of traditions and examples of contemporary practice;
- identify key concepts and terms and build pupils' art, craft, design vocabulary;
- set a variety of types of homework, e.g. gathering information, practice of techniques, writing up methods, etc.

CONCLUSION

There are many variables in teaching which make it difficult to find a perfect formula for planning SoW. A lesson idea which works well with one group can be a disaster with another. The weather, time of day, incidents from previous lessons, etc., can all affect the pupils' receptiveness to your aims.

Some student teachers find that they have been too elaborate in their planning or have complex aims which they cannot communicate to pupils. In an effort to provide breadth, you may try to cover too many strands of the PoS without allowing time to establish connections between them. You may feel that every piece of work needs to be accompanied by reference to a maker or that all work should start with drawings from observation in order to fully comply with the four strands of the attainment target: exploring and developing ideas, experimenting with and using media, reviewing and adapting work as it progresses, investigating others' work and applying knowledge and understanding. As a result a veil of predictability descends as the formula tightens and is perpetuated!

Variety of input and outcome is one way of guaranteeing the involvement of the largest numbers of pupils. It should be possible when planning your SoW to change the focus from time to time, sometimes concentrating specifically on technique, sometimes investigating traditions and changing cultural forms through discussion and written tasks, sometimes providing opportunities for pupils to make discoveries of their own and at other times directly instructing them and transferring knowledge.

In Chapter 3 it is noted that pupils have many different ways of learning. Consequently you need to find different ways of presenting instructions or stimuli and adapt your teaching method and style accordingly. This does not, however, remove the need for the security provided by consistent routines, rewards and sanctions.

LESSON PLANNING

Much of the advice in the previous section on SoW also applies to lesson planning. When planning lessons you should set specific learning objectives for each lesson, a skeletal outline of this should form part of the SoW. However, individual lessons are

Table 4.1.1 Scheme of work

SCHEME OF WORK	Class: Teachers name: No. in class:　　　age:　　　KS:

Aims (related to NC PoS)

(objectives are listed in each lesson plan)

STRUCTURE OF THE PROJECT/FRAMEWORK OF LESSONS

		NC Reference
Day/Week	one	
Day/Week	two	
Day/Week	three	
Day/Week	four	
Day/Week	five	

RESOURCES

ASSESSMENT STRATEGIES

OTHER NOTES (health and safety, etc.)

best planned in response to and after careful reflection on what has been achieved and can be developed from previous lessons. It can prove useful to devise a proforma for planning lessons. This encourages you to develop effective working habits and balanced, purposeful lessons. The layout and content of your proforma may need revision after a few lessons in order to reflect and assist your emerging teaching style and methods.

Task 4.1.10 Defining aims and objectives

- Write down your definitions for the terms below:
  aim,
  objective.
- Compare and discuss your definitions with other student teachers.
- Check them against dictionary definitions.

Consider the following check list before planning lessons:

SUBJECT KNOWLEDGE

- Do your lesson objectives meet your overall aim?
- Are the concepts, resources and tasks suitable for the age and range of achievement of the pupils?
- How does the lesson make use of pupils' existing skills, knowledge and experience?
- How does the lesson build on the achievements and challenges of earlier lessons?
- At the end of the lesson, what new knowledge, understanding and skills do you hope pupils have gained (i.e. the learning objectives)?
- What new vocabulary do you want pupils to learn and how can it be communicated and reinforced?

TEACHING AND CLASS MANAGEMENT

- What teaching methods can best meet your objectives? e.g. whole class, group work, pair work, discussions, talk/demonstration, etc.?
- What is the most appropriate way to arrange the tables/equipment, etc.?
- How do you establish or support existing routines for distributing materials, work, etc.?
- How do you sustain pupils' interest e.g. by balancing discussion with practical activities, taking pupils off task to discuss progress, changing activities, address different ways of learning, reviewing and summarising achievements in order to set future targets?

Table 4.1.2 Constructing a lesson

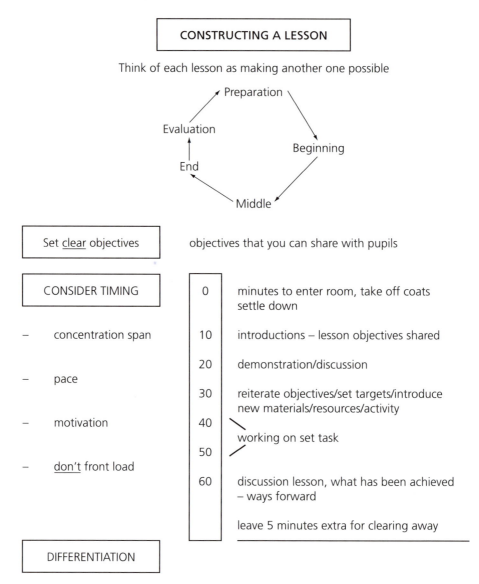

- What strategies do you use to help pupils compare methods, approaches and outcomes with each other and the work of others?
- How do you ensure that all pupils benefit from the lesson? How do you differentiate individual needs? In what ways do you cater for the SEN pupils? What extension tasks are provided for pupils who successfully complete work before the rest of the class?
- How do you communicate concepts and tasks to pupils for whom English is an additional language (EAL)?
- How and at what point in the lesson do you set homework?

Table 4.1.3 Lesson plan (a)

LESSON PLAN

OBJECTIVES

By the end of the lesson, students/pupils will be able to:

State ....

Describe ....

List ....

Identify ....

Prioritise ....

Solve ....

Demonstrate an understanding of ....

Objectives within a lesson

1   to learning ⟶

2   to behaviour ⟶

Work collaboratively/in pairs/cooperatively

Work independently

Sustain effort – concentrate (adults, concentration span is 20 minutes maximum!)

Maintain an orderly environment

Table 4.1.4 Lesson plan (b)

LESSON PLAN

CLASS		Yr		KS		
Date: time:				length of lesson:		

OBJECTIVES

      *
      *
      *

LANGUAGE:
KEY CONCEPTS:

RESOURCES:

PLAN:

* a clear sequence of actions to be undertaken by you e.g. instructions, demonstration
* a clear, timed sequence of activities
* organisation of room, distribution, collection of materials, children's work
* a note of the questions to be asked group and pupil evaluations

EVALUATE:

(Assessment) How will you identify if learning objectives have been achieved?

Criteria for success

NC STIMULUS	ACTIVITY	EVALUATION

ASSESSMENT

(Checklist continued from p. 73.)

How do you ensure:

- pupils know if they have achieved the learning objectives?
- feedback is provided: group/individual?
- they understand the criteria used for assessing progress and achievement?

HOMEWORK

Homework can extend pupils' Art & Design education by many weeks over their schooling. Reasons for setting homework are to:

- consolidate the skills, knowledge and understanding taught during lessons;
- practise techniques and media;
- resource and gather information (e.g. objects, cuttings, quotations, etc.);
- assess and set targets;
- investigate histories, issues, artists, craftspeople and designers, etc.

Task 4.1.11 Homework survey

Consult a copy of your department's homework policies.
    Discuss with other teachers the various ways and reasons for setting homework. Add to the list provided above.

4.2 EQUAL OPPORTUNITIES: AN OPPORTUNITY FOR THE ART & DESIGN CURRICULUM

OBJECTIVES

By the end of this unit you should be able to:

- examine equal opportunities and identify ways to develop strategies which acknowledge difference and plurality;
- adopt approaches for developing and interpreting visual images and artefacts which recognise that interpretations may be changed, by reference to gender, race and class, to give widely different meanings.

INTRODUCTION

> The range of what we think and do
> is limited by what we fail to notice.
> And because we fail to notice
> that we fail to notice
> there is little we can do
> to change
> until we notice
> how failing to notice
> shapes our thoughts and deeds.
>
> (R. D. Laing)

VITAL LIES, SIMPLE TRUTHS: THE PSYCHOLOGY OF SELF-DECEPTION

Equal opportunities policies attempt to foster the conditions whereby pupils can fulfil their potential irrespective of their gender, race, ethnicity, religion, nationality, social class, sexual orientation or level of ability. Any serious engagement with visual and material culture must tackle issues of representation and identify ways in which meanings are constructed. In order to engage with these you need to acknowledge that meaning is not fixed but constantly changing and needs to be negotiated within a range of forces: historical, social, cultural, political, and economic.

Equal opportunities does not imply the assimilation of 'other' cultures or groups by blurring distinctions, in other words encouraging minorities to radically alter their identity in order to conform to the majority. Neither does it involve ignoring the background of pupils by assuming or pretending that they share the same cultural traditions and history: rather, it means acknowledging all the groups who make up contemporary society in curriculum planning and in the content of PoS. To this end teachers of Art & Design need to develop a broad subject knowledge informed by an understanding of changing traditions and ensure that this is reflected in their SoW. This is happening increasingly as more teachers and pupils develop plural or multiple identities through travel and migration. The Art & Design curriculum presents many opportunities for teachers and pupils to understand, appreciate and inform the diversity and hybridity of traditions and shifting practices.

The writers of NC Art are concerned to make the Order applicable to as wide a range of pupils and communities as possible. Initial concern regarding a bias towards European traditions was modified for NC Art (DFE 1995) in order to ascribe equal value to cultures 'Western and non-Western'. Meecham (1996) aptly describes the 1995 Orders as 'an almost multi-cultural approach, with an eye to the National Heritage' (p. 75). However, the PoS for KS3 identifies a list of named periods in the Western tradition as examples: 'Classical, Medieval, Renaissance, post-Renaissance through to the nineteenth and twentieth centuries' (DFE 1995: 6). Although this was not meant to suggest a chronological approach, it has resulted in teachers feeling obliged to include the given examples to the exclusion of more inclusive work, whether contemporary or drawing on a range of cultural traditions. The work towards the NC 2000 (QCA 1999) has revised this list, adding that 'pupils investigate the diverse roles and intentions of artists, makers and designers in the community, from past and present and from familiar and unfamiliar cultures'. Neither document mentions race, gender or social circumstance. Children with SEN are only provided for by means of general 'access statements', none being specific to Art & Design. These statements are directed towards those children with marked needs, that is the 2 per cent with Statements, rather than the 18 per cent without. No strategies are offered for teaching gifted pupils although 'exceptional performance' is recognised in the Attainment Target Levels. In fact, the Order is so skeletal in its new form, that while a broad, informed, differentiated, intercultural, and issue-based approach is possible, it is not specifically required. However, it is your responsibility to ensure that the curriculum you develop is inclusive and gives full consideration to equal opportunities.

When attempting to plan a curriculum which embodies a full range of traditions it is necessary to locate your own cultural bias or frame of reference in order to see how this 'fits' with that of the NC. To enable you to 'position' yourself in a different group or setting you may find it helpful to complete the following task.

Task 4.2.1 Navigating the National Curriculum

1   Identify from the NC for Art a statement which requires additional qualification or terms of reference to be fully understood. Note down the mental routes and associations you used to provide that understanding.
2   Repeat this exercise, this time imagining your background, culture, education or place of origin is completely different: try to identify what alternative routes or means you would use to arrive at a meaning.

The last exercise suggests that one of the many differences between pupils is their prior experience and 'frames of reference'. In many schools this presents a number of challenges, for example:

- pupils have attended feeder primary schools which are at different stages of developing a broad and balanced PoS in line with the NC;
- pupils may not have received a comparable experience of NC Art and/or use a different frame of reference for identifying and practising art, craft and design, e.g. those who have recently arrived in this country.

The common requirements for PoS state: 'A small number of pupils may be working at an earlier or later key stage' (DFE 1995).

EQUAL OPPORTUNITIES: CHECKLIST FOR PLANNING

Gender

Foster mutual respect and achievement between the sexes:

- Investigate the roles of men and women as makers, spectators, participants, consumers and critics of art, craft and design.
- Investigate the cultural/economic reasons for the predominance of one sex in some traditions/practices, e.g. the 'great masters' of Western Art, the batik craftswomen of Nigeria.
- Investigate the causes of bias and stereotyping within art, craft and design practice.
- Introduce pupils to activities which may once have been the preserve of one sex.
- Highlight women's achievements, through your choice of references, book, exhibitions, collections, etc.

Race/culture/religion/class

- Investigate the different status and definitions and relationships between art, craft and design in different communities.
- Investigate the different roles and functions art, craft and design have for different communities.
- Aim to plan SoW which reflect cultural and social diversity both within and outside the UK and question orthodoxies and hierarchies.
- Identify the main cultural and religious traditions represented by the groups/ individuals in your school: recognise the contribution pupils can make to the department's range of references and resources.
- Incorporate pupils' backgrounds and frames of reference in SoW.
- Be aware and respectful of pupils' religious beliefs and customs.
- Seek examples of achievements from a broad range of activities and social groups, e.g. include vernacular styles, folk traditions, popular culture.

5 Resource-Based Learning (RBL)

Lesley Burgess

> The most important resource of any school is its teachers: the most
> important quality of any teacher is resourcefulness.
>
> (Robinson 1982: 56)

This chapter introduces you to the central role of RBL in Art & Design education. It
encourages you to reflect critically on the resources used by yourself and others in
order to determine whether they are appropriate and, if not, how they might be
developed. You are encouraged to think carefully about the ways in which outdated
attitudes and values can be perpetuated through the uncritical use of existing
resources. You are alerted to ways in which you can, all too easily, reproduce traditional power structures by working with existing exemplars of 'good' practice and/or
promoting cultural heritage uncritically and without reference to context.

The SCAA (1997b) *Survey and Analysis of Published Resources for Art (5–19)* found
that resources are selected on the basis of:

- teacher's interests, motivation and training;
- what happens to be accessible;
- finances available.

Specific areas identified as difficult to resource were: 'non-Western Art, sculpture,
women artists and craftspeople, designers, design education/design history, local artists, contemporary art/artists. Other areas mentioned were: photography, environmental art and architecture. These gaps in provision provide the focus for this chapter
with the exceptions of sculpture, craft and design which are covered in other parts of
this book.

OBJECTIVES

By the end of this chapter you should be able to:

- resource schemes of work (SoW) giving careful consideration to the impact that your selection, use and presentation of resources can have on pupils' reception, enjoyment and understanding of the subject;
- identify a wide range of resources available for teaching and learning in art, craft and design;
- recognise the central role of new technologies in accessing information and locating resources;
- reflect critically on the range of resources currently used in schools and HE, including those accessed through WWW and the Internet, and consider their suitability for different pupils;
- identify opportunities to work in partnership with outside agencies, including galleries and museums, in order to develop the range of resources currently at your disposal;
- appreciate the need to constantly update and reconsider resources to ensure that they reflect contemporary practice as well as changing traditions.

5.1 DEFINITIONS

OBJECTIVES

By the end of this unit you should be able to understand:

- why the term RBL was introduced in the 1970s;
- its reception in Art & Design education at a time when it was being denigrated in some other areas of the curriculum;
- the changing definition of RBL in response to curriculum developments and new technologies.

> So much pupil learning in schools is alienated in the sense that it consists of other people's knowledge purveyed in transmissional mode. Pupils have no share in the knowledge nor any control over the learning processes. In addition, it is difficult to see the relevance of such learning for their own interests.
>
> (Woods 1996: 127)

Woods (1996) suggests that RBL was promoted in the 1970s in an attempt to provide the antidote to 'alienated' learning. He identifies its genesis in child-centred progressivism which promoted: learning by discovery, making teaching relevant to pupils' concerns, integrated knowledge, democratic decision making. Critics of RBL claimed that, in practice, it became: a restraining rather than a liberating culture, a coping rather than a teaching strategy; operating in the interests of social control rather than pupils' interests (*ibid*.). However, despite this criticism and RBL's limited reception in other areas of the curriculum, it has had a significant impact in Art & Design. Rather than dismiss RBL as problematic, many Art & Design educators successfully related it to the pedagogic principles underpinning critical practice and intelligent decision-making.

> The concept of 'Resource-Based Learning' (a significant feature of GCSE art and design), developed from the successful teaching and learning methods which were based on extensive provision of an appropriate range of learning resources. In the main, these consisted of collections of objects and reference materials used both as stimulus for practical work and 'research' as work progressed.
>
> (Binch and Robertson 1994: 112)

The term RBL was initially used in Art & Design to signify the promotion of work from 'direct experience' using primary sources. This signalled a clear rejection of the ubiquitous practice of inviting pupils to respond creatively to a given topic using only their imagination or images (usually gleaned from the Sunday newspaper supplements). Many educators acknowledged the need to move away from such restricted practices. They recognised sheer self-expression required no artistic form and mere transcriptions of photographic reproductions required no understanding of content. They responded by insisting that the main stimulus for all pupils' work must be close observation of collections of 'real' objects. This gave rise to the 'big still life', large classroom installations of objects and artefacts selected in response to a given theme: natural forms, reflections, celebrations.

Kennedy (1995a) explains why he considered it to be an essential part of the curriculum:

> Students are encouraged to work from real things for two main reasons. The first is a matter of quality control – there is less of a credibility gap between concept and realisation in representational terms if students work from direct observation rather than a second hand image. The second concerns empowerment – the student is in control of the basic elements when working directly and can determine angle, view, scale, rendering etc. Working from secondhand imagery frequently means that the student's own intentions have to be compromised by compositional decisions made by others. Generally, in one scenario the student is active and learns more, in the other the role is more passive with the student potentially disempowered.
>
> (p. 9)

However, RBL was not without its problems, the extent to which some teachers rigidly adhered to this 'perceptual' approach proved limiting; pupils fed on an exclusive diet of close observation tended to develop a narrow definition of art. Critical and contextual resources used to support this approach drew exclusively upon the nineteenth- and early twentieth-century Western still-life tradition. The perceptual approach fostered the development of the 'school art' orthodoxy that Hughes claimed was 'static, safe and predictable . . . divorced from contemporary ideas in the spheres of art practice, critical theory, art history or museology' (Hughes 1998a: 41). By restricting the range of practices, emphasising 'basic skills' and techniques and reducing drawing to accurate recordings of things seen, it avoided the difficulties of assessing work which was issue based, innovative, idiosyncratic or from cultural origins other than Western.

Simpson (1987) insisted:

> The emphasis has been on direct observation and recording – the object being the 'environment' . . . drawings of twigs, shells, sections of any fruit or vegetable you care to name, sculls, bits of bikes' lamps and those tedious pencil drawings of crushed coke cans and toothpaste tubes; all now definitely passé.
>
> (p. 255)

By the mid-1980s use of the term had expanded to include galleries, museums and other outside agencies, artists, craftspeople and designers in education. Art & Design teachers began to collect resources to fulfil the GCSE requirement to develop pupils' critical and contextual skills. The proposed introduction of the National Curriculum Art Order (NC) (DES 1992) with its requirement 'to study the work of others' resulted in a rush to locate suitable resources. In deliberations about NC Art, RBL extended beyond concrete examples (object, material, human) to include abstract concepts such as time and space. At an NSEAD conference in 1992 the then HMI for Art, Colin Robinson, claimed that: 'the most valuable resource any art and design teacher has is time'.

TIME

On average one hour per week is allocated for Art & Design at KS3. However, given that each pupil will probably be ill for at least one lesson, and out for school trips, choir, orchestra, steel drums, sports day and rubella injections you can expect any one child to attend thirty timetabled hours each year. At KS4 time allocation is typically 2.5 hours per week, but with reductions for work experience, field trips and mock exams the likely total is seventy-five hours per year. Many Art & Design teachers find it necessary to provide extra-curricular opportunities for pupils to develop their studies. Prentice (1995: Chapter 2) insists that student teachers need time to become familiar with material, tools, equipment and processes; the same claim can be made for pupils. Time restrictions are often cited as the reason why some teachers offer a limited range of practices; they claim to provide depth in the basics rather than introduce a broad range superficially. Similarly, restrictions in equipment, materials room size and available storage space are cited as the reasons why pupils do not develop work in three dimensions or ICT. It is important to take these issues into consideration when developing SoW. However, the opportunity provided by your ITE course to visit a number of Art & Design departments introduces you to different ways to overcome such hurdles.

ROOMS/EQUIPMENT/HARDWARE RESOURCES

> The visual resources available to the art teacher today are many and varied. The extent to which they are used in any one school will depend on a great many circumstances, some economical, some geographical, but the most important factor will be the energy and the initiative of the art teacher concerned . . . The stultifying effects of resourceless departments must be rectified as a matter of urgency.
>
> (Taylor 1986: 211)

Surveys by OFSTED (1998), RSA (1998), reveal that there are strong contrasts from school to school in the range of activities undertaken and the resources available. Some schools have a wealth of equipment: airbrushes, cameras (video, digital, still) computers (monitors, scanners, colour printers, Internet, etc.), looms, silkscreens and kilns; others have to make do with hog-hair brushes, powder paints, recycled boxes and a few blunt lino-cutting tools. It is useful to determine how often different resources are used. You should explore the ways teachers in your Higher Education Institute (HEI) and placement schools organise resources; are they stacked in neatly labelled boxes or do they litter the department in a cacophony of conflicting colours and textures? Do pupils have open access, or are they closely guarded for fear they might disappear? Are they catalogued? If so, what headings are used? Find out where you can obtain or borrow resources. Some local education authorities have extensive loan collections, as do a number of galleries and museums. Most regions have 'scrap banks': collections of recycled /industrial waste, off-cuts of paper, cloth, wire, metal, plastic, all made available for educational use at give-away prices. Your ITE year is an opportunity to identify what is available and to develop a comprehensive bank of critical and contextual resources.

More recently the term RBL has been reintroduced in mainstream education and has taken on a new, extended meaning – the pedagogic use of new technologies or ICT. RBL can now be defined as 'diverse course delivery methods using non-traditional modes'. It takes place in learning resource centres, quiet study spaces with access to printed and computer-based resources (CD-ROM, Internet, Intranet). In FE and HE these are already well established and teachers from all curriculum areas are expected to be actively involved with 'learning assistants' ensuring that course design, support, assessment, review and evaluation are available on the Internet and the Intranet. The terms 'library' and 'librarian' suddenly seem quaint and old fash-ioned. RBL has shifted from the periphery to the heart of the learning process. Tomlinson (1996), Kennedy (1997) and Dearing (1997) have argued for accessibility, inclusivity, flexibility, transferable skills; greater opportunities for students to manage their own learning. The implications for teaching in secondary school are rapidly becoming apparent.

Task 5.1.1 Audit of resources

Make a list of resources available in your teaching practice school(s). (Every department should have an inventory of non-consumable resources.) Over a number of weeks record the following; a simple tick system is sufficient:

- how often each resource is used;
- the year group using it;
- the teacher responsible;
- your use of the resource with your classes.

Discuss your findings with teachers in the department. Can you identify a pattern of usage?

Compare and contrast your findings with other student teachers. Access recent OFSTED reports (*http://www.ofsted.gov.uk/ofsted.htm*) and identify whether your findings reflect the local and/or national picture.

FURTHER READING

SCAA (1997) *Survey and Analysis of Published Resources for Art (5–19)*, London: SCAA.

QCA (1997) *A Survey of Teachers' Reference to the Work of Artists, Craftspeople and Designers in Teaching Art*, London: QCA.

OFSTED (1998) *The Arts Inspected*, Oxford: Heinemann.

RSA (1998) *The Effects and Effectiveness of Arts Education in Schools*, Slough: NFER.

Department of National Heritage (DNH) (1996) *Setting the Scene; The Arts and Young People*, London: HMSO.

5.2 OWNERSHIP

OBJECTIVES

By the end of this unit you should be able to:

- consider who or what decides or dictates the types of resources you use;
- understand whether selection is based upon your subject specialism, interests and enthusiasm or is prescribed by others;
- ensure your use of resources provides an inclusive education informed by cultural and social issues.

Task 5.2.1 The naming game

Name 6 artists, give their dates and identify their art form. Take no more than 2 minutes to complete this task.
Once a name has been written down you must not change it or cross it out.
Do not confer with anyone else.

Education has been named as one of the major ideological state apparatuses – that is, not just a place of learning, but an institution where, as in the family, we are taught our places within a hierarchical system of class, gender and race relations.

(Pollock 1996a: 54)

The important role that you play in constructing pupils' understanding of 'what counts' in Art & Design cannot be overlooked. Your choice and use of resources provides pupils with a clear statement about what is to be valued. Before you address issues of marginalisation and invisibility you have to question your 'own ready implication in a discourse of mastery' (Usher and Edwards 1994: 81); to recognise that, often unwittingly, you can reinforce orthodoxies. Even when they think they are being 'inclusive', research (Stanworth 1987; Spender 1989) has proved that teachers continue subtly, albeit unintentionally, to reinforce 'bias' and 'otherness'.

Long-established metanarratives are powerful forces which are difficult to disrupt. This is as true in Art and Design as it is in other areas of the curriculum. However it is worthy of note that in successive versions of the NC Art Order, unlike any other NC Order, teachers have been actively encouraged to 'explore different codes and conventions', 'recognise ways in which works of art, craft and design reflect the time and place in which they were made' (DES 1995), 'investigate the diverse role and intentions of artists, makers and designers in the community, from past and present and from familiar and unfamiliar cultures' (QCA 1999). Far from being an exhortation to sustain a 'school art orthodoxy', these statements insist pupils should engage with a range of social and cultural issues.

Task 5.2.2 Reviewing personal choices

Look back at the list of 6 artists produced in Task 5.2.1. Analyse them using the following questions as prompts:

- Does your selection represent: exhibitions you have recently visited, artists with a high media profile, artists who have been influential in the development of your own practice, your contemporaries?
- If you had been asked to list 6 artists, craftspeople or designers, would your list be different?
- If you were completing this list in a contemporary art gallery would this alter your selection?
- Is there a gender balance?
- Is cultural diversity considered?
- Have you included yourself?

Throughout this unit you are encouraged to think carefully about the way the critical and contextual resources can conceal a hidden curriculum. You are asked to query the 'modernist' trap which reproduces traditional hegemonic power structures and reinforces dominant Western patriarchy. Hughes (1998a) calls for a total reconceptualisation of the Art & Design curriculum. Like Efland (1996) he insists that merely 'tinkering on the edges' is insufficient. Perhaps such demands are needed to effect the smallest changes in practice. However, as a student teacher you need to think carefully about your role in developing pedagogies. If Art & Design education is to evolve; to reflect new ideas and ways of working, then you must see yourself as both an advocate for change and an agent of change. It is important that you consider what a 'reconceptualised' curriculum might comprise and what resources you need to support it. However, until you understand the existing system tread carefully. To 'boldly go where no-one has gone before' would be foolhardy. The approach you are asked to adopt is one of interventions (Pollock 1988) or what Spivak calls 'strategic essentialism' and Hall refers to as 'arbitrary closure' (in Shohat and Stam 1995). Through interventions we can destabilise, question and identify why invisibilities have come about and through strategic essentialism/arbitrary closure, understand that although we need to adopt a position in order to practise, in time it is likely to be replaced. You need to recognise that permanently prescribing resources and their usage in school is a redundant project. Some of the resources we use with conviction today, are likely to be spurious/irrelevant in the future. Shohat and Stam (1998) insist that any study of visual culture should be a 'provocation', to cause new questions to be asked, 'to interrogate the conventional sequencing of realism/ modernism/postmodernism' (p. 31). Spivak insists, 'it is the questions we ask that produce the field of inquiry and not some body of materials which determines what questions need to be posed to it' (Rogoff 1998: 16).

While it is impossible to start from scratch it is important that you do not only adopt resources produced by others. You need to intervene in 'packaged pedagogies' by challenging the authority of stereotypical resources, turning statements into questions and by introducing artists, objects and artefacts which refuse to comply.

Task 5.2.3 Visual resource sheets

You probably have acquired or produced visual reference sheets and reproductions early on in your course and are using them in your introductions, demonstrations and displays. How are you selecting these resources?

Are they there to produce exemplars, canonic or otherwise, for pupils to emulate?
Do they include diverse cultural perspectives?
Have you selected them to act as a stimulus for discussion?
What questions might they provoke?
Do you provide contextual information to complement choices made for formal reasons?
At what stage in a SoW should you bring in critical and contextual resources?
What is the function of your references in each SoW?
Are you producing these resources to promote learning as a sign of your effort and enthusiasm, a form of spectacle?
Would it be more beneficial if pupils investigated an issue of interest and produced their own referenced presentations?

PUPIL INTERVENTIONS – PUPILS AS A RESOURCE: GATHERERS AND COLLECTORS

> Every student transferring to secondary education at the age of 11 brings with them a unique canon of visual literacy ranging from, in most cases a subliminal expertise of sophisticated televisual constructs (absorbed osmosis-fashioned from many thousands of hours of home viewing) to conscious and articulate appraisals of what one year ten student termed TCCC Twentieth Century Cultural Clutter.
>
> (Kennedy 1995a: 8)

Ownership of lesson content merits careful consideration. Edwards and Furlong (1978) suggest that RBL is conditional upon teachers' prevailing knowledge; that it rarely takes into account pupils' interests. The NC 2000 infers that the responsibility is not just the teachers': 'Pupils should be taught to research, select and organise a range of visual evidence and information'. Loeb (1984: 6) notes that when pupils are encouraged to become researchers and are 'given permission' to become contributors to, as well as clients of, the education system, a rich and sometimes unexpected resource becomes available.

Willis (1990a) insists we need to broaden education's traditional, restricted notion of art and design in order to embrace 'moving culture'; to recognise that children are already engaged in imaginative, expressive and decorative activities that are grounded in the needs and functions of everyday life.

Task 5.2.4 Moving culture

Willis (1990a: 68) includes the following question in a list of seven at the end of *Moving Culture*:

Are the creative activities and everyday aesthetics of young people's culture fully recognised in your own thinking and in the practices of institutions of which you may be part?

In your tutor group consider:

- the extent to which 'the everyday aesthetics' of young people are acknowledged in your resources;
- where the 'work of others' might be usefully extended to include aspects of young people's culture;
- how this inclusion might change or extend possible outcomes.

In groups of 3 or 4 devise a SoW for year 10 that uses 'grounded aesthetics' as its main stimulus and invites pupils to contribute to resources: objects, artefacts, images, books, articles, comics, etc., and encourages them to reflect on the ways content and context construe meaning.

As an art and design graduate you are familiar with the impact context can have on the reading and reception of objects and artefacts. You need to consider how this applies to pupils' contributions. An object, image or statement which one pupil values for highly personal reasons (memory, association) may be perceived by another as promoting stereotypes (race, gender, social class, age, sexuality). Troyna and Hatcher (1993) have suggested that while both racism and antiracism feature in pupils' culture 'the existence in every class of children who have a clear anti-racist commitment is potentially the most powerful resource' (Troyna and Hatcher 1993: 203). Pupils can provide perceptive and sometimes unexpected insights into a wide range of issues from ecology to exploitation and, given the right environment, will voice their concerns about censorship, marginalisation and prevailing hierarchies. This encourages both teachers and pupils to question their value systems and to develop classroom relationships on the basis of equality of treatment and an ability to acknowledge difference and value the viewpoints of others.

Pupil interventions, both material and theoretical, should be acknowledged as an important resource for teaching and learning in Art & Design. However, just as pupil participation is significant, non-participation can also be significant, it can conceal a desire not to be exposed (hooks 1994a). Stanley warns us that 'pupils may well have good reasons to be suspicious of an invitation to share their cultural heritage with the teacher and with other pupils not otherwise exhibiting much tolerance, let alone respect, for minority cultures' (Stanley 1986: 177). He reminds you that you do not start with a cultural 'tabula rasa' and you may be perceived as patronising or overbearing (*ibid.*). Williamson (1982) and Buckingham and Sefton-Green (1994) identify 'subtle resistances' by pupils who are unwilling to examine issues which impinge on their own identity. Such pupils contribute only what they think fits the status quo rather than their own beliefs and values.

> The decision to proceed or not to proceed, to intervene or not to intervene, to take up issues or to leave them unchallenged are decisions the teacher is compelled to make as an authority – both real and imagined – in the class. This is not to say that students cannot intervene or challenge, but that students' actions have very different consequences because of their very different relation to power.
>
> (Todd 1997: 72)

PLUGGING THE GAPS

> There are significant gaps in provision in the content of resources for art. Resources relating to cultures and traditions from the more remote past, in contemporary society, by women artists and from non-Western traditions are difficult to locate and hard to access. The gaps in provision match the main areas in art which are overlooked by many teachers in their planning and teaching of the programme of study.
>
> (SCAA 1997b: 12)

Any attempt to try to ensure that no 'gaps' exist is clearly impossible. SCAA claimed 'there appears to be plentiful resources to support teaching about fine art of the nineteenth and twentieth centuries, possibly reflecting common perceptions of what art is and what most people value' (*ibid.*). To disregard the past in favour of the present is not the solution. Similarly, any suggestion that race, gender, class can be viewed as discrete entities is erroneous.

CULTURAL DIVERSITY

Brandt (1991) expressed his disquiet about the way Art & Design education seemed to be promoting cultural diversity through the use of cultural artefacts. He claimed that they were defined by their place of origin; there was no reference to the way 'images and objects presented were not value-free but value laden; not politics-free but politics laden, not context-free but context-bound'. He insisted education is not simply about making pupils aware of diversity or providing exotic or celebratory examples; it must also confront negative stereotyping. Brandt reminds us that art is not divorced from society, 'it is not simply an aesthetic expression, but also a socio-political and cultural statement which has added to a legacy of heritage of representation and speaks loudly to contemporary society'. Brandt insists we move beyond providing cultural artefacts or adopting 'ethnic techniques' by ensuring that they are grounded in the right cultural contexts.

Task 5.2.5 Resource development plan

Constraints in terms of resources is the first excuse teachers make in defence of these omissions so you must try and plug these gaps from the outset.

You have already produced an audit of the resources available in your placement school; do the same for your HE institution and your own collection.

Consider these in relation to the canon of 'plentiful resources' outlined above.

What are you making visible?

Are there any gaps?

What do you need to form a more comprehensive and inclusive set of resources and reference packs?

On the basis of these omissions produce a resource development plan for the year. Don't feel that you have to cover each aspect separately but find ways of combining them.

What further resources can you recommend to your host department and their school library?

Multiculturalism has been replaced and/or subsumed by a succession of terms, 'new internationalism', 'global art', 'world art', 'cross-cultural studies', 'cultural hybridity'. More recently the notions of cultural diversity and interculturalism have encouraged educators to recognise that most earlier definitions are inadequate and limiting. Perminder Kaur, a British artist who is Sikh, reiterates this point when she claims:

> It's very difficult to make statements about particular things. Issues about race and colour are very complex now. In my work there is no longer direct polarisation between two distinct cultures. The work still contains questions concerning identity but on a more subtle level.

> (Proctor 1996: 11)

Proctor insists that Kaur's work does not suggest that 'home is "no-place", but rather "an-other" place, somewhere just over the next hill, in the next land . . . and any belief that we have found it is just an illusion' (*ibid*.). Meera Chauda (Plate 14) addresses similar issues in her work when she introduces traditional Hindu images into collages which ironically represent her place in contemporary British society.

The gap between theory and practice, between rhetoric and reality remains wide. It exists on a number of levels. Perhaps the most significant here is the difference between what teachers suggest should happen in the classroom and what can be observed. Aware of the complexity of the hidden curriculum, and the unspoken messages given to others by our choice of resources, it is tempting to play safe rather than take risks. To get it right is time consuming and involves research. Resources to support an inclusive curriculum are rapidly increasing. The following organisations and publications are worthy of note:

Resources

The Institute of International Visual Arts (inIVA) is an arts organisation which promotes the work of artists, academics and curators from a plurality of cultures and cultural perspectives. It has a reference library and archive which includes reference sources such as books, exhibitions,catalogues, slides, CD-ROMs, video and audio tapes. It is an excellent source of information for conferences, residencies, useful Web sites and on-line lectures (www.iniva.org). Its education packs on landscape and portraiture introduce British artists from a number of cultural backgrounds confirming a broad and rich notion of British art practices. They position Hogarth and Bacon alongside Sonia Boyce, Chila Kumari Burman and Vong Phaophanit with Andy Goldsworthy.

The African and Asian Visual Art Archive (AAVAA) run by Sonia Boyce and David A. Bailey from the University of East London E13 0BG is an important resource centre for slides and publications, dissertations, audio and video tapes. Tel: 0181 548 9146.

Cahan, S. and Kocur, Z. (eds) (1996) *Contemporary Art and Multicultural Education*, New York and London: Routledge. This provides theoretical foundations, good colour plates, lesson plans and a useful annotated bibliography. Its focus on the USA should not be seen as a deterrent in this context. Deals with interrelationship between contemporary art, multi-culturalism and social class. Promotes issue-based work within an expanding field. Includes contemporary concerns such as ageism, Aids, gay rights, and racism.

Shohat, E. and Stam, R. (1995) 'The politics of multiculturalism in the postmodern age', 'Art & cultural difference: hybrids & clusters', *Art & Design* Magazine, 43. A collection of theoretical papers supported by visuals.

Mason, R. (1995) *Art Education and Multiculturalism*, London: Croom Helm. This includes case studies of multicultural programmes in education, discussion re the interrelationship between aesthetics and social learning, art criticism and anthropology.

Loeb, H., Slight, P. and Stanley, N. (1993) *Designs We Live By*, Corsham: NSEAD. This provides cross-cultural resources for design work including West African, Celtic, Japanese and Islamic examples.

Third Text: an international journal covering 'third world perspectives on contemporary Art and Culture' provides a critical forum for discussion and appraisal of artists marginalised by racial, sexual and cultural differences. For further information write to: Third Text, PO Box 3509, London NW6 3PQ.

Areean, R. *The Other Story*, London: Hayward Gallery.

Lloyd, F. (1999) *Contemporary Arab Women's Art: Dialogues of the Present*, London: WAL.

198 Gallery is 'a centre for the production and exhibition and interpretation and participation in contemporary art representing a range of diasporan identities'. 198 Railton Road, Brixton (*http://www.198gallery.co.uk*).

GENDER

Since the mid–1970s there has been a legal requirement for equality of opportunity of treatment of boys and girls . . . One approach to this is to present pupils with examples of work, of artists, craftworkers and designers of both sexes. In this way, both boys and girls can grow up knowing that the full spectrum of media techniques and skills is open to them.

(DES 1991a: 60)

Ignorance does not just mean not knowing women's names or being able to identify pictures, sculptures, photographs, films or videos by women. It is much more complex. It is about an invisibility of meaning that arises from the indifference and indeed hostility of the culture to where these works come from, what they address and why they have something to offer that realigns our understanding of the world in general. If I call the work of women 'different' I immediately fall prey to the deadly paradox: to name what makes it interesting to study art by artists who are women is to condemn the artists to be less than artists: women.

(Pollock 1996b: xiii)

While Borzello claims that feminist art criticism has 'just touched the national curriculum with its fingertips' (Deepwell 1995: 22), Meecham insists the NC 'does nothing to undo the structures that hold in place received opinion which argues that women's art is derivative and not innovative enough to gain a place in the canon defined in a masculine culture . . . a hard paradigm to shift' (Meecham 1996: 74). Although critical and contextual resources for Art & Design are no longer totally dominated by references to male artists an imbalance certainly exists. Within the fields of art production and art education, you are still confronted by entrenched hierarchies 'the overwhelming masculinities of historically privileged knowledges' (Grosz 1995: 45) which require critique and questioning rather than (in)difference. In the face of such established elitism, attempts to insert women retrospectively into a visual canon constructed by white males has never been sufficient.

Janson's *History of Art*, the most widely used text book, didn't mention a single woman artist until Janson died. Then his son revised it, including a big 19 out of 2,300.

(Guerilla Girls 1995: 26)

Task 5.2.6 Checking for visibility

Look through the textbooks used on your course. Concentrate on books dealing with art history and cultural studies over the last 40 years. What proportion of the artists cited are female?

Ask pupils in your classes to name the women artists they used to inform their own production.

(Allen (1996: 86) has analysed successive NC Orders to uncover bias.)

The feminist critique of art history (Nochlin 1991 and 1999; Pollock 1988; Chadwick 1990) has made most teachers aware of Artemisia Gentileschi, Rosa Bonheur, Berthe Morisot, Mary Cassatt, Käthe Kollwitz and Frida Kahlo. More recently the

Turner Prize has introduced names such as: Helen Chadwick, Paula Rego, Rachel Whiteread, Shirazeh Houshiary, Mona Hatoum, Cornelia Parker, Angela Bulloch, Cathy de Monchaux (Plate 21), Gillian Wearing and Sam Taylor-Wood. Through the Sensations exhibition at the Royal Academy 1997 and via newspaper journalism young British artists (yBa), Sarah Lucas, Abigail Lane and Tracey Emin have become well-known figures. However, it is media notoriety rather than consideration of artistic acumen that has helped to develop their reputations. Indeed, even McRobbie (1998: 55) contends Tracey Emin's *Tent* 'owes more to the girls just wanna have fun humour of *More* magazine than it does to her feminist elders Cindy Sherman or Mary Kelly'.

While most teachers are aware of the need to move beyond the limited definitions and debates promoted by 1970s feminists, many are ill-informed about more recent feminist interventions and readings. Pollock (1996a) highlights the false dichotomy between past and contemporary 'generations' and calls instead for 'constructed correspondence'. She also reiterates the importance of 'geographies' which are cultural and social as well as political. There is no homogeneous community of women artists, critics or academics. Petersen and Wilson (1976) reveal how women artists were frequently referred to as wives or lovers rather than artists in their own right. Chadwick (1990) explains how some women artists have received considerable public and critical attention but cautions against:

> the dangers of confusing tokenism with equal representation, or the momentary embrace of selective feminist strategies with the continuing subordination of art by and about women to what is, in the words of Pollock 'the gender free Art of men'.
>
> (p. 349)

Hoorn's research reveals how the impetus for inclusion of women in the writing of art history has not necessarily altered the view of women as inferior artists. Her analysis of art texts shows the subtle ways in which women's artmaking is devalued through limited discussions and reference to women's personal characteristics rather than contributions to artistic practice (in White 1998).

This serves as a reminder that you need to scrutinise all texts for such practices and encourage pupils to do the same. All too often, the media either trivialises or pathologises the work of women artists. Journalist Hunter Davies describes Jenny Saville in the following way: 'She doesn't look like the artist, more like a lower-sixth-former, so young, so small, so conventionally dressed' (Rowley 1996).

Jenny Saville's work is a useful example. As a figure painter she is seen as part of a long tradition often compared with Freud and Bacon and a British painterly school. Comparison with artists such as Jo Spence, Cindy Sherman and Orlan confers a matriarchal lineage and invites a different interpretation, one which confronts traditional constructs of female subjectivity. Rowley (1996) compares more typical representations of 'the supine female object body', where the female model is observed by the male artist, to the self-examination that Saville undertakes, the way 'scale' and 'gaze' and 'perspective' are used as interventions which work with, yet against, traditional modes.

This requires that you, as a teacher, ask: 'Why am I teaching this?' 'Does it still apply?' 'Is it relevant?' 'Am I presenting pupils with fixed interpretations or am I asking them to consider different readings?'

Task 5.2.7 More than tokenism?

Look through your GCSE SoW. Identify how women artists have been represented in your supporting resources.

Is representation based only on materials, skills and techniques or on other issues?

Is gender important in their work?

What other constructs inform women artists' practice: race, religion, social class?

Do, or can, their artworks act as 'interventions', questioning traditional hegemonic practices?

Can they be recognised as part of a matriarchal lineage or are they grouped alongside male artists as part of the same genre?

With other members of your tutor group develop a SoW which encourages pupils to question traditional representations using the work of women artists as interventions.

Consider how pupils might be encouraged to use these findings to inform their own production in art, craft and design.

Freedman offers a useful way forward when she promotes the exploration of 'sites of contestation and frayed boundaries' (Freedman 1994: 48). Rather than women (and other marginalised groups) being injected into the curriculum and their stories adjusted to conform with the modernist model of history, you should consider areas of contestation or disagreement. Introduce pupils to different critical positions and help them to understand that interpretation is open, how each one is a construct and why some have been privileged over others.

Pollock (1996a) insists that:

> It is vital to show that the present is historically shaped. Sexual difference and sexual divisions in society are not natural but historical and that is why they can be challenged and changed. The past as tradition – in Art History it becomes the Canon – is used to justify the present status quo. Validated by time, the canons of great art brook no discussion or serious consideration. Feminist interventions have to disrupt canonicity and tradition by representing the past not as a flow or development, but as conflict, political, struggles on the battlefields of representations of power in the structured relations we call class gender and race.
>
> (p. 12)

The easily accessible resources and publications dealing with classical and modernist genealogies (the canon) need no promotion, they can be found in any bookshop on any high street. Gretton (forthcoming) suggests that they can usefully provide the starting point for raising issues of power relations, marginalisation and stereotypes.

However, deconstructing the canon in this way can been seen to reinforce its hierarchical position.

Wolff (1990) reveals that just as women are more or less invisible in mainstream histories of modernism, the same pattern can be seen to be developing in postmodernism (p. 6). Various working parties set up to inform the development of NC 2000 advocated a postmodern approach to art education. However, there is a growing suspicion that postmodernism has failed to resolve issues of marginalisation. By claiming to embrace all previously marginalised groups within its wider remit and including them within its discourses postmodernism can be seen to have subsumed and thus silenced them (Burgess and Reay 1999).

Panting (1999) points out how 'debating gender has become fraught with anxiety and is considered anachronistic and unnecessary' (1999: 20). She reveals how work that deals with gender is itself on the periphery; how it is no longer synonymous with feminism but has been subsumed into the broader picture of identity politics.

Task 5.2.8 Identity politics

Repeat Task 5.2.6 but this time examine the work of male artists and subject their work to the same scrutiny.

Consider how the representation of men has changed in art and the media over the last 40 years.

Not only do we have to grasp that art is a part of social production, but we also have to realise that it is itself productive, that it actively produces meanings . . . it is one of the social practices through which particular views of the world, definitions and identities for us to live are constructed, reproduced and even defined.

(Pollock 1982: 9)

Resources

The Women's Art Library (WAL) *www.womensart.org.uk* is a voluntary-aided organisation constituted as an educational charity since 1982 to promote women artists working in any medium in any part of the world. It is the only resource of its kind in the UK. Situated in Fulham Palace, London, it houses over 140,000 slides of art works divided into three main sections: historical, documentary, contemporary and 5,000 files of press cuttings. In addition it has a comprehensive library of books, videos, monographs and exhibition catalogues and a growing collection of artists' books. It houses a comprehensive 'Women of Colour Index' set up by artist Rita Keegan. It receives information from other key feminist, women's libraries, such as the Feminist Archive in Bristol, the Glasgow Women's Library and the Fawcett Library at London Guildhall University. In 1994 WAL produced a slide pack of forty women artists. This provides a useful selection, from Artemisia Gentileschi's *Judith Beheading Holofernes* (1618) to Sonia Boyce's *She Ain't Holding Them Up, She's Holding On* (1986). It identifies recurrent themes such as identity, memory, motherhood and war. It highlights the range of media used by recent women artists including examples as various as

Zarina Bhimji's 1992 installation of suspended and children's kurtas to Tess Jaray's public art floor tiling for the Centenary Square in Birmingham.

There are two Journals dedicated to Women's Art: *MAKE* by WAL and *n.paradox* the international feminist art journal

Broude, N. and Garrard, M. D. (1994) *The Power of Feminist Art*, New York: Harry N Abrams. Chronological series of articles on feminist practice in the USA since 1960s supported by quality reproductions.

Dalton, P. (1999) *Dismantling Art Education: Modernism, Art Education and Critical Feminism*, Milton Keynes: Open University Press, looks at the way art education has always been implicated in producing gendered identities.

Hiett, C. and Orbach, S. (1996) *Venus Re-Defined*. A resource pack produced to accompany a display of sculptures by Matisse, Renoir and Rodin at the Tate, Liverpool. Drawing on feminist critique themes are identified. The pack aims to involve pupils in discussion about the work and surrounding discourses. The pack is evaluated in *JADE* 1996, 15 (3).

Hyde, S. (1997) *Exhibiting Gender*, Manchester: Manchester University Press, examines ways in which women's art and representations of women are collected and 'represent-ed' in galleries and museums. It presents full colour reproductions of pairs of unidentified works, one by a woman and one by a man and asks, 'Can you tell which one is the woman's work?' Issues raised are discussed further in the text.

Women Artists, Packs 1 and 2 available from Philip Green Educational Ltd, Tel: 01527 854711. Each pack contains sixteen good quality reproductions including Chila Burman, Shirazeh Houshiary, Ana Maria Pacheco, Paula Rego.

MAKE Issue 81, Nov. 1998, 20 years of women's art. This is a special edition charting the development of women's art from 1978–98 in an accessible series of illustrated articles.

IRIS (*www.staffs.ac.uk/ariadne/index.htlm*) The Women's Photography Project: celebrates the contribution made by women practitioners to the development of photographic theory and practice.

FRIDA KAHLO: (*www.cascade.net/kahlo.html*).

VARO (*www.netdreams.com/registry/*) an electronic registry of contemporary international women's art. Includes work by Audrey Flack, Barbara Kruger, Sophie Calle, Rachel Whiteread.

If you are interested in the way gender identities are constructed the following books will be of interest; they are sociological texts, therefore they make no direct reference to Art & Design education. Both books reinforce the notion that gender identities are multiple not singular, not necessarily fixed or constant, and that they are subject to a range of powerful influences.

Mac an Ghaill, M. (1996) 'The making of men', in P. Woods, *Contemporary Issues in Teaching and Learning*, London: Routledge, (in which the author discusses the many different types of masculinities he observed in one comprehensive school).

Mirza, H. (1992) *Young, Female and Black*, London: Routledge, (deals with the way personal agency as well as socialisation theory contribute to constructs of femininity).

CONTEMPORARY ART

> Contemporary art introduces many aspects of popular culture: photography, video and computers. It raises relevant issues, it uses contemporary materials and technologies, it erodes traditional boundaries because it does not always fit neatly into traditional categories of painting, sculpture or print and art, craft, and design.
>
> (Burgess and Holman 1993: 9)

Examples of artwork referenced in this chapter have been deliberately selected from contemporary practitioners. It is important that you introduce pupils to artists who share their pupils' lives and times: but remember you cannot introduce pupils to exciting installation pieces and then expect them to be prepared to respond enthusiastically using only powder paint. You must carefully consider ways in which such work can impact on pupils' own production. There are numerous examples of contemporary artists who deal with natural, found, reclaimed objects and materials (Bill Woodrow, David Mach, Andy Goldsworthy, Rona Pondick, Meret Oppenheim, Annette Messager). Their practices translate into classroom practice. Others (Helen Chadwick, Cornelia Parker, Mariko Mori) use ephemeral, degradable pieces, space consuming installation works or highly sophisticated expensive technologies which make application to the classroom all but impossible. When using contemporary art to inform pupils' work it is important to explore the ways in which it can reinterpret the old and/or introduce new issues and ideas rather than encouraging only transcription or pastiche. Remember, contemporary conceptual clichés can be as limiting as modernist 'masterpieces'.

The work of contemporary artists such as Joseph Beuys, Barbara Kruger, Mary Kelly, Jenny Holtzer, Hans Haccke and Christian Boltanski can seem incomprehensible without reference to its political content. The work of Robert Gober, Tracey Emin, Judy Chicago, Hannah Wilke, Andrés Serrano or the Chapman Brothers of an overtly sexual or abject nature raises interesting issues of censorship and freedom of expression (work distanced by time is often not regarded as so problematic). However, the extent to which such challenging work can be discussed with older pupils depends on the ethos of the school and the attitudes and values of staff and parents. As a student teacher, you must always consult your mentor first. The school environment should be one where, within reason, young people are encouraged to look critically at challenging images, in order to understand the ways in which they represent negative stereotypes and call into question morals, attitudes and values. Art can be seen to have a privileged place in the curriculum, one where it is possible to explore such ideas. Artists play games with intentions in order to encourage engagement with social, political, ecological or moral issues, sometimes provocatively, sometimes deliberately confrontational. The painting of Myra Hindley by Marcus Harvey (1995) was reviled and refused by many; for others it raised significant issues about the role of art in contemporary society.

Contemporary art has become a popular topic for reportage. Along with the Turner Prize, which receives annual rebuke from the media, contemporary shows exhibiting the most radical and unexpected use of materials attract the most coverage.

It is through this avenue that pupils' impressions of what comprises contemporary practice are often formed and reinforced. Such articles provide a cheap and potentially informative resource for teaching and learning. Also scrutinise TV viewing guides and ask media resources staff to tape programmes relating to material and visual culture.

Task 5.2.9 Collecting 'critical' comments: cuttings

With an 'A' level or GNVQ group complete the following:
    Ask students to bring in any articles or reviews about contemporary art. Ask them to compare reviews in different newspapers and magazines to investigate the ways in which the media construct attitudes and values.
    Invite pupils to write reviews for their own and others' work.
    Set up cuttings files (lamination helps preserve otherwise degradable paper).

Resources

Information on contemporary art is not difficult to obtain, although books and catalogues with good quality reproductions are unavoidably expensive, often beyond the limitations of capitations. Contemporary galleries such as the Tate, Ikon, Laing, Serpentine, Camden Arts Centre, Whitechapel and the ICA have excellent bookshops with catalogues and publications available by mail order. Making your own slides is one way of building resources, providing you keep within the copyright laws! A less well-exploited alternative is the wide variety of magazines currently available; the list is endless but the following are recommended in addition to those already cited: *Art Monthly, Contemporary Visual Arts, Frieze*.

The Department of History of Art & Design at Manchester Metropolitan University, 0161–247 1930, produces and sells slide sets and videos. They deal with design, architecture and fine art, cultural diversity, popular culture. Many are based on exhibitions and original research. It houses the Design Council collection of slide sets which are available on loan, and sells slide sets produced by galleries and museums.

Adams, J., Meecham, P. and Orbach, C. (eds) (1998) *Working with Modern British Art, Tate Collection @ Tate Liverpool*. Produced to support visits and includes sixteen colour prints, a curator's description of the display and a critical essay. It provides a useful framework/model for producing resource packs.

The following provide good colour reproductions of contemporary art:

Art & Design Magazine published six times a year by Academy Group Ltd.

Archer, M. (1996) *Installation Art*, London: Thames & Hudson.

Blurring the Boundaries: Installation Art 1969–1996, San Diego, CA: Museum of Contemporary Art.

Button, V. (1998) *The Turner Prize*, London: Tate Gallery.

Morgan, S. and Morris, F. (1995) *Rites of Passage: Art for the End of the Century*, London: Tate Gallery.

Price, D. (1998) *The New Neurotic Realists*, London: The Saatchi Gallery.

Sensations Exhibition Catalogue (1997), London: Royal Academy of Arts.

5.3 PARTNERSHIPS WITH OUTSIDE AGENCIES

Giroux (1994) insists that:

> Teachers cannot locate teaching in one space, they need to engage with other educators in a variety of sites in order to expand the meaning and places where pedagogy is undertaken.
>
> (p. 166)

OBJECTIVES

By the end of this unit you should be able to:

- identify some of the important national and local contributors to resource-based learning;
- examine how resource packs support visits and follow-up work;
- identify the role of the professional artists in schools, LEAs and local arts boards.

GALLERIES AND MUSEUMS

In a special edition of the American journal *Art Education* entitled *Art Museum/School Collaboration*, Mayer (1998) suggests that in gallery education there is a shift away from object-centred values towards people-centred values. She claims that:

> the most significant change in art historical theory that semiotics offers educators concerns the 'receiver' – the viewer. Central to semiotics are theories of reception . . . the goal of semiotic analysis of works of art is not to produce interpretations of those works, but to investigate the processes viewers use to make sense of visual art.
>
> (p. 19)

The idea that the viewer has agency and is not just a passive recipient is significant and calls into question more traditional ways of gallery education. Mayer believes that by engaging pupils in this way, 'they become empowered as active, equal thinking members of an expanding community of inquiry' (*ibid*.) as critically literate.

Task 5.3.1 Gallery education

Most galleries and museums, whether small local galleries or large national institutions are willing to allow student teachers to observe gallery educators/artists/art historians working with school groups.

Arrange to go to a number of gallery talks for different age groups. (Always phone first.)

Enquire whether the content/delivery has been negotiated with the class teacher.

Observe different approaches to interpretation.

Identify what determines the approach: the type of work; the age of the audience; the specialism of the leader.

What types of discussion take place? Are questions open, closed, framed?

Is the talk descriptive, analytical, challenging, provocative? Does it raise issues, introduce or invite different readings?

This is a useful exercise, as Reeve (1995: 80) points out, 'at least 80 per cent of museums and galleries do not have any educators' therefore if you are to work with gallery and museum exhibits you will need to understand how working in this situation can differ from the classroom.

Anderson refers to 'an expanded concept of museums' insisting that no culture can be adequately represented in museums by material culture alone (Anderson 1997: xiv). He suggests that the museum walls should not be seen as a boundary. He calls for a concept of museums that includes their natural and cultural environment: one which reflects not just the physical and economic environment of the museum but also how the institution can depend on non-material culture: 'shared values, ideologies, oral traditions, rituals, ethical standards and beliefs that give meaning and symbolic significance to our material world' (*ibid*.). Given this definition no one can claim not to have a 'museum' equivalent in close proximity to the school: a local cemetery, church, mosque or synagogue, historic buildings, local festivals, football grounds, visiting and permanent funfairs, all have their cultural significance and value. Anderson claims that museum education develops 'cultural literacy', the capacity to understand, respect and interact with people from different cultural backgrounds; it develops a sense of our own identity (*ibid.*: 7). He suggests that museums can demonstrate ethical leadership; that their values, codes of morality and expectations of behaviour (embraced by this concept of cultural literacy) help to shape those of the public. Whereas in the past most museums claimed to provide a neutral space, excluding controversy, this is now changing and the museum can be seen as a site for different voices, a place that encourages questioning and enquiry: 'children need a gallery environment that allows open and exploratory learning and encourages them to question and challenge' (*ibid.*: 23). Feminist critics, whilst acknowledging some change, do not believe it was ever neutral and suggest most exhibition spaces have a long way to go before they can claim to operate as dynamic and reflexive cultural forms (Duncan 1995; Duffin 1995).

Thirty per cent of museums have libraries and study collections; the potential for teachers and older students to use these for independent on-site and on-line learning has yet to be exploited to the full. The development of digital technology links between the museum and education sector has yet to realise its potential, but given the rapid nature of development in this field it is one that teachers can ill afford to ignore.

Galleries and museums are a rich resource for art and design education; both permanent and temporary exhibitions at such venues are accompanied by catalogues and teachers' packs. Many offer talks and practical workshops run by art historians, critics or artists. Some are involved in outreach projects involving artists in community projects including school placements. The benefits of establishing links with local exhibition spaces and collections should not be overlooked. They provide an opportunity to involve pupils in discussions about the type, origins, quality and significance of the work on display, to make comparisons with other exhibition spaces, and to consider issues of patronage, funding, curatorial decisions, target audience; issues which can all too easily be overlooked when working with a large national institution where there is a temptation to accept what is written as the 'correct' interpretation.

Selwood *et al.* (1994: 37) claim that secondary teachers equate the accumulation of cultural capital with education. They are critical of the way teachers bring charabancs of pupils from the north, spending hours on the motorway, in order to briefly visit a national collection based in London just to be able to 'tick off items on the cultural agenda' rather than gain an intimate knowledge of less famous local resources. They reveal that this attitude is commonplace citing the fact that a third of all visitors to the best-known Western museums never get as far as the galleries. 'Their object in visiting is to buy souvenirs, savour the ambience, have a cup of coffee and say they have been. Their visit is simply a symbolic gesture' (*ibid.*).

Resource packs

Hughes (1998b) suggests that student teachers entering the teaching profession all too quickly forget their graduate training which becomes 'rapidly overlain with the orthodoxy of the school curriculum' (p. 44). As a result of the pressures from an overfull traditional, subject-based, mandated curriculum, Art & Design in all its practical and theoretical complexity has to be delivered to most pupils in less than an hour. In order to cope teachers use commercial books and resource packs: expediency is taking over from philosophy and teachers are in danger of becoming ahistorical.

You need to approach resource packs with a degree of scepticism. It is all too easy to presume that because the content has been selected for publication by reputable publishing firms it is sound. It is usually 'safe', selected to appeal to the widest audience possible. Most commercial packs, whether slide packs, postcard packs, video tapes or CD-ROMs, deal with information about established art; artists and cultures. Many deal with content, context and history but are less well developed in terms of pedagogy. Some promote what is disparagingly called a 'Blue Peter' approach to education. Bancroft (1995) examined a number of packs selected from advertising material and catalogues sent to her school. Her evaluations identified that most packs promoted the limited canon of DWEMs. Although 9 per cent of the images were by black artists this was often accompanied by patronising (mis)information with work from other cultures placed in a 'timeless past'. The artists in the packs may be, in the main, dead, but issues such as 'death' are studiously avoided. All too often packs are a sanitised collection of favourite images chosen to cause no offence but which fit neatly into a 'safe' curriculum and are easily subjected to a traditional, formalist analysis. Textual information is in the form of biographical notes; questions are 'closed' suggesting that there is a correct response.

Unpacking Teachers' Packs (Clive and Geggie 1998) is an evaluation of teachers' packs from ten London galleries and museums by Engage, the national association for gallery education.

Task 5.3.2 Unpacking teachers' packs

Contact a number of local and/or national museums and galleries and request a recent teachers' pack. Ask to be placed on their mailing lists to receive posters advertising exhibitions and invitations to free teachers' evenings.

With other student teachers consider the following for each pack:

- Who is the target audience – is it teachers, pupils or both?
- Does the pack provide information and/or raise questions?
- Are fixed curatorial views presented or are different readings given?
- What is provided: text, images, historical, contextual information, further reading, etc.?
- How is it presented?
- How could it be developed to make it appropriate for use with pupils at KS3?
- Does the pack provide useful INSET material or is it limited to a taster tempting you to visit the exihibition?
- Are connections made with the NC, GCSE, A level and GNVQ criteria? Is this necessary?

ARTISTS, VISITS AND REGIONAL ARTS BOARDS (RAB)

Artists are often provided subsidised studios on the understanding that they engage in community work including projects with local schools. Information about such schemes is readily available from local education authorities (LEAs), gallery education departments and libraries. However, the first port of call should be your Regional Arts Board (RAB) which is in a position to advise you about educational projects involving artists. Joint applications between artists and teachers which have a clear endorsement from the school's senior management team are most favourably received. RAB also support larger programmes involving a number of schools, including cross-phase and cross-arts initiatives or identified regional educational priorities in the visual arts. For example, environmental projects, site-specific or urban regeneration projects bring together professional artists from a range of specialisms to work collaboratively with schools, town planners and environmentalists and RABs can also give up-to-date information and advice about funding.

Task 5.3.3 Networking

Contact the RAB (*http://www.arts.org.uk*) and ask them to send you their most recent newsletter or bulletin. Find out about recent educational projects, how they were set up, funded and evaluated.

Ask to be placed on a mailing list informing you of local and national projects.

The role and value of the professional artist as a resource for teaching and learning has given rise to interesting debates, especially in secondary schools where the teacher has a degree in art or design and may practise as an artist part-time. The potential benefits (for artist, teacher and pupils) of such collaborations are enormous but so are the disadvantages if it goes wrong. The placement of an artist in education is seen as one answer to the problem of the separation of art and artist from the public, enabling young people to observe artists' working processes, work collaboratively with them, engage them in discussion, etc. However, it must never be seen as a panacea. It is important to help pupils understand the complexity of contemporary practice; introducing them to the work of one artist can give them a limited view of contemporary art. Pupils need to be introduced to the diversity of practice and a well-planned residency in school enables them to extend their definition, understanding and skills in art, craft or design. However, it should also involve an element of curriculum development so that, once the artist has left, the concepts and processes can be revisited and continue to inform the curriculum.

Resources

Dahl (1990), Manser and Wilmot (1995), Sharp and Dust (1998), and Oddie (1998) provide comprehensive information on artists' placements in education including: contacts, selection of the artist, funding, planning, INSET possibilities through to evaluation of the project. Taylor (1991), Dickson (1995), Clive and Binch (1994) provide examples of exemplary practice. Burgess (1995) stresses the need to ensure that any placement involves curriculum development.

PUBLIC ART, ARCHITECTURE AND THE BUILT ENVIRONMENT

The usual public art story, as we know, is of pieces plonked in fore-courts and pedestrianised streets in ill-conceived bids to brighten up (or compensate for) botched bits of urban design. Couched mostly in the language of modernism, and therefore intrinsically siteless and nomadic, public art often

> succeeds merely in transforming places which were once public into an annex of the contemporary art gallery – that forbidding, empty space where only the initiated feel comfortable.
>
> (Usherwood 1998: 46)

Gablik (1995), like Usherwood (1998) questions the meaning of art in a postmodern age. She suggests that the focus should shift from culturally sanctioned spaces, such as museums and galleries to social, natural and community spaces. Similarly, Lacy (1995) and Neperud (1995) contend that educators need to move away from art on gallery walls to seeing the natural and built environment as spaces of social responsibility and aesthetic improvement. If these are important sites for artistic production then they should be recognised as resources for teaching and learning. Mark Fisher MP (1994) insists: 'We are not going to get good critical appreciation of the built environment without having it understood at school level'.

Avery (1994) insists that built environment education must encourage pupils to be socially responsible and consider the needs of gender, race and class in relation to the environment. She believes that the introduction of literature which focuses on women and the built environment can inspire pupils to investigate critically their surroundings and recognise how it impacts on others. She provides an extensive list of relevant references including Women and Built Environment (WEB) and Matrix, a feminist architectural practice. Adams (1989) stresses the importance of fostering aesthetic responses to the environment. She expresses her concern that learning which emphasises socio-economic and political aspects too often neglects the subjective view and individual response (see Plates 5 and 6). She quotes Jonathan Raban to reinforce this perception: 'the city as we know it, the soft city of illusion, myth and nightmare is as real as the hard city one can find on maps, in statistics, monographs on urban sociology and architecture' (p. 192). Both Adams and Avery agree that in order to encourage attitudinal changes, support from outside agencies is important.

Building on her earlier research, Adams (1997) presents a convincing case for 'public art' education. She suggests that projects to change the school environment echo the function of many public art projects, that: 'it is site-specific, concerned with environmental improvement. It develops a sense of place and aims to create a sense of identity and local distinctiveness' (p. 236). Thurber (1997) asks: 'What sources and resources, what "sites to behold", do you have in nearby communities that are waiting to become rich sites for inquiry in Art for your students?' (p. 39). One can safely assert that the majority of school and college grounds are often a sadly neglected resource. They can be best described as flat, featureless places, consisting of vast areas of tarmac and close-mown grass (Ker 1997: 61). Redesigning a 'close-mown grass' lawn into a landscaped garden or the school's own sculpture park involving local artists, town planners, designers and architects is an ambitious project for student teachers. However, the use of particular locations as a stimulus for site-specific plans and maquettes is worth considering. This is excellent preparation for more ambitious projects once you have secured a teaching post.

Task 5.3.4 Redesigning the school environment

Survey the school environment and identify possible areas to redesign. Locate plans of the school and its grounds. If the school does not have copies then the local town hall or library should.

Devise a SoW which encourages pupils to investigate the school environment, to look at it critically with a view to improving it. Geographical, environmental, health and safety issues and cost implications should be taken into consideration.

Ensure aesthetic response and social and political understanding are all included.

Resources

Your RAB and/or LEA has information about local artists currently working on environmental projects both independently and with pupils. They can provide information about previous projects in the area that they have sponsored. Find out how these projects were set up and funded; identify opportunities for you to get involved in the future.

The Architecture in Education Project is an initiative of the Architecture Unit at the Arts Council of England (ACE) in collaboration with Design Museum and the University of Greenwich School of Architecture. As part of the project the Design Museum produced an architectural handling collection. The project is currently working on a national outreach programme. It aims to deliver training and teaching modules to secondary schools in England from September 1999 to July 2002 and provide a national frameworking for the advocacy of architecture and the built environment.

English Heritage produce a series of user-friendly booklets including *Using Listed Buildings*. They have regional offices throughout the UK and for a small charge provide conducted tours of listed buildings: 0171–973 3442.

The Royal Institute of British Architecture (RIBA) 1998 relaunched its architecture gallery with a programme of exhibitions designed to attract school parties and families alongside architecture students and design professionals. Contact the education coordinator: 0171–307 3682.

Public Art Journal is a features/review magazine that brings together art education and the built environment.

The Building Experience Trust has developed a national schools programme to promote work on the built environment. Workshops are presented by trained animateurs. Contact: 01223 65378.

www.publicart.sunderland.com provides useful information and images about regional development projects involving artists, the community and schools.

5.4 CHANGING DEFINITIONS – NEW TECHNOLOGIES ICT

The increasing importance of 'new technology' in the development of digital imaging, multimedia and video is widely recognised but, as yet it is unclear how those of us responsible for visual education should respond. For many of us, our own understanding of the technology and its potential is limited and we are reluctant to replace tried and trusted methods and subject methods with unknown content.

(Binch 1997: 6)

OBJECTIVES

By the end of this unit you should be able to:

- recognise how new technologies are changing the way educational resources are accessed and used;
- appreciate the continuities and discontinuities between still and digital imagery;
- identify venues, sites and publications which enable you to develop RBL.

PHOTOGRAPHY AND NEW TECHNOLOGIES

It is with caution that I make photography a special case. It is perhaps, more correctly identified as one of the many media within contemporary art practice. Dewdney (1996) recognises that, although the art, photography, technology continuum provides developing historical narratives, technology will not be limited by its predecessors:

> New technology has a culture of very rapid transmission. On the one hand it is a technological development which builds upon and supersedes others before it like the magic lantern, photography, film, telegraphy, radio and television. On the other hand, it is a cluster of technologies which converge to produce a hybrid of machines which can combine visual, text and auditory information transmission, extremely large information storage with manipulation and authoring programmes. For art, media and design, the quickening pace of new information technology heralds the dissolution, not only of previous distinct forms and practices of design, photography, animation, film and fine art but also the categories and disciplines of knowledge upon which those practices rest.
>
> (p. 86)

Photography once a distinct, albeit small, area of Art & Design education has all but been subsumed by the digital revolution. However, whether chemical or digital, photography still plays an important role in the lives of young people: family albums, newspapers, books, posters, postcards, advertisements, compact disc covers and so on. It is all too easy to 'respect' the photographic image as a factual record; often it carries no mark of its maker to remind us that it is a construct. When using photographic images in the classroom (still, digital or moving) you should always consider the way the producer has intervened with the 'real', how images are created, mediated and circulated. Stanley (1996) insists 'the very ubiquity of manipulated images may, if not studied seriously, inoculate us against a close and continual suspicious inspection of their contents, intent and manipulative strategy' (p. 4). You should help pupils to understand that photography does considerably more than record or confirm an event. Whether displayed on a gallery wall or on the back of a bus, photography should be viewed critically, especially if it is to inform practice. Buchanan (1995) promotes the notion of critical literacy, a combination of visual literacy and critical studies which he believes can counter the false separation between theory and practice. Pupils should be encouraged to be 'critically literate', to uncover the ways photographic imagery can represent and position the subject, build pathos, indicate status, aesthetise the politics of Third-World poverty or reinforce gender stereotypes, both male and female. Works by Richard Billingham, Adrian Piper, Ingrid Pollard, Martha Rosler, Cindy Sherman, Yinka Shonibare, Jo Spence, and Sam Taylor-Wood, present us with questioning, occasionally disturbing, sometimes ambiguous, occasionally playful, images which challenge typical expectations of the photographic image. An image can be interpreted in many ways. Artists often construct images that suggest unlimited and unstable meanings. In contrast advertisers often produce images aimed at a particular audience (Morley 1992). Ferguson (1995) insists that advertisements not only sell products, they also sell politics. Jones (1996) states that at secondary level the range of work in ICT applications needs to, 'be made more responsive to pupils' backgrounds and interests ... There are opportunities for well planned Art/IT projects which have for example, a social or ethical dimension' (p. 1).

Task 5.4.1 More than just a pretty picture

> Any representation can be 'problematised for the ways in which it both produces and covers over forms of cultural self-representation that are constructed within dominant historical, hierarchical, and representational systems'.
>
> (Giroux 1994: 103)

With other members of your tutor group, devise a year 12 SoW which encourages pupils to consider the way photographic images 'frame' the way people represent ourselves and others.

Use work from two or more of the above artists plus images selected by students to contextualise making.

The advances in digital photography, including the improved quality and reduced cost of the computer print-out, heralds the demise of the single-lens reflex camera and its environmentally unfriendly chemical processes. However, Stanley (1996) calls for caution when he asks educators to ponder the following question: 'do we, on the basis of the development of digitised imagery to date, have reason to rethink our views about the significance of photography?' (p. 98) He suggests you consider the relative merits of an interactive package with that of a family photo album. He warns us to be, 'wary of special pleading by those who offer the future which all too often bears an uncanny resemblance to the past' (*ibid.*). Similarly Tate (1996) insists we need to 'evaluate the somewhat messianic claims we keep hearing about how we ought to reconstruct the curriculum to take into account the need of the information super highway' (p. 2).

Resources

The journals: *Aperture, Creative Review, Creative Camera* and *Doubletake, 20:20* (the national magazine for photography and media education) provide up-to-date information on exhibitions and also suppliers.

Isherwood, S. and Stanley, N. (eds) (1994) *Creating Vision*, London: ACE. Ends with a useful selection of resources and contacts.

Boyd, F., Dewdney, A., and Lister, M. (1996) From Silver to Silicon CD-ROM. London: Artec. Explores the nature of photography in the digital age.

Creative Camera (March 1999) includes a guide to UK galleries. This identifies 154 photographic galleries throughout the UK – the majority are art galleries or museums which devote a significant proportion of their programme to photography. Forty are 'dedicated' photography spaces. My personal selection includes:

Dazed & Confused, 112 Old Street London EC1V 9BD (0171–336 0766) Exhibits challenging contemporary photography.

Lighthouse Media Centre, Chubb Buildings, Fryer Street, Wolverhampton (01902 716055) *www.light-house.co.uk*. A dedicated space for lens-based and digital media.

The Pavillion, 2 Woodhouse Square, Leeds LS3 1DA, (01132 431749). Photographic arts centre for women.

Viewpoint, Old Fire Station, The Crescent, Salford M5 4NZ (0161 737 1040). Exhibits and commissions a wide range of photo-based work.

The Photographers' Gallery, 5 & 8 Great Newport Street London, N1 (0171 831 1772). Exhibits a wide range of image-based works including historical and contemporary.

National Museum of Photography, Film and Television, Pictureville, Bradford, BD1 1NQ *www.nmsi.ac.uk/nmpft* (01274 727488). Exhibits both historical and contemporary works.

The Victoria and Albert Museum holds the national collection of the art of photography in the nineteenth and twentieth centuries with 'Education Boxes' designed for use with visiting groups.

(CON)FUSION: THE WAY FORWARD?

> It is essential that we try and form a contemporary view of art and IT
> developments for ourselves, using our subject expertise and working towards
> an understanding of where art, craft, design and technology are taking each
> other.
>
> (Jones 1996: 5)

Fusion: Art & IT in Practice (NCET 1998), a resource produced 'to support NC Art at
KS3 and beyond' suggests that teachers need to form a 'fusion' between traditional
practice and new opportunities. However, when subject to closer scrutiny it appears
to be in danger of just reproducing the dominant orthodoxy electronically. It fails to
raise issues or refer to artists who are currently developing work which explores how
ICT can be productive rather than just reproductive. Reference to the work of
'others' is restricted to the established canon with one notable exception: Barbara
Kruger. As Meecham (1999) points out: 'Currently documents like *Fusion* and the
AVP *Art Computer* [1998], software CD-ROMs and videos – modernism's technologically validated publishing arm – still reproduce the old hierarchies but faster'
(p. 81).

Fusion makes no claims towards providing exemplary materials; its authors suggest
that it provides a stimulus for discussion about how ICT can enhance the creative
process. However, the document is supported by HMI, funded by the DfEE and
distributed free to all state schools; therefore it is perceived as the validated approach.
If you accept this easy 'solution' you will have missed an opportunity to use new
technologies to move beyond the existing school art orthodoxy:

> The fragmented, often contradictory, multidisciplinary and intercultural
> references to knowledge that students interact with through visual
> technology may have more to do with student understanding of the subject
> than does a curriculum based on the structure of a discipline. Such
> postmodern visual experience should not be made to fit into modernist
> curriculum frameworks. Instead, the interpretative, didactic, even seductive
> power of the imagery should be given attention in school.
>
> (Freedman 1997: 7)

Freedman reminds you that computer 'images speak to us'. Unlike other resources
which can be carefully selected to reinforce learning, the images accessed by pupils via
the computer are not so easy for teachers to control. Pupil access to unsuitable
materials can be denied (material deemed unsuitable because of its racist, porno-
graphic or violent subject matter). There are other important issues in relation to new
technologies which should not be ignored. It is your responsibility to ensure that you
promote a 'screen culture' that refuses social isolation: recognise your responsibility
as a teacher to devise projects which extend communication not isolation. Look too
at the gendered use of ICT (Spender 1995): research to date shows usage in the
classroom is dominated by males.

CD-ROMS

De Montfort University, Leicester has established a National Collection of CD-ROMS for Art & Design education. Since 1996 it has been collecting and evaluating CD-ROMs in partnership with the British Education and Communications Technology Agency (BECTa) and the Qualifications and Curriculum Authority (QCA). The purpose of the collection is to provide information about the range of CD-ROMs available to support Art & Design education including evaluations by teachers who have tested them with pupils to ascertain their usefulness and relevance as a classroom resource. Edited versions of the evaluations can be found on Virtual Teacher Centre (*http://www.vtc.ngfl.gov.uk*).

CD-ROMs are a powerful tool in any art department for supporting work at key stages 3 and 4. However, while there are many CD-ROMs available, their coverage is currently focused largely on Western art (BECTa 1998: 37). The National Collection confirms this statement; however, it is anticipated that as teacher evaluations of the collection are fed back to publishers via the centre, greater reference will be made to non-European art, women, photography, architecture, contemporary artists, designers and craftspeople.

The list of CD-ROMs available for Art & Design education is growing rapidly, and earlier limitations are being resolved. Improved technical reproduction and download facilities raise issues of copyright. There are a number of ways of 'grabbing', saving and downloading images from CD-ROMs; before you do, check the terms of purchase and find out how the school operates under copyright law.

Resources

My personal selection would include:

Art 20th Century: Multimedia Dictionary of Modern Art, Thames & Hudson.

Art Connections: Cultural Links (1997) Bradford Art Galleries and Museums, includes seventy works from their collections covering past, contemporary, Western and non-Western art. Produced specifically for use in schools it relates to the NC. Text is in both English and Urdu.

Arts Council of England CD-ROM celebrates what is new and challenging in contemporary visual art: Gillian Ayres, Damien Hirst, Rachel Whiteread and Catherine Yass.

Contemporary Artists in Scotland (1998) produced by AXIS at Leeds Metropolitan University.

Keith Piper: Relocating the Remains (1997) inIVA, accompanies a book. It documents Piper's work from 1980–1997 in which he powerfully interprets the iniquities and struggles of the black diaspora. (See Plate 17).

Urban Feedback inspired by the chaotic energy of modern cities and produced by artists Sophie Greenfield and Giles Rollestone, available from Research Publishing, 78 Liverpool Street, London, N1 0QD.

> **Task 5.4.2 Working with CD-ROMs**
>
> Select and view an Art and Design CD-ROM. Devise an activity for a year 7 group which encourages them to translate the information on the disk into their own practice in the classroom. Remember many classrooms have only one computer therefore you will need to consider access. Group work and clearly focused tasks encouraging pupils to skip over irrelevant information can be useful. It is important that you differentiate between information and understanding.
>
> How does the CD-ROM relate to the KS3 NC orders?
>
> Is the language used appropriate?
>
> Does it encourage critical enquiry or provide answers?
>
> Does it promote or value a particular way of working?
>
> What hidden curriculum does it contain?
>
> In what ways might it be used to inform making?
>
> What is it providing that more conventional teaching cannot?
>
> What can you learn from teaching methods deployed by successful interactive CD-ROMs?
>
> What are the disadvantages of CD-ROMs as a learning resource?

INTERNET AND WORLD WIDE WEB

There is an endless supply of digital art on the Net; from digitised images of cave paintings from Vallon-Pont-d'Arc in France to the latest techno art at the Pixel Pushers Gallery. Web site *http://www.world-arts-resources.com* provides information on current exhibitions and access to high-quality images. Downloading images from the Net is time consuming and the quality is dependent on numerous variables from original site to the colour adjustment on your printer. Clearly it is a poor substitute for the real thing. Many artists are now producing work directly for the Web or adapting work to the new media. The Arts Ed Net Getty Foundation (*http://www.artsednet.getty.edu/*) provides a 'gateway' to art-related sites on the Web, an extensive list of museums around the world. This should be visited and useful sites bookmarked for use by pupils. The interactivity of ICT makes possible new relationships between HE-schools, schools-schools, schools-galleries and museums. The National Grid for Learning (*www.vtc.ngfl.gov.uk/resource/cits/art*) connects schools to a wide range of learning resources, including libraries, galleries and museums.

The value of the Internet and WWW as resources for Art & Design education is indisputable. However, trawling through cyberspace for resources is time consuming and distractions from the initial task almost inevitable. Whilst it is important that you develop the skills necessary to surf the Net and utilise search engines to locate interesting and sometimes unexpected resources there needs to be a valid reason before you use valuable time during lessons teaching pupils how to do the same. All you need to do is to identify relevant sites and store them under bookmarks thus providing pupils with ready access. Pupils should be encouraged to note down any relevant Web site addresses they come across at home or in IT and other lessons and include these in an art directory (bookmarked) and store pertinent images in folders (virtual sketch-books). It is important that you realise that some pupils have highly developed skills in ICT. They will have acquired these at home, in cybercafés and in other areas of the curriculum. Sefton-Green (1998) believes that young people who regularly 'surf the Net' at their own pace may well find the regimented structure of a teacher-led curriculum tedious (p. 12). Indeed pupils may know much more than you do about new technologies. Respect this expertise and recognise it as a valuable resource.

Schoolart (*www.schoolart.co.uk*) is an 'art education publication and teacher centered service' available on the Internet. It claims to be: '. . . an architecture that constructs virtual spaces in which good ideas about teaching can be clearly expressed . . . (a site where) innovation and originality in art education thinking can be set alongside down-to-earth advice for . . . inexperienced teachers and students.' Clearly this is a publicity statement targeting student teachers. Before paying a subscription to this or any other site, you are encouraged to access demonstration units on a Web browser first.

Resources

There is a never-ending supply of useful Web sites. My selection to date includes:

ADAM: (*adam.ac.uk/*). Art, Design, Architecture & Media information gateway: a searchable catalogue of hundreds of Internet resources.

Anne Baker's project: (*www.flatearth.co.uk/annescafe*)

ART on FILE: (*www.artonfile.com/*). Images of Art, Architecture and Design. ART on FILE's mission is to document important new developments in the built environment. Many projects combine art, architecture, urban planning and landscape design.

ARTEC: (*www.artec.org.uk/*). Digital arts and multimedia, 'laboratory for artists and creative professionals to explore this new territory'.

Channel: (*www.channel.org.uk*). Showcases Net art works with links to a wide range of work on-line.

Computer ARTWORKS: (*artworks.co.uk*). Contains Latham's virtual sculpture.

CTI: Computers in Teaching Initiative (*www.bton.ac.uk/ctiad/*). This provides an on-line resource guide and outline newsletter reviewing new developments and CD-ROMs.

ACTLAB: (*http://home.actlab.utex.edu/*) is at the boundaries where technology, art and culture collide. Worried about computers as the sewing machines of the twenty-first century ACT-LAB calls instead for 'room to stretch and do risky things'.

Dia center for the Arts: (*www.diacenter.org/*). This contains arts projects for the Web, long-term installations, information re art education in the USA.

Franklin Furnace Gallery: (*www.franklinfurnace.org/cyber.html*). Avant-garde performance and interventions include Group Material and Guerrilla Girls.

Institute of Contemporary Art (ICA): (*www.ica.org.uk*).

Lisson Gallery: (*www.lisson.co.uk*).

LUX Centre: video and digital art work by contemporary artists using electronic technology (*www.lux.org.uk/*).

The Metropolitan Museum of Worldwide Arts Resource: (*www.wwar.com/*).

MOMA: Museum of Modern Art, New York (*www.moma.org*).

New Media Encyclopedia: (*www.newmedia.arts.org/*)

TEST: (*www.test.org.uk*). A digital research facility providing public access for anyone wanting to explore the creative potential of digital media.

Twenty-four Hour Museum: (*www.24hourmuseum.org.uk*).

The Walker Art Centre: (*www.walkerart.org/*).

White Cube: a project room for contemporary art (*www.whitecube.com*).

RBL: THE WAY FORWARD?

In the foreword to the Green Paper, *Teachers Meeting the Challenge of Change*, (Blair 1998) suggests that teachers have suffered 'decades of drift'. He insists that the profession needs to 'engender a strong culture of professional development'. It calls for a contractual duty for all teachers to keep their skills up to date. Your FPD requirements as a newly qualified teacher will be identified in your Career Entry Profile (CEP); a statutory 'induction' programme enabling you to achieve these is now an entitlement for all those entering the teaching profession. The CEP serves to remind you that initial teacher education is a launching pad, that teaching needs constant refuelling and refreshing. During your ITE course you need to identify and establish contacts with resources in the field of Art & Design that will enable you to identify your FPD.

Task 5.4.3 Identifying FPD resources

Identify the Art & Design INSET/FPD provision available in your placement schools.

Discuss with your mentors resources available locally and nationally. Make a record of these.

Contact AAH, Engage and NSEAD (visit Web site) to find out what is currently being offered.

Find out about proposed conferences and publications via the above or through your LEA or RAB.

Get your name on as many mailing lists as possible.

Identify provision for FPD accreditation, links with a modular MA in Art & Design Education, Museums & Galleries, Critical Studies and Art History.

It is no simple task to develop the type of RBL that engages with the issues raised in this chapter. It demands that you appreciate difference, understand context, recognise agency (your own pupils and others) make critical comparative judgements on the basis of evidence and empathy, not received opinion. Giroux (1994) insists that teachers must be researchers and work with each other in producing curricula. Lippard (1995) identifies teachers as agents of exchange as well as change agents. She believes that we have to shift the emphasis away from bringing 'great' art to people for their veneration towards working with people to create meanings both in and through art:

> We are laying out the ingredients but still looking for the recipe. Once there are more cooks, everyone will use the ingredients differently . . . Critical consciousness is a process of recognising both limitations and possibilities. We need to collaborate with small and large, social, political, specialized groups of people already informed on and immersed in the issues . . . At the same time we have to collaborate with those whose backgrounds and maybe foregrounds are unfamiliar to us, rejecting insidious notions of "diversity" that simply neutralise difference.
>
> (p. 114)

Resources

AAH is the professional association for art historians. Its annual conference engages with current debates in the discipline, and across disciplines. The schools sub-committee organises events for students and teachers (*www.aah.org*).

Engage is a professional body which promotes greater understanding and enjoyment of the visual arts by engaging with the public, artists, galleries and educators. Address: 1 Herbal Hill, London, EC1R 5EF.

NSEAD is one of the most significant resources for Art and Design. Many of the articles cited in this book come from *JADE*, the society's journal which provides an international forum for the dissemination of ideas, practical development and research findings. The important role that NSEAD plays in producing and selling publications for art and design education should not be overlooked. In addition it provides INSET and organises conferences. Detailed information about its activities can be found on its Web site (*www.nsead.org*).

Standards site includes a discussion forum, allowing teachers to share with each other their experience in implementing strategies for raising standards in school – includes video clips. Not art and design specific, nevertheless it has the potential to promote curriculum development projects to a wide audience both in and beyond the subject field. It is important that as student teachers you recognise the importance of advocacy and ensure that visual and aesthetic literacy is promoted alongside numeracy and literacy. It includes best practice, schemes of work and lesson plans for teachers to access and customise and encourages teachers to share teaching materials (*www.standards.dfee.gov.uk*).

6 Practice in PGCE Art & Design

Nicholas Addison and
Lesley Burgess

This chapter is divided into two sections, the first examining your teaching experience (TE) and the second your Higher Education (HE) course studies. Some units evidently belong to both, indeed the two components are indivisibly related. It is only the opportunities and resources offered by these respective sites that determine differences: their common concern is pedagogy. The way the two components are sequenced in time varies from course to course, but the HE component is often front- and end-loaded to allow for a sustained period in school. The way these components have been divided in this chapter means it does not run in strict chronological order but looks at different issues and practices.

OBJECTIVES

By the end of this chapter you should be able to understand:

- the trajectory of your TE, its different components and cumulative effect;
- the function of course studies to enable you to develop and transform subject knowledge into effective pedagogical practice;
- the means by which the partnership model enables you to plan, develop, evaluate and record:

  1 your own studies and practice of art, craft and design;
  2 a broad theoretical and philosophical base on which to build educational practice;
  3 specific school experiences for developing Standards (DfEE 4/98) in subject knowledge and its application, classroom management and assessment.

6.1 TEACHING EXPERIENCE (TE): A PARTNERSHIP MODEL

OBJECTIVES

By the end of this unit you should be able to:

- understand how to use focused observations to gather knowledge of teaching and learning;
- understand the importance of planning and reflection for effective teaching;
- consider ways of developing competence in teaching in relation to the Standards.

The partnership between schools and HE in the education of teachers provides an ideal situation in which to relate theory to practice and vice versa. The interrelated elements of the partnership are made coherent and dynamic through a process of personal reflection in combination with theoretical study and collaborative action (Schon 1987). This provides a basis for the development of your competence in teaching and enables you to contribute across a wide range of Art & Design educational practice. It also promotes your further professional development (see Chapter 13).

The partnership model itself is complex. Consequently responsibilty lies with you, as a postgraduate student to negotiate and manage this partnership alongside your HE and school tutors.

INDUCTION TO TEACHING EXPERIENCE

Before taking responsibility for teaching you are required to undertake a number of activities and tasks in your placement schools directly related to subject application, including classroom observation and shadowing, collaborative teaching and lesson analysis. Documents (proformas and observation schedules) supporting these activities should be provided by your partnership, but the following tasks indicate something of the range you should cover.

OBSERVATION TASKS

During induction you are involved in all or most of the following activities as part of a programme of structured/supported classroom work:

- classroom observation;
- assisting with teaching;
- collecting information about the previous projects undertaken by the pupils you are to teach;
- finalising TE timetable;
- planning lessons;
- target setting, etc.;
- identifying the department's philosophy and its relationship to NC Art;
- pupil/teacher shadowing;
- area–based study.

Task 6.1.1 Induction exercise

During the induction periods at your TE schools you should find out and record the following information:

- the names of colleagues with whom you are working;
- lesson times;
- sizes of classes/groups;
- the composition of groups: mixed ability, etc.;
- available materials and equipment (including health and safety requirements);
- available resources: visual material, videos, books, slides, etc.;
- the types of courses on offer in Art and Design;
- departmental and school policies: equal opportunities, discipline, differentiation, assessment, homework etc;
- examples of programmes of study (PoS), schemes of work (SoW), etc.;
- your teaching timetable;
- groups you are expected to observe/assist/teach;
- provision for, and use of, Information Communication Technology (ICT) in the school and your department;
- Special Educational Needs (SEN) provision within the school and your department.

TE, SCHOOL 1

Progressively during the TE you are introduced to teaching through systematic observation, planning, practice and reflection. As you gain competence you work with individuals and small groups and participate in team teaching with experienced colleagues. Throughout your TE you are supported by school tutors, probably your subject mentor and a teacher responsible for professional studies, who monitor and review your progress in relation to the training Standards (DfEE 4/98). Your HE tutor visits you and, with school colleagues, observes and assesses your teaching, evaluating your development and setting targets.

During your TE you may be preoccupied with surviving the mechanics of teaching, however, it is essential that you are ready to apply theory to practice and use both to develop a personal educational philosophy. It is also important for you to establish yourself as a professional colleague across the partnership, someone who enters into and contributes to the life of the whole school. To assist you in planning and evaluating this experience you are required to keep a teaching file.

THE TE FILE: GUIDELINES AND CRITERIA FOR ASSESSMENT

The TE file is your primary means of recording your developing practice and is a vital resource for your teaching. This file should contain records of your schemes of work, lesson plans and evaluations. It is important that you adopt a systematic approach, filing each teaching group or SoW in a logical order. An A4 loose-leaf file with dividers is ideal for this purpose. From your file, your tutors and colleagues should be able to gain insights into the nature of your teaching and the quality of the learning experiences you provide.

SoW and lesson plans should contain three sections:

Intentions: aims and objectives

You are advised to produce a cover sheet clearly explaining the following:

- *Why* you have chosen a given topic.
- *What* learning outcomes you expect.
- *How* you can enable pupils to achieve them.

Preparation

Make lists of:

- Materials and equipment: the things pupils work with – paper, printing rollers, inks, etc.
- Resources: the things you use to support your teaching – slides, videos, reproductions, objects, visits, texts, etc.
- Health and safety regulations.

Reflections

Following each lesson you are required to record your evaluation of your:

- performance;
- pupils' responses and the extent to which they have realised the learning objectives;
- implications for the following lesson and differentiating need.

Your TE reflections and notes should clearly show the ways in which you have made connections between your workshop experiences and your teaching, and the requirements of the NC for Art, GCSE and post-16 courses. It should reflect your resourcefulness and include relevant visual material: photographs, illustrations, drawings, diagrams, samples, reproductions, catalogues, trigger sheets, etc.

Planning is the key to effective teaching. You will feel confident if you have:

- understood the educational aims and learning objectives of a SoW;
- researched and resourced all relevant aspects of subject knowledge and its application to classroom activities;
- related knowledge to the particular pupils you are to teach, including their prior knowledge and special educational needs;
- fully prepared yourself with questions, key words, visual aids and materials;
- considered the arrangement of the room and the different activities of particular lessons, their pace and their impact on the following lesson;
- considered what methods of assessment to employ.

Once this is done, and only then, are you ready to face the question of how to manage classes and how to interact with and motivate pupils. These are elements of teaching that you can only learn through experience in the classroom.

GUIDELINES FOR TEACHING

General conduct

1 Remember not only to learn the names of your mentors and departmental colleagues but also those of the headteacher, SENCO and support staff, ICT coordinators, librarians, etc.

2 Be punctual. If you are unwell and unable to attend school inform the secretary by telephone before morning school and leave a message for your HE tutor.

3 Remain in school all day (unless special prior arrangements have been made with your tutors).

4 Observe school regulations, e.g. about smoking, use of common rooms, dress code, health and safety, fire drill, etc.

Subject knowledge

5 Develop and promote an understanding of the intrinsic and extrinsic significance of Art & Design in the school curriculum and the wider community (Unit 12.2), advocating the importance of visual and aesthetic literacy across the curriculum (Unit 14.1).

6 Support your lessons with systematic plans and carefully selected resources, ensuring coverage of the NC Art PoS and examination syllabuses. Always take

into consideration the variety of needs of pupils relating work to their age and conceptual understanding by using appropriate language and activities.

7   Think of yourself not only as a teacher concerned with practical activities but as someone who can promote understanding and appreciation of art and design through critical, contextual and historical studies. Stimulate intellectual curiosity and communicate your enthusiasm for art, craft and design.

Planning

8   Prepare thoroughly: prepare yourself to meet and to work with your pupils. Find out about their skills, attitudes and previous experience. Attend to practical details, e.g. health and safety, distribution of materials and equipment, storage of work, organisation of space and time.

9   Have constantly in your mind the quality of your pupils' experience of art, craft and design in your lessons. Identify clear learning objectives; develop SoW that challenge pupils and ensure high levels of pupil interest and motivation.

10   There are important decisions to make. Some of these are made by you; others by your pupils, for example: timing/pacing, starting points, media and scale of work, the degree of teacher direction or pupil autonomy.

11   Recognise the need for continuity and progression. Find out about pupils' previous learning in Art & Design. Plan sequences rather than isolated lessons. Conceive every lesson as making another lesson possible.

12   It is important to anticipate what your pupils should be able to gain from your presence. Try to imagine what the teaching/learning event might be like before it happens. Record your anticipations and compare them with your post-lesson evaluations.

Teaching and class management

13   Reinforce the school/department's code of conduct. Recognise the importance of establishing and maintaining a good standard of discipline through well-focused teaching and positive reinforcement.

14   Learn the names of your pupils as efficiently as possible (a diagram of seating arrangements may be helpful). Keep a register.

15   Maintain classrooms, studios and workshops in a clean and safe condition to ensure they operate successfully. You are responsible for ensuring that pupils clean and clear all work surfaces, tools and equipment at the end of every lesson. If you are not conversant with a particular tool, material, or item of equipment, it is your responsibility to ask for assistance and try them out before introducing them to pupils. At no time should pupils use electrical tools or other potentially dangerous equipment without the support of a qualified teacher: remember you are not 'in loco parentis'.

16   Identify pupils who have SEN including those not yet fluent in English and ensure that they are given positive and targeted support. Remember also to identify pupils who are very able; they too need support through differentiated, challenging work and extension activities.

Monitoring assessment, recording, reporting and accountability

17   After each lesson you should reflect upon your teaching and the pupils' learning. To what extent do you meet your aims and objectives? Become conscious of yourself as an effective cause in what happened: make a written record of your evaluation; assess and record each pupil's progress systematically providing targets for learning.

Subsequent TE

During subsequent TEs you can build on the experience and expertise acquired in your first school to develop high levels of teaching skills and classroom management. Placement in other schools also provides you with the opportunity to develop an understanding of how different contexts affect learning.

Regular feedback and evaluation of your progress, coordinated by your subject mentor, takes place in each TE. This leads to a report assessing your teaching and identifying targets for further developing your practice in relation to the Standards. You should endeavour to achieve these targets while consolidating and developing your existing strengths. Experienced teachers continue to observe and support you in your teaching. In addition your HE tutor visits you to monitor your development. You are given notice if you are chosen as one of the representative sample of students to be observed by external examiners at the end of the TE. A final report measures your achievements against the Standards for the Award of Qualified Teacher Status.

SPECIAL EDUCATIONAL NEEDS (SEN)

By the end of this section you should be able to:

- understand what constitutes a special educational need;
- acknowledge the statutory provision for SEN;
- recognise your responsibility to implement the SEN code of practice.

Understanding and meeting the educational needs of all pupils is one of the standards expected of newly qualified teachers. You are expected to teach and support a wide range of pupils enabling them to realise their full potential, including pupils who have:

- mild learning difficulties;
- moderate learning difficulties;
- specific learning difficulties;
- emotional and behavioural difficulties;
- physical or sensory impairment;
- exceptional ability;
- have yet to develop fluency in English.

Your growing understanding of pupils' individual educational needs affects the way you plan, resource, deliver, monitor and evaluate your SoW and communicate with pupils. This process is called differentiation and applies to all pupils including those with exceptional ability. Arrangements for the education of pupils with SEN vary between schools and LEAs. Your placement school has a policy for SEN provision based on the *Statutory Code of Practice* (DfEE 1995). In addition many Art & Design departments devise their own subject-specific policies and practice. Bearing in mind these complementary documents, negotiate your responsibilities with your subject mentor and develop effective methods and strategies for their implementation. Some pupils will have been identified as SEN in your school and the documentation, procedures and shared strategies necessary for their support put in place. Occasionally specific learning difficulties arise through art, craft and design activity which may not have been noted elsewhere or which pupils have learned to disguise: you should discuss any such concerns with experienced teachers and/or the Special Educational Needs Coordinator (SENCO).

You can discover pupil differences by carefully observing and monitoring all aspects of their learning and behaviour in the classroom. To do this effectively you should devise your own efficient methods of continuous assessment to record individual needs. However, it is important that you remember to acknowledge those strengths and abilities which fall outside the criteria by which development in Art & Design is usually assessed, particularly social and precognitive skills. You should also provide pupils with the opportunity to assess their own progress which in turn informs your evaluation of their needs and development.

Pupils are assessed and provided with additional support according to their level or 'stage'. This may lead to a 'statement' which entitles them to specialist support. Support is limited and often focused on pupils' literacy and numeracy needs, therefore it cannot be guaranteed in Art & Design. The NC for Art states that some pupils may be working at an earlier or later key stage, to enable them to progress and demonstrate achievement in:

- developing a visual and tactile memory;
- learning about visual and spatial ideas;
- using visual and spatial thinking in creative ways;
- making connections between things, especially those connections which are impeded by their disability;
- recognising art as a means of communication.

Appropriate provision should be made for pupils who need to:

- learn through sensory experiences, including visual and tactile demonstrations about art, craft and design;
- use relevant media, technological aids and/or adapted equipment in practical work, in order to develop their strengths, overcome difficulties and gain new abilities;
- practise movements of hand and body and manipulative skills, learning to relate to their expressive use of gesture to expression in art;
- work objectively to analyse what they see or for the more subjective purposes of personal expression or imaginative use of media;
- work in smaller steps, exploring and comparing visual and tactile qualities of media in depth, observing closely and in detail, as the work progresses;
- respond to art, craft and design in ways other than using speech, including using computers, technological aids, signing, symbols or lip reading;
- use non-sighted methods of reading, such as Braille, or non-visual or non-aural ways of acquiring information, if appropriate.

(QCA 1999)

SEN provision varies widely between different types of school and therefore you are encouraged to discuss and compare experiences with other students. This should add a political and social dimension to your understanding. The literature on SEN in Art & Design is limited. However, Robinson (1982: 95–110) devotes a chapter to SEN across the arts. The main focus of his argument is to ensure provision for gifted or talented pupils who may be found in any constituency: at 'every social level,' amongst the physically able and disabled. Where he considers the needs of pupils with learning difficulties, defined as 'less able and disturbed', he concentrates on the therapeutic possibilities of the arts, an approach we feel is at odds with the educational function of Art & Design in the curriculum if not in schools.

It is often argued that Art & Design provides for SEN implicitly; it is seen to be a part of the 'nature' of active learning that traditionally informs practice:

> Art education can make a valuable contribution to pupils with special educational needs (physical, sensory, emotional and behavioural). Like all other pupils, they can derive much benefit from working with a range of materials, media and processes which can help to develop positive attitudes not only to themselves but also to other people.
>
> (DES 1991a: 59, 11.6)

As the Art & Design curriculum moves to more cognitive, constructivist models this assumption is open to question. Rather than accept SEN provision as a naturalised aspect of teaching we believe that it is best achieved through a conscious decision to differentiate in relation to individual need. We looked at the implications for learning arising from this in Chapter 3, exploring strategies to enable differentiated learning.

HEALTH AND SAFETY

It is particularly important that you are aware of your responsibilities regarding health and safety, and ensure that pupils develop safe working practices. The DfEE (1995) has published *A Guide to Safe Practice in Art & Design*. It is essential that you are familiar with the content of this publication. It offers advice on all aspects of safe practice in Art & Design, including the legal framework regarding health and safety responsibilities and good practice guidance.

The Department for Education and Employment has identified that safe working practices are dependent upon:

- commitment and a sound health and safety policy;
- common sense, good management and organisation;
- general awareness of requirements and shared responsibility;
- properly planned and maintained accommodation;
- appropriate techniques, use of tools and materials;
- use of adequate safety devices;
- a knowledge and awareness of potential hazards.

Classroom practice in Art & Design need not be unduly restricted because of fears for health and safety. Normal practices are safe, within the accepted bounds of risk-taking. Pupils should be trained to work sensibly and safely, and to acquire positive attitudes towards safe practice. Teachers (including student teachers) should give a clear lead by their own planning and personal example.

As a student teacher in your TE schools, you should have knowledge of the health and safety legislation as it relates to schools. You need to ensure that you are reliably safe in the practical activities that you do yourself and in those that you ask pupils to do. See a copy of the safety handbook in your school and ensure you are aware of and understand its contents. All schools have a health and safety representative whom you can consult.

Health and safety and special educational needs

Account should be taken of:

- pupils' ability to understand instructions, follow them and understand any dangers involved;
- pupils' ability to communicate any difficulty or discomfort;
- any physical disability which might affect pupils' ability to perform a task safely;
- any medical conditions which may be adversely affected by exposure to equipment or material.

It is important for teachers to know about the medical conditions of any pupils which may give rise to risk in certain activities, e.g. respiratory problems like asthma:

clay dust, aerosol sprays, screen wash; skin allergies such as dermatitis: dust, glues and pastes; epilepsy: rapid optical sensations in CD-Roms and other computer programs.

Ask for this type of information from each class teacher and seek further information from the SENCO.

Remember to count out and in, equipment which has the potential to be abused, e.g. knives and scissors, adhesives and solvents.

WAYS IN WHICH THE SCHOOL MENTOR CAN SUPPORT YOU

It is the subject mentor's responsibility to organise, supervise and assess your TE in conjunction with your HE tutor. The following list indicates the ways in which these responsibilities manifest themselves throughout the year and provides issues to discuss in your debriefing meetings.

During Induction your subject mentor:

1   introduces you to the other members of the department and support staff;
2   arranges working and storage areas, pigeonholes, etc., to enable you to plan, evaluate and record your teaching effectively;
3   provides you with copies of the department's policies, systems and handbooks;
4   prepares a timetable of classes for your TE covering a range of age, ability, motivation, etc.;
5   provides class lists for the groups you teach and liaises with form tutors, year heads, etc.;
6   arranges a regular time for a weekly debriefing tutorial;
7   supports your planning by providing suggestions about SoW and their relationship to PoS, individual lessons and their sequence within a SoW; providing advice on resources, approaches, methods and materials, addressing the attainment target levels within the NC Art; providing occasions for joint marking, assessing and responding to pupils' work – this should take place between you and the relevant class teacher.

During Teaching Experience your subject mentor:

1   ensures you receive your entitlement as indicated in the course handbook;
2   discusses your TE file: preparation of SoW, record of lessons observed, reflections on and evaluation of lessons, and monitoring of assessment of pupils' progress;
3   observes your teaching and afterwards discusses the lesson in detail and on a number of occasions provides written comments on the lessons observed, including setting specific targets;
4   gives you increasing responsibility for devising your own SoW (in line with the department's PoS) and adding to, plus developing, teaching resources;
5   takes an active interest in reading and commenting on your HE tasks, enquiries and assignments;

6   takes an active interest in the ongoing development of your Career Entry Profile (CEP) and assessment record file;

7   helps you to develop your teaching Standards (DfEE 4/98);

8   completes reports in time for you to read and respond to them before the final date for submission.

Feedback to you could take the following forms:

- a weekly meeting in a suitable place for discussion between you, your mentor and/or other teachers involved;
- a discussion between you, your school and HE tutors about your overall progress to set targets for further development with reference to your TE file and notes about the lessons observed.

Good practice occurs when:

- you take an active role in the feedback sessions;
- you are able to evaluate your own teaching;
- your lesson's aims or means of assessing pupil progress and achievement are used to structure feedback;
- feedback is thorough, comprehensive and, where appropriate, diagnostic;
- there is a balance of praise, criticism and suggestions for alternative strategies;
- points made in feedback are given an order of priority, targeting three for immediate action;
- you receive written comments on the lessons observed.

FURTHER READING

Capel, S. *et al.* (1999) *Learning to Teach in the Secondary School* (2nd edn), London: Routledge.

Pearson, A. and Aloysius, C. (1994) *The Big Foot: Museums and Children with Learning Difficulties*, London: British Museum.

Robinson, K. (1982) *The Arts in Schools: Principles, Practices and Provisions*, London: Calouste Gulbenkian.

6.2 DEVELOPING PEDAGOGY: IN PARTNERSHIPS

OBJECTIVES

By the end of this unit you should be able to:

- identify strategies to develop your subject knowledge and application;
- develop ways to record your teaching experience and curriculum studies.

WORKSHOPS IN ART & DESIGN EDUCATION (see also Chapter 2)

Developing subject knowledge and application through workshops

Teachers of Art & Design are expected to have a wide range of conceptual, critical and technical skills in order to address the broad-based curriculum required by the NC Order for Art & Design and the examination boards. During the PGCE year you should participate in as many workshops as possible, developing existing skills and learning new ones. Your course may not differentiate between technical and concept-based workshops. However, occasionally it is instructive to consider them discretely so that particular needs can be addressed. All workshops should have a critical dimension. These workshops can be sited at a variety of venues, including: colleges, schools, studios, galleries, museums and the natural and built environment.

It is an effective use of human resources for you to organise workshops with members of your tutor group. The content of these is negotiable, building on the experience, knowledge and expertise of the group members. You can, in teams of three to four, plan, organise and present workshops of this kind. These teams can usefully produce a supporting teaching/resource pack including some or all of the following:

- aims and objectives of workshop;
- a visualisation of processes, diagrams, photographs, etc.;
- critical, historical and contextual support material;
- relationship to the NC for Art and public examination syllabuses;
- implications for classroom management and organisation;
- names, addresses and contact numbers for resources including suppliers and galleries and museums;
- helpful hints and handy tips, checklists, etc.;
- useful books, videos, etc.;
- evaluation.

The resulting packs can be shared with your tutor group or collated into a central resource.

CONCEPT-BASED WORKSHOPS

Through concept-based workshops you are encouraged to investigate issues and ideas using a full range of skills, critical, conceptual and technical, shifting the emphasis away from a limited technical and formalist training towards a more reflexive education. You engage with concepts which help you to define what art is, how and why it is a significant area of social production and how you can contribute to debates on pedagogy by reflecting on your observations and practice in schools. They provide a structure within which you can engage in practical activities to draw upon and extend your experience and expertise. You are asked to identify and investigate different ways of recording experience using the processes of art, craft and design, encouraging you to respond by moving from familiar towards less familiar ways of working. You can examine and exploit materials and processes in new combinations as you question traditional approaches. You are invited to locate practice in its historical, social and cultural contexts. Central to this activity is your awareness of the function and potency of metaphor and analogy; the equivalents by which you are able to embody and communicate your experiences in visual and other forms. Subsequently, you can develop strategies to make connections between concepts, skills, processes, products, visual and verbal modes of communication, the NC Art and external examinations. In addition you are encouraged to consider how to overcome constraints such as limited time, space and materials so that you can plan effectively for learning in the classroom. Fundamental to the concept-based workshop is the way it promotes praxis, the linking of theory and practice.

Task 6.2.1 Audit of technical skills and needs

Identify the technical skills and processes with which you

- are familiar;
- feel confident to teach;
- are required to teach;

and those for which you

- require basic training;
- require a revision workshop.

  Prioritise your needs from this audit and devise a programme to meet your requirements. Remember that in addition to timetabled workshops you should find ways to learn from the expertise of fellow students and colleagues and utilise the facilities and INSET at your TE schools.

Criteria for evaluating workshops

Engagement in concept-based workshops enables you to:

- acquire, develop, refine and apply a wide range of skills, knowledge and understanding;
- demonstrate a broad base of practical skills across art, craft and design (including ICT);
- acquire and use skills of reflection and critical analysis;
- develop a personal response;
- locate ideas in a wider context: social, cultural, historical;
- show evidence of reading to inform theoretical understanding;

- make connections between professional practice and ways of teaching and learning in art, craft and design;
- recognise the implication for lesson planning – content, methods, resources;
- make explicit links with the National Curriculum.

The above criteria can be addressed through interrelated modes of communication and expression including:

- visual;
- tactile;
- verbal/written and spoken;
- multimedia, including the aural and kinaesthetic;
- ICT.

Task 6.2.2 Concept-based workshop self-assessment

Informed by these criteria, devise a self-assessment proforma to help you identify your strengths and targets for further development.

APPROACHES TO ICT CAPABILITY IN ART & DESIGN

The PoS for Art (DFE 1995) state that: 'Pupils should be given opportunities, where appropriate, to apply and develop their Information and Communications Technology (ICT) capability in their study of art, craft and design'.

The new digital technologies are determining the development of ICT, whether lens-based, audio or multimedia. It is therefore essential that you develop your computer literacy in line with required standards detailed in the Standards for the Award of Qualified Teacher Status (DfEE 4/98).

ICT forms a fundamental part of professional practice: it is central to design and increasingly an area for exploration in contemporary art. Many of you already possess highly specialised skills in this area, transferring them to the classroom is therefore your primary task. Fusion (NCET 1998) suggests ways in which traditional practice can be harnessed by ICT. However you are advised to take a critical approach to ICT, discriminating between what it can do well, how it can assist learning and where it duplicates inefficiently other methods and skills.

Task 6.2.3 Developing practice through ICT

Consider ways in which digital technologies can create new forms and modes of investigation.

Develop a SoW for a KS3 class that incorporates these forms/methods.

Throughout the course, both at your HE institution and in the partnership schools, you have the opportunity to develop your ICT skills, some of which may have to take place in twilight sessions, depending upon your level of capability. By the end of the course you have to provide evidence of your ability to use ICT to support subject pedagogy.

Task 6.2.4 ICT: beginner, user or expert?

Define your present ability in relation to the following categories:

- *Beginners* – little or no ICT experience at all;
- *Users* – a fair ICT experience, but no insight into pedagogical uses;
- *Expert users* – much ICT experience, but no insight into pedagogical uses;
- *Leaders* – much ICT experience and significant insight into pedagogical uses.

('Leaders' are invited to contribute to the training of *Beginners* and *Users*.)
Monitor your development in relation to these categories as the year progresses.
Devise a training programme to develop your skills and identify opportunities on your timetable to use ICT with pupils.

How ICT capability can be developed in Art & Design

Communicating and handling information:

- by using painting and drawing software you can enable pupils to electronically acquire their own and others' images by scanning, digitising, exploring, investigating and developing them through desktop publishing and multimedia;
- by providing access to a wide variety of visual resources through the Internet and CD-ROM, including multimedia art galleries, multimedia presentations, search engines and visual databases; helping pupils develop their knowledge and understanding of the methods and techniques used to produce artefacts, their function and their historical and cultural contexts.

Modelling:

- by using painting and drawing software and three-dimensional modelling packages, enabling pupils to explore situations and visualise the outcomes of different approaches, methods and techniques.

It is a requirement of the ITT NC Standards (DfEE 4/98) that all students provide evidence that they can use ICT effectively in their subject teaching.

Table 6.2.1 Some routes for developing your ICT capacity

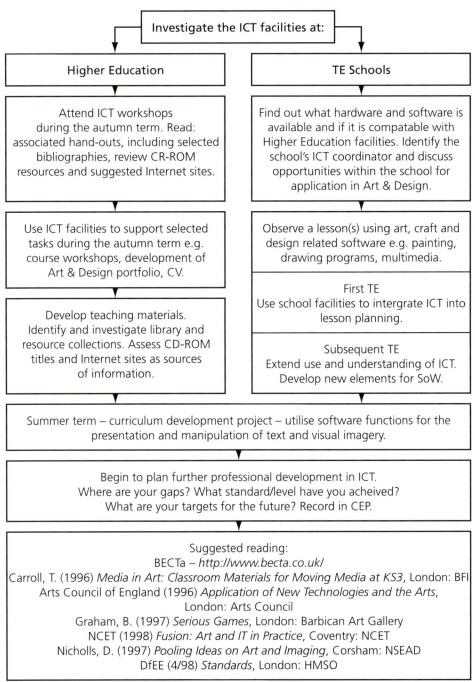

Investigate the ICT facilities at:

Higher Education

Attend ICT workshops during the autumn term. Read: associated hand-outs, including selected bibliographies, review CR-ROM resources and suggested Internet sites.

Use ICT facilities to support selected tasks during the autumn term e.g. course workshops, development of Art & Design portfolio, CV.

Develop teaching materials. Identify and investigate library and resource collections. Assess CD-ROM titles and Internet sites as sources of information.

TE Schools

Find out what hardware and software is available and if it is compatable with Higher Education facilities. Identify the school's ICT coordinator and discuss opportunities within the school for application in Art & Design.

Observe a lesson(s) using art, craft and design related software e.g. painting, drawing programs, multimedia.

First TE
Use school facilities to intergrate ICT into lesson planning.

Subsequent TE
Extend use and understanding of ICT. Develop new elements for SoW.

Summer term – curriculum development project – utilise software functions for the presentation and manipulation of text and visual imagery.

Begin to plan further professional development in ICT. Where are your gaps? What standard/level have you acheived? What are your targets for the future? Record in CEP.

Suggested reading:
BECTa – *http://www.becta.co.uk/*
Carroll, T. (1996) *Media in Art: Classroom Materials for Moving Media at KS3*, London: BFI
Arts Council of England (1996) *Application of New Technologies and the Arts*, London: Arts Council
Graham, B. (1997) *Serious Games*, London: Barbican Art Gallery
NCET (1998) *Fusion: Art and IT in Practice*, Coventry: NCET
Nicholls, D. (1997) *Pooling Ideas on Art and Imaging*, Corsham: NSEAD
DfEE (4/98) *Standards*, London: HMSO

THE ART & DESIGN COURSE PORTFOLIO, RECORDING YOUR TE

The development of an Art & Design course portfolio should be seen as a continuous activity which provides a valuable record of the PGCE course and resources for your teaching. The contents should clearly communicate your understanding of Art & Design education indicating both the breadth of your experience and your personal philosophy. It should also reflect the highest professional standards of presentation: great importance is attached to it by interviewing panels for teaching posts (see Plates 5, 6, 7). It is essential that during your teaching experience you collect records and examples of work using ICT, photocopying, photography or by selecting actual samples of pupils' work. These form the core component of your course portfolio and Career Entry Profile (CEP). Remember always to obtain permission when borrowing pupils' work and ensure that you respect arrangements to return it by an agreed date.

The following list gives some indication of the content of a portfolio:

- Links with primary schools;
- gallery/museum projects;
- course workshops;
- resources and visual aids;
- SoW undertaken on TE including samples of lesson plans;
- pupil work in progress;
- a range of pupil responses and outcomes;
- evidence that you have assessed and recorded pupils' progress effectively;
- evaluations of your SoW;
- involvement in ICT;
- areas of interest/or concerns that extend across SoW, e.g. sketchbooks, visual literacy, differentiation, SEN, equal opportunities, language;
- curriculum development projects;
- classroom organisation and display;
- assessment proformas;
- resource sheets: trigger, reference, technical;
- school-based research and aspects of professional studies;
- links with Career Entry Profile;
- examples of your own work.

Some sheets will cover a single issue in depth, others will include related themes and issues. When designing portfolio sheets you should:

- clearly present ideas in visual form;
- be concise;
- use terms consistently;
- ensure that texts and images complement rather than duplicate one another;
- show a balance between information, descriptive records, analyses and critical evaluation;

- consider different audiences; e.g. interviewing panels, non-specialists, pupils.

The portfolio should provide evidence that you have successfully covered the Standards (DfEE 4/98).

7 Assessment and Examinations in Art & Design

Andy Ash and Kate Schofield with Amanda Starkey

Assessment and reporting is an area of contention in Art & Design. Finding a common policy is a troublesome issue for teachers, with clashes between ideologies and practice fuelling the debate. The intervention of the 1988 Education Reform Act (ERA) in England and Wales provided statutory assessment procedures for the first time. This chapter discusses these developments and considers ways in which teachers can understand and make use of assessment procedures to develop ways of recording and reporting.

> Unit 7.1 provides a general overview looking at terminology and methods. Unit 7.2 focuses more specifically on classroom-based practice and the requirements of the National Curriculum (NC). This should help you to make judgements in relation to the attainment target levels.
> Unit 7.3 discusses external assessment and examinations in Art & Design, with reference to GCSE, A/AS Level and GNVQs.
> Unit 7.4 examines assessment procedures in GNVQs in Art & Design.

7.1 OVERVIEW OF ASSESSMENT: PRINCIPLES AND PRACTICE

INTRODUCTION

Assessment and examinations in Art & Design have now become a central issue in teaching and learning. Since the introduction of GCSE 1986, consolidated by the 1988 ERA, fundamental changes have occurred in the secondary art and design curriculum. The introduction of the NC Art Order (DES 1992) encouraged teachers to make explicit decisions about the assessment of pupils and to formulate comprehensive systems. The ERA established guidelines which still influence the way pupils are monitored, assessed and examined.

The purposes of assessment are:

- for the differentiation of individual pupils' learning needs;
- to provide information about pupils' progress;
- to enable pupils to understand their own learning and progression;
- to enable pupils to appreciate and relate their work to the work of others (artists, designers and craftspeople);
- for teachers to evaluate and thus develop their schemes of work (SoW);
- for teachers to make judgements about pupils' achievements in relation to attainment target levels;
- to provide information for parents/guardians and pupils;
- to enable comparisons to be made about all pupils and schools across the country.

OBJECTIVES

By the end of this unit you should be able to:

- understand the different terms used in Art & Design assessment;
- consider the various methods which can be applied within the Art & Design curriculum;
- look at a range of systems and procedures for assessment at KS3.

SOME DEFINITIONS

Assessment is a general term embracing all methods customarily used to appraise the performance of an individual or a group. It may refer to a broad appraisal including many sources of evidence and aspects of pupils' knowledge, understanding, skills and attitudes, or a particular occasion or instrument. An assessment instrument may be any method or procedure, formal or informal, for providing information about pupils' learning: individual discussion, homework, project outcomes, mock examination work, group critiques. You should consider existing assessment policies operating in schools alongside your own developing understanding of the its purposes and value.

> The notion that one programme of assessment could fulfil four functions (formative, diagnostic, summative and evaluative) has been shown to be false: different purposes require different models of assessment, and different relationships between pupils and teacher. It may be possible to design one assessment system which measures performance for accountability and selection purposes, whilst at the same time supporting the teaching and learning process but no one has yet done so.
>
> (Gipps and Stobart 1993: 3)

In your Art & Design education to date (primary through to HE) reflect on the ways you have been assessed.

- Make a list of the methods of assessment you have encountered;
- discuss with others which methods you feel have been the most and least successful in aiding your learning;
- compile two lists to show ways in which Art & Design assessment has been, for you:

   a a negative experience;
   b a positive experience;

- share and discuss your findings with other student teachers.

TYPES OF ASSESSMENT

Formative assessment is usually continuous throughout the process of a particular learning activity. It has been recognised by most Art & Design educators that formative assessment is particularly constructive. It is concerned with providing ongoing feedback during the process of making rather than assessing and grading an isolated finished product. Formative assessment is an integral part in the development of projects which supports learning; it encourages and guides the pupils' work forward. If pupils do not receive regular feedback on their work they quickly lose motivation and become unsure of their own assessment of success or failure.

Summative assessment occurs at the end of a project, SoW or course of study and therefore focuses on finished outcomes or more holistically on a body of work. Summative assessment can be a culmination of evaluations made over a period of time, for example GCSE course work.

Ipsative assessment gauges the development of an individual from one moment in time to another (usually the present). It is concerned with the evaluation of personal achievement rather than an individual's relationship to national or local norms. It often takes the form of pupil self-assessment providing an opportunity for pupils to appraise themselves in a non-competitive climate. Additionally pupils' self-assessment provides teachers with insights into pupils' understanding of their own progress (see Figure 7.1.1). This can then be used for diagnostic assessment.

Diagnostic assessment approaches pupils' work and behaviour as evidence for the analysis of their ability in a given field (it is often used to discover learning needs). It can be used constructively as a vehicle for discussion between teacher and pupil where both parties consider progress and set targets for future development. Negotiated assessment in the form of constructive criticism promotes learning and a degree of pupil ownership in the assessment process. Diagnostic assessment can be a powerful motivating factor for pupils.

At the end of every project you need to think about what you have learnt and achieved, and write an evaluation. This is a very important part of your project so you need to take time to do it thoroughly. Here are some questions which might help when writing your evaluation.

What have you learnt during this project and does this correspond with the learning objectives?
What have you produced?
How did you produce this work?
What materials did you use?
Which artists/craftspeople/designers have you referred to?
In what way did they inform your work?
Do you think your final work communicates your ideas effectively?
How could you develop it?

Primary colours, secondary colours, complementary colours, tessellation, star, hexagon, arabesque, curved line, leaves, flowers, calligraphy, repeat pattern, symmetry, compass, ruler, tissue paper, Islamic art, paint, felt tips.

(PGCE student 1998)

Figure 7.1.1 Self-assessment for year 8 pupils

Task 7.1.2 Evaluating forms of assessment

Collect examples of all four types of assessment. Evaluate their strengths and weaknesses.
To what extent do you think the self-assessment form (Figure 7.1.1) is ipsative? What subtractions and additions are required to make it so?

ELEMENTS OF ASSESSMENT

Monitoring and tracking

It is important that as a teacher you monitor pupils' progress on a regular basis. In order to facilitate effective planning it is imperative that pupils are constantly assessed to establish achievement and learning: this provides evidence to evaluate the success of your teaching. A proven method of monitoring pupils' progress is to use a tracking sheet – see Figure 7.1.2. The following example shows how, lesson by lesson, you can monitor each pupil by making a quick assessment during or at the end of a session. At stages during a SoW this cumulative assessment provides patterns of learning and behaviour which indicate pupils' responses and can help you to differentiate needs in relation to different types of activity. For example, Pupil Q you will note, has not completed any homework. Your action might be to speak to the pupil and discover what the issues are and why this is the case. You may want to contact the form tutor and/or the parents.

Task 7.1.3 Tracking sheet

Examine the tracking sheet example given, look at Pupil S. What action should you take? Now look at Pupil E. What action should you take? Finally look at Pupil K. What action should you take?

What sort of additional aspects would you want to track/identify? Devise your own criteria and symbols relevant to your teaching.

Use this tracking sheet to monitor a selected KS3 class for 6 weeks. At the end of this time you will be able to see clearly how your pupils have been performing, learning, progressing, behaving.

Achievement

Achievement refers to the overall accomplishment of a pupil including personal factors.

Case study 1

Refer to Pupil F on the tracking sheet. You can see that the symbol 'off task' occurs frequently. This pupil is not achieving the SoW objectives. Differentiation in terms of tasks and teacher input needs to be considered. The implications suggest that actions needed for this case may be differentiated resources, more appropriate work sheets or additional support from you. You will need to identify the pupil's needs for there may be a combination of reasons. It may be that this pupil is an 'invisible child' as described by Pye (1988).

Case study 2

Very able pupils (Pupil D) who, from the tracking sheet appear to be 'coasting' are shown by the frequently occurring symbol. You need to consider whether they require more challenging or differentiated tasks to be developed for them. Providing extension work in your initial lesson planning is a way of ensuring that they are continuously challenged from the outset. Alternatively you might wish to engage in discussion to enable these pupils to take more responsibility for their work and become involved in setting their own learning targets.

You need to select appropriate times to identify the issues raised by the tracking sheet and to decide on appropriate action. By constantly building this process into your practice, this reflective approach supports and strengthens your preparation and planning, leading ultimately to more effective teaching. It may be useful to collect visual evidence on a weekly basis in tandem with the tracking sheet. This can provide a useful focus for a meeting with your mentor.

DESIGN AND MAKE – frame project – using medieval patterns and designs

	Name Orange/Purple	1	H/W	2	H/W	3	H/W	4	H/W	5	H/W	Comments
1	Pupil A	⊠	O	□	A	□⊘	A	□	O	□	O	Producing a successful piece of work.
2	Pupil B	O	O	□	B	⊘	B	⊘	B	⊘	A	Report: last 2 weeks has made a good start.
3	Pupil C	□	A	□	A	□	B	□⊘	A	O	A	Making good progress.
4	Pupil D	□⊘	A	□⊘	A+	~□	A	~□	A	~	A	Works well/good idea development.
5	Pupil E	O	O	~□	O	O	O	~□	O	O	O	Missed a lot of lessons sp/studies.
6	Pupil F	⊠	C	∫□	C	⊠	O	⊠	C+	⊠	C+	Confidence but work is not improving.
7	Pupil G	∫□	C	∫□	B+	∫□	B	□	B+	□	A	Works very slowly but carefully on task.
8	Pupil H	⊠	C	~⊠	C	□	B	~□	C	∫□	C	Needs to concentrate. Keep to task set.
9	Pupil I	~□	C	□⊘	C	□	B	□	O	□	B	Started project well. Keeps on task.
10	Pupil J	~⊠	B	□	A	□⊘	O	□	B	□⊘	A	Worked very well this term/motivated.
11	Pupil K	□D⊘	A+	□D⊘	A	□D⊘	A+	□⊘	A+	□D⊘	A	Moved away from other pupils. Good start.
12	Pupil L	□	B	□	B	⊘	B+	⊘	B	□	B	Is progressing well. Developing design.
13	Pupil M	∫□	B+	∫□	O	□⊘	B+	□	A	□⊘	A	Slow, but the work is of a high quality.
14	Pupil N	□	C	∫□	C	∫□	C	□	B	□	O	Refining and developing work. Slow start.
15	Pupil O	~□	A	∫□	A	∫□	B	□	B+	□⊘	B+	Progressing well from a slow start.
16	Pupil P	□	A	□⊘	B	□⊘	O	O	B	□	B	Growing in confidence and ability.
17	Pupil Q	∫□	O	∫⊠	O	O	O	□	O	∫□	O	Tends to copy but is gaining confidence.
18	Pupil R	∫⊠	B	□	O	□⊘	O	□	B	□⊘	B	Very motivated and is kept on task.
19	Pupil S	∫D⊠	C	D⊠	B	D□	B	∫D□	C	D⊠	O	Limited attention span. Easily distracted.
20	Pupil T	□	B	□⊘	A	□	B	□	B	□⊘	B	Technical ability improving every lesson.
21	Pupil U	⊘	A	⊘	B+	⊘	B+	□⊘	A	□⊘	B+	Working extremely well. Keen on project.

AIMS OF LESSONS/HOMEWORK SET

1. Intro. to medieval patterns. Start designing individual patterns.
   H/W – Observational drawing in response to patterns and texture.
2. Finish design. Enlarge pattern and repeat on to frame shape.
   H/W – View from a window – observation and recording. Tone/shade.
3. Start to decorate frame layering papier mâché.
   H/W – Collage using newspaper or magazines. Fruit or vegetables.
4. Continue to add layers, building up the surface of the frame.
   H/W – Referring to skills learnt last term, draw hard and soft objects.
5. Complete project.
   H/W – evaluation worksheet.

KEY
- ☐ On Task
- ⊠ Off Task
- ⊘ Excellent
- ⊗ Unsettled
- D Disruptive
- ∫ Limited
- ~ Coasting
- O Absent

Figure 7.1.2 Tracking sheet

Grading and marking

Up to the end of KS3 the responsibility for marking and grading pupils' Art & Design work largely rests with teachers although they must use the levels provided by the NC. It is only once pupils reach KS4 (GCSE) that external criteria are imposed by the examination boards.

Task 7.1.4 Investigating assessment procedures

Collect examples of your school's assessment procedures (forms, policies, grading schemes, etc.).

- Identify how these are being applied in the Art & Design department. Examine how individual teachers interpret them differently.
- Evaluate the strengths and weaknesses of these and consider how you would modify them.

FURTHER READING

Allison, B. (1986) 'Some aspects of assessment in art & design', in M. Ross, (ed.) *Assessment in Arts Education – A Necessary Discipline or a Loss of Happiness?*, Oxford: Pergamon Press.

Boughton, D., Eisner, E. and Ligtvoet, J. (eds) (1996) *Evaluating and Assessing the Visual Arts*, New York: Teachers College Press.

Gipps, C. (1994) *Beyond Testing: Towards a Theory of Educational Assessment*, London: Falmer Press.

Kennedy, M. (1995b) 'Approaching assessment', in R. Prentice (ed.) *Teaching Art and Design: Addressing Issues and Identifying Directions*, London: Cassell.

7.2 ASSESSMENT IN ART & DESIGN

OBJECTIVES

By the end of this unit you should be:

- informed about assessment within the NC;
- able to consider different methods of using recording systems for assessment.

GOOD PRACTICE

With the growing emphasis on assessment and accountability, the collecting of evidence demands that Art & Design teachers pay more attention to the organisation of their time. Successful Art & Design teaching is based on observations and knowledge and understanding of the ways pupils learn as individuals. This means providing activities and an environment where pupils:

- combine practical making with theoretical understanding: make connections between knowledge, understanding and skills; evaluate the work of artists, craftspeople and designers and relate it to their own;
- work collaboratively: help each other while making; negotiate roles and responsibilities in the group; are open and receptive to issues and ideas within the group; are able to understand and make choices about tools, techniques and materials appropriate to their needs;
- discuss: dialogue with the teacher to aid progress and understanding; develop an art-specific vocabulary; speak confidently about the work of other artists, craftspeople and designers;
- reflect on their learning in feedback and critique sessions: encourage explanation of work processes, skills and concepts; challenge and respond to each others' work.

ASSESSMENT IN THE NC

The NC Programmes of Study (PoS) provide the basis for planning, teaching and learning and lead on to assessment during each key stage setting the knowledge, understanding and skills that should be taught to pupils as their minimum statutory entitlement. Art & Design departments use the NC PoS to provide a basis from which to develop schemes/units/projects of work for teaching and everyday assessment. These schemes specify objectives for teaching and pupil learning and ensure that over the period of the key stage pupils meet the legal requirements.

Decisions about how to mark and record progress in relation to these objectives are matters for schools to consider in the context of the needs and achievements of their pupils (SCAA 1996: 2).

ATTAINMENT TARGET FOR NC ART

Art is a foundation subject in KS 1, 2, 3, but is optional at KS4 (GCSE) and post-16. Art & Design PoS have one attainment target (AT): knowledge, understanding and skills, with four strands which outline the process of working in the subject:

1   Recording and exploring ideas;
2   Investigating and making art, craft and design;

3   Evaluating and developing that work;
4   Applying knowledge and understanding.

It is recommended that teachers adopt an integrated approach to teaching within the four strands. Although SoW do not necessarily address all these strands, the PoS must. In the past, when there were two ATs (AT 1 Investigating and Making and AT 2 Knowledge and Understanding), teachers tended to follow the suggestions given in the Art Non-Statutory Guidance booklet, (DES 1992) to weight the ATs two to one in favour of AT 1. This ensured that the practical element comprised two thirds of the PoS.

The Attainment Target Levels are designed to clearly indicate the expectations for Art & Design for the majority of pupils by the end of KS3. Teachers are expected to make summative judgements about their pupils' attainment and the way they relate to the standards of performance at KS3. To make a rounded judgement teachers will use:

- their knowledge of a pupils' overall performance, based on the information gained during the day-to-day assessment of any work across a range of contexts;
- their knowledge of the expectations set out in the NC Orders;
- their knowledge of what pupils have been taught.

Task 7.2.1 The function of assessment

Important questions to consider during this activity are:

- How is evidence recorded and stored?
- Who is it for?
- What is it for?

The SCAA (1996), now the QCA, produced terms in their booklet *Consistency in Teachers' Assessment – Exemplification of Standards at KS3* which gives Art & Design teachers a series of statements exemplified by visual material. The booklet is for reference when you intend to:

- consider the standards set out in the revised Order;
- work with other teachers to reach a shared understanding of assessment procedures;
- make judgements at the end of the key stage.

These statements encompass the broad spectrum of pupil attainment, i.e. pupils who are:

A   working towards;
B   achieving;
C   working beyond – the expectations 'that the majority of pupils should characteristically demonstrate by the end of the KS';

D   demonstrating exceptional performance.

(The use of the categories A, B, C, and D are optional for schools to provide summary data and are not intended for reporting to pupils or their parents.)

These are provided as a guide to differentiate between those pupils who do not reach the expectation, described here as 'working towards'. Others surpass the expectation and demonstrate that they are 'working beyond'. The description for exceptional performance is likely to apply to only a small number of pupils.

Task 7.2.2 Making judgements

Collect examples of pupils' work which you consider demonstrate the following:

A   working towards;
B   achieving;
C   working beyond;
D   demonstrating exceptional performance.

Relate the same work to the level descriptions. Do these methods of assessment correlate?

DAY-TO-DAY ASSESSMENT: PRACTICAL WAYS TO RECORD AND ASSESS PUPILS' PROGRESS

In assessing their own work pupils have to be reflective. They need to develop and use an art vocabulary as part of the requirements of the NC. Self-assessment is a vital vehicle for this process. This involvement actively promotes a sense of ownership, empowerment and responsibility. If you remain distant, aloof and appear judgemental you reinforce the traditional barriers between pupils and teachers.

Openness and accountability are increasingly important and we advise that you adopt a model of dialogue. You should engage in discussion with pupils so that together you can develop, evaluate projects and set targets for learning. In order for this to happen you need to enable pupils to:

- articulate clearly;
- develop a specialist art vocabulary;
- to move from description to interpretation;
- apply their analytical skills;
- extend their range of concepts/experiences.

RECORDING SYSTEMS

Recording systems enable pupils to document progress, set targets and take an active role in their learning. Some of the tried and tested methods are:

Diaries
These can be a sequential and chronological, recording the process of work in progress, reviewing work completed, setting deadlines, tasks, and identifying needs. A diary may also record achievement and progress.

Sketchbooks
These are a compilation and development of ideas, combining visual material, texts, objects, plans for future work – a process-based book.

Portfolios
A portfolio is a formal compilation of work with a more self-conscious arrangement, generally with a higher level of presentation. They may include preparatory studies and outcomes rather than process-based work. They are mainly used for work which demonstrates evidence of achieving learning objectives. Often the portfolio is presented to others as a filtered collection of Art & Design work representing range and depth of study.

Log book
A Log book is a way of monitoring achievement, keeping records and dating tasks. It is also a way of documenting units of work completed.

Scrap book
Scrap books are an informal collection of useful source material, showing ideas and the ways in which pupils have engaged in initial levels of research.

Photographs
These can be seen as an independent element to record moments and events (e.g. installations), individual pieces or objects or can be used as part of any of the above.

ICT
The term ICT covers CD-ROMs, video and audio tapes, CAD and computers. Through using various software packages, machines and processes in collaboration, material can be stored, retrieved, manipulated and used to generate art, craft and design work.

Task 7.2.3 Recording

In your practice, which of the above recording systems do you employ? Which ones do you under-use and why?
Consider additional methods of recording: add to the list.

FURTHER READING

DES (1992) 'Art non-statutory guidance booklet', in *Art in the National Curriculum*, London: HMSO.

SCAA (1996) *Consistency in Teachers' Assessment – Exemplification of Standards, KS3*, London: SCAA.

7.3 ART & DESIGN EXAMINATIONS

INTRODUCTION

Units 7.1 and 7.2 examined the principles and definitions of assessment, focusing on the day-to-day methods employed in the classroom. In this unit the focus is on the external examination systems, how they have developed and relate to the NC.

Compared with other subjects in the curriculum there are noticeable differences in the approach to Art & Design assessment and examination of pupils' work. External Art & Design examination syllabuses have been designed to provide an extension for pupils' work at KS3. The syllabus requirements for GCSE (KS4) are intended to build upon the philosophy and practice of the NC. This is particularly highlighted in the construction of the assessment objectives against which all pupils are marked.

Art & Design A Levels have been radically restructured following the Dearing Report (1997). The review of post-16 qualifications attempted to broaden the curriculum for 17 and 18 year olds which had historically been considered too narrow. This could signal the end of the restricted specialisation which has been the basis of the system to date. With new A Levels and refined GNVQs post-16 students have a more flexible learning path preparing them for the demands of work in the twenty-first century.

OBJECTIVES

By the end of this unit you should:

- be aware of external assessment and examinations;
- understand the role of the teacher in external examinations;
- be confident to apply written criteria to visual work;
- have been introduced to the assessment of GCSE, A Level and GNVQ examinations.

BACKGROUND

For many Art & Design teachers the idea of examining art work runs counter to a long-held philosophy and imbued ideas about the nature of creativity, self-expression and the idiosyncratic nature of Art & Design and making. However Art & Design teachers are aware of and clearly recognise the value of assessment strategies as a formal part of the school curriculum. At worst an exam set by examination boards might be perceived as an artificial imposition on a course of study, but nevertheless it is considered a way in which status among subjects in the school curriculum and NC is determined. Ross (1993) stated: 'The status system embodied in the mundane curriculum stresses impartiality, inequality, secularity, acquisitiveness, discrimination, epicureanism. All these are anathema to the arts and to the true artist' (p. 92). In his view Art & Design was somehow outside the testing or examination procedure and did not sit comfortably within the school curriculum. He continued:

> If teachers do participate in public examinations they run the risk of allowing their work to be wrestled from its legitimate roots, yet if they do not they seem to push the arts further out along the educational limb, accepting the more the arts become exceptions to the rules of schooling the less relevant they are likely to appear.
>
> (*ibid.*)

Around the same time, Allison (1986) promoted the idea that examinations were a necessary enterprise. Stevenson (1983) too, began to see that if Art & Design was to be taken seriously then it should stand up to the same rigours as the other subjects in the school curriculum. He put forward a case for accountability: 'In the absence of other means of evaluation and assessment . . . we may see external examinations' results assuming ever increasing importance as the means through which schools give their account to the wider community' (p. 298).

Since then the government has intervened about what should and must be taught in schools and consolidated them with the implementation of the NC in 1992. Art & Design as a NC subject standing alongside, for example Modern Languages, Geography and Religious Studies, has now to be taken seriously as a compulsory activity for every pupil up to and including KS3, although recent government initiatives on literacy and numeracy have eroded that entitlement. If Art & Design as a NC subject is to be taken seriously as an intellectual activity then systematic testing and examination is one way of making visible the art-making process to interested parties, such as parents, governors and employers, as well as to pupils themselves.

Following the Task Group on Assessment and Testing (TGAT) report (DES 1987) and the introduction and implementation of the General Certificate of Secondary Education (GCSE), all secondary age pupils had the opportunity to take the same examination. GCSE was an amalgamation of the previous General Certificate of Education (GCE) O Level and Certificate of Secondary Education (CSE).

THE PROCESS INVOLVED IN EXTERNAL ASSESSMENT

For many years O Level prioritised the piece submitted for the final summative examination rather than evidence of course work. Pupils' course work and test pieces were posted to the examination board's headquarters and marked by a team of examiners 'in house'. These examiners were a group of qualified and very experienced teachers who were specialists in their given subjects. Examining or 'marking' as it was known, was done according to and by using given visual exemplars. Now, however, GCSE tends to be examined through exhibition where pupils' course and examination work are assessed by their teachers holistically and moderated by the board.

Task 7.3.1 GCSE: continuity and progression

Reflect upon your experience of taking the GCE/GCSE examination in Art & Design. Consider the following issues:

- Was there an emphasis on process or product?
- How well and in what ways did it prepare you for your future Art & Design education career?

THE GCSE EXAMINATION

The word examination refers to a formal procedure whereby pupils' achievement is assessed over a specified period of time. It is designed to ensure comparability of results between different test administrators, for example, examination boards, and between different examination occasions. The GCSE in Art & Design comprises both an assessment of pupils' coursework done over a period of time, usually two years, along with an externally-prescribed controlled or timed test (ten hours). These two elements of course work and test make up the GCSE examination and all three examination boards adhere to the same principles. GCSE is assessed holistically; the marks from both the course work and test are added together to form the overall mark. This is then accorded a grade by the examination board. From 1998 the course work counts for 60 per cent of the mark whilst the timed or controlled test makes up the remaining 40 per cent. This is the same for all pupils and for all Art & Design syllabuses offered by the examination boards. They all have to comply with a code of practice which in turn is monitored by a government body, the Qualifications and Curriculum Authority (QCA). In theory, whichever GCSE examination syllabus is taken by candidates, parity of marking is ensured through a system of monitoring by the QCA.

The examination boards vary the way their syllabuses are designed and laid out. In practice, for example, some require candidates to submit four pieces of course work while others demand three. Examination question papers for the timed test also vary in their length and presentation. The number of questions set and the amount of written stimulus or starting points depend on which examination board is selected by schools. Generally teachers investigate all syllabuses and then make a choice based on the one most closely allied to the type of work they already do in their particular department. The way in which teachers are required to mark their pupils' work also varies between one board and another. This said, all candidates are assessed according to written criteria which are common to all three.

THE GCSE CRITERIA

During the 1990s much work was done to develop and provide written criteria for assessing GCSE work and, from 1998, these were put in place and made statutory. From one viewpoint the criteria are intended to encapsulate the essence and philosophy of NC Art, the idea being that all secondary pupils who had followed it through the first three key stages would naturally be able to develop these ways of working to meet the criteria for GCSE (KS4). All candidates entering for the GCSE examination have to address the following criteria in both their course work and test units.

ASSESSMENT OBJECTIVES

The following assessment objectives are not in any order of priority, and are approximately equally weighted.

In each of the syllabuses candidates are required to demonstrate their ability to:

i   record responses to direct experience, observation and imagination;

ii   develop ideas for their work, investigating visual and other sources of information;

iii   explore and use a range of media for working in two and/or three dimensions;

iv   review, modify and refine work as it progresses and realise intentions;

v   identify the distinctive characteristics of art, craft and design and relate them to the context in which the work was created, making connections with their own work;

vi   make critical judgements about art, craft and design, using a specialist vocabulary.

Although in theory it seems a good idea to gain nationwide comparability, teachers have found some difficulty in using a written set of criteria to assess Art & Design work when the subject is primarily material and visual. Attempting to define and understand exactly what is meant by the above has inevitably caused some initial worry and difficulty.

Task 7.3.2 Reviewing GCSE criteria

What do you think is meant by criteria v and vi?
How would you integrate these into your SoW?

Past practice has meant that a number of pupils may not be familiar with some of the these criteria. Although pupils may be good at working within the broad aegis of say criteria i, ii and iii they may flounder when trying to score marks in v and vi.

If pupils are taught to engage with the way their work relates to each criterion they are provided with a strong foundation for future practice, learning and development.

ENDORSEMENTS

An endorsement is a specialism within a broad Art & Design syllabus. All three examination boards offer either an unendorsed syllabus where candidates follow a broad course of work, or an endorsed syllabus, which has a specialist approach such as textiles, graphics, 3D (three dimensional), photography, drawing and painting.

MARKING GCSE

Since the inception of GCSE the responsibility for the assessment and marking of pupils' course and test work has been with the teacher. The Art & Design teacher is therefore seen to be in the best position to judge pupils' level of attainment because they are aware of the formative, preparatory and developmental work at all stages. In the past, if a candidate for some reason performed badly in an examination there was no means of awarding a fair mark.

POST-16 EDUCATION

This book is published at a time when educational provision for post-16 education has seen major changes. Pupils taking their GCSEs in the summer of 2000 should prepare for new systems and examinations to be in place. The Dearing Report (1997) recommended three distinct pathways for post-16 education: academic (A Level), general vocational (GNVQ) and work-based (NVQ). It was considered that a broad-based approach was the way forward. Students could opt to follow any particular route. Central to this idea, was a new argument for breadth. Dearing recommended that post-16 education should largely be in the form of academic education for those who wish to continue with school subjects (30 per cent of pupils), and for the rest it should be restricted to key skills. The debate continued by proposing closer links between the academic and vocational educational routes. A multi-dimensional approach has been suggested which emphasises modularity. Thus, the flexibility offered to students in selecting pathways allows for sideways movement (from academic to vocational and vice versa) enabling a broader and more adaptable programme of study to suit individual needs for future employment or further study.

ADVANCED SUBSIDIARY AND ADVANCED LEVEL EXAMINATIONS

During the first year of post-16 education students take a number of AS Levels (up to five). In the second year they select from these two to four subjects to take through to A Level. In Art & Design the aims and criteria for AS and A Level are the same. The subject criteria set out the knowledge, understanding, skills and assessment objectives common to all AS and A Level specifications. Specifications in Art & Design should promote the development of:

- intellectual, imaginative, creative, and intuitive powers;
- investigative, analytical, experimental, technical and expressive skills, aesthetic understanding and critical judgement;
- an understanding of the interrelationship between art, craft and design and an awareness of the contexts in which they operate;
- knowledge and understanding of art, craft and design in contemporary society and in other times and cultures.

The content of the proposed syllabuses integrates critical, practical and theoretical work in Art & Design.

Task 7.3.3 GCSE and A Level: identifying difference

Examine the new AS and A Level examination syllabuses for Art & Design used by your schools.
What differences can you find between GCSE and A Level?
Do they indicate continuity and progression?

Built into the AS, A Level are key skills which are:

- communication;
- application of number;
- ICT;
- improving own learning and performance;
- working with others;
- problem solving.

Task 7.3.4 Key skills

Devise, plan and write a SoW, unit of work or project in which all the key skills are covered. Make a list of the different ways in which you can collect evidence and assess students' understanding of the above.

FURTHER READING

Barrett, M. (1990) 'Guidelines for assessment in art & design education, 5–18 years', *Journal of Art and Design Education* 9 (3).

Hodgson A. and Spours, K. (eds) (1997) *Dearing and Beyond: 14–19 Qualifications, Frameworks and Systems*, London: Kogan Page.

Ross, M. (1993) *Assessing Achievement in the Arts*, Milton Keynes: Open University Press.

Steers, J. (1994a) 'Art & design: assessment and public examinations', *Journal of Art and Design Education*, 13 (3).

7.4 GNVQs IN ART & DESIGN EDUCATION

Amanda Starkey

INTRODUCTION

Since their introduction in 1993, General National Vocational Qualifications (GNVQs) have attracted a certain amount of controversy and have undergone many rapid changes and revisions in a relatively short time. However, as the qualification has evolved GNVQs have provided a valid and much needed alternative route through post-16 education for students who might previously have been excluded from further education (FE) and therefore progression to higher education (HE).

Initial misconceptions and criticisms about GNVQs were largely concerned with:

1   Restrictions over course content
On first reading, the course specifications seem lengthy and prescriptive. However, when approached creatively they provide a useful framework for devising projects geared to the strengths of the centre and the needs of the students.

2   Workload
Although the GNVQ workload for assessment appears to be heavy it is open to interpretation. It is easy to let the assessment mechanisms, the various forms and check lists, obscure what students have actually produced. GNVQ programmes seek to encourage students to take a far greater responsibility for their own learning than in GCSE and A Level programmes and to practise self-assessment from an early stage. It is central to the GNVQ philosophy that the pace and direction of candidates' progress is, wherever possible, their responsibility, negotiated with tutors. The method of assessment is based largely on course-work evidence with some external testing. While end-of-unit tests have been controversial, most committed students do not find them an obstacle to success. Many detractors feel that because GNVQ course work can be continually revised and resubmitted following feedback from tutors it is therefore easier to pass than equivalent qualifications. But, if anything, the rigour of the assessment and the breadth of the syllabus ensure quality and confirm understanding. Recent research found that GNVQ lessons had more varied outcomes than A Level and that the quality of lessons sampled were generally better, concluding that: 'GNVQ teaching is more intellectually varied than A Level, yet is holding its own in traditional A Level territory' (Guardian Education 1998).

3   The language used in course specifications
The language used has been criticised for being complex with teachers and students alike finding it off-putting. This has been addressed over the past few years by the publication of increasingly user-friendly language in specifications and the most recent is clear and accessible (September 2000).

4   Parity of status
It has been suggested that GNVQ Advanced Level (equivalent to two GCE A Levels) is not adequate for those students wishing to continue to HE because it lacks the intellectual rigour of A Level. It took some time to educate admissions tutors and universities to a point where they understood the different but equivalent skills attained by GNVQ candidates. However, since 1994 there has been a dramatic rise in both applications from and acceptance of GNVQ advanced students into HE; in 1994, GNVQ students (in all disciplines) represented less than 0.2 per cent of overall HE applicants but application from this group had grown by 1997 to 6.5 per cent, some 22,853 applicants. In 1997, 94 per cent of GNVQ applicants were given offers, a rate higher than A Level students. Acceptance is still much higher in the post-1992 universities and applicants are most successful on more vocationally- and practically-based courses (EDEXCEL 1998).

STRUCTURE AND GENERAL AIMS OF THE GNVQ QUALIFICATION

GNVQs have been designed to provide a broad education as a foundation for both training leading to employment, and for further and higher education. This is achieved by ensuring that students develop the general skills, knowledge and understanding that underpin a range of occupations or professions, and by incorporating a number of additional general skills including application of number, communication and information technology into every GNVQ.

(EDEXCEL 1995: Foreword)

There are three levels of entry to the GNVQ programme, Foundation, Intermediate and Advanced, and these link to GCSEs and A Levels as shown in Table 7.4.1.

Until recently the GNVQ tended to be a post-16 qualification, students entering at a level determined by their results at GCSE. Typically a Foundation student has a very low average GCSE grade (F/G) with weakness in Maths and English. A proven interest in Art & Design is sufficient grounds for entry on to this course. It is really intended as an alternative entry qualification to the Intermediate course and therefore has no direct equivalence to GCSE grades. Typically an Intermediate student has an average GCSE grade below the entry requirements for A Level courses (i.e. D/E), again, with probable weakness in Maths and/or English. Often students take an additional subject e.g. a retake of Maths and/or English GCSE. Typically Art & Design is a GCSE subject in which the student has enjoyed some success (grade C or above) or a merit/distinction grade in Foundation GNVQ. The Intermediate qualification is the equivalent of four or five GCSEs. Typically an Advanced student has an average

GCSE grade suitable for entry on to an A Level programme (i.e. C or above) including Art & Design or a merit/distinction grade in Intermediate GNVQ. The Advanced qualification is the equivalent of two A Levels and an additional A Level course can be taken alongside.

Table 7.4.1 Post-16 qualification framework: a comparison between traditional GCE/GCSE and GNVQ levels

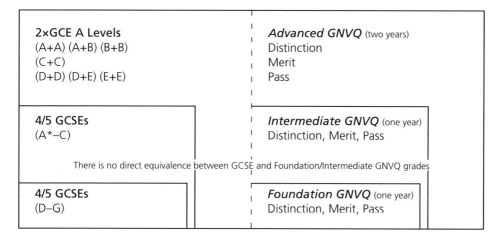

2×GCE A Levels (A+A) (A+B) (B+B) (C+C) (D+D) (D+E) (E+E)	*Advanced GNVQ* (two years) Distinction Merit Pass
4/5 GCSEs (A*–C)	*Intermediate GNVQ* (one year) Distinction, Merit, Pass
There is no direct equivalence between GCSE and Foundation/Intermediate GNVQ grades	
4/5 GCSEs (D–G)	*Foundation GNVQ* (one year) Distinction, Merit, Pass

Source: Adapted from Table 1, 'GNVQs: the context, the rhetoric and the reality', *J. Steers* (1996) *Journal of Art and Design Education* 15 (2): 204.

The main characteristics of the qualification are as follows:

1 All GNVQs are unit based and students demonstrate cognitive and practical skills, underpinned by knowledge and understanding of Art & Design as a vocational area through active learning supplemented with classroom learning, etc. Active learning means work that is carried out by the student as a result of individual interpretation of a given task to meet a personal need rather than a task controlled by the teacher from start to finish. For example, in setting a jewellery-design task you should expect the student to draw up a research plan which independently identifies:

- places to visit to look at existing jewellery design;
- sources of information on the history of jewellery and jewellery from cultures other than their own;
- sites for contextual information, e.g. libraries, ICT, discussion with experts in industry;
- the need to develop ideas, e.g. through drawing from first-hand sources;
- opportunities to explore and test conventional and alternative materials through experimentation and developing prototypes.

At the start of a course, you are advised to set projects of a more structured nature (i.e. strongly directed). Such prescription can gradually be removed with successive projects. Students are assessed in relation to their degree of independence, ideally the course should result in each student working on a personalised assignment and assessment plan.

The student can then negotiate deadlines and use of time with you as tutor, e.g. they might visit a gallery or the library rather than attend a timetabled lesson. With your agreement, the student decides the pace and order of work with your support and guidance. You might instigate whole class sessions to cover essential common ground interjected into the students' personal time plan: normally, these occur no more then once or twice a week. These sessions involve class discussion and presentation of students' work. They provide a forum for critical evaluation and the sharing of good practice. A programme of one-to-one tutorials is integral to the course. They form part of the assessment procedure since oral evidence (from the student) and witness testimony (from you in written form, often in the form of signed check lists) can be used to assess understanding and application of process.

GNVQ unit outlines state what is required in terms of learning outcomes but there is flexibility in the order and thematic content of the learning programme. As they progress, students acquire credits, one per unit, which are transferable. Once registered on a GNVQ programme, a student has a personal candidate number which is 'live' for five years. This number can, if necessary, be transferred from one centre to another following a break in study; e.g. if a student moves home. The student can enrol at a new college with their existing candidate number, even if there is a gap of several years, and any units previously passed will automatically be credited at the new enrolment centre.

2 At all three levels there is an emphasis on work–related aspects of Art & Design with the aim that conceptual and practical skills are developed in a way that has relevance to the workplace as well as further study at a higher level. At each level students enquire into working practices in the Art & Design industry. They are expected to be realistically aware of developments and expectations in industry.

Task 7.4.1 Links with industry

Develop a year 12 project which is product based i.e. students working for a particular market such as a public sculpture, a magazine, a garment or a fashion accessory. The project must involve contact with a suitable practitioner who provides an initial stimulus for the students through either a studio visit or a school-based slide talk. There should be scope for research into real-life constraints, techniques and procedures, perhaps investigating team structures or legalities relating to health and safety or copyright to form an assessment plan. The aim should be to provide simulation of a real situation and the assessment plan should include consideration of the viability of outcomes. The importance of resourcefulness and initiative should be stressed.

COURSE STRUCTURE

The structure of each course is designed to be as flexible as possible. Work carried out for each unit has a particular focus such as '2D visual language', 'working to a design brief' or 'investigating working in art, craft and design'. Achievement of a full award at any level requires the student to complete a set number of units in full, including:

1   Mandatory compulsory units focusing on general skills and concepts in Art & Design which aim for breadth in techniques and knowledge, e.g. learning to manipulate a range of mark-making materials through practical experimentation and study of the work of others.

2   Optional selected units from a range of possibilities intended to promote depth in skills and concepts through more specialist focus tailored to the student's particular needs or interests, e.g. to focus exclusively on a chosen material such as ceramics rather than a range of 3D materials.

3   Key skills are assessed as an integral part of the award and this has the advantage of providing students with opportunities to apply skills in numeracy, IT and communication. GNVQ students often lack confidence in precisely these areas, particularly on Foundation and Intermediate courses and the integration of key skills is often of great benefit as the activities are rarely seen as Maths or English. This can help to overcome previous barriers to learning and achievement. The approach to the application of key skills is relevant to working situations and is practical rather than abstract.

For a typical Advanced project see Figures 7.4.1, 2, 3, 4, Unit 7. 'Working to Design Briefs' in which students work as a team on set and costume designs for a college play. The specifications allow any design brief to be carried out, i.e. the teachers and/or the students can choose any theme, discipline or materials as a focus. This project generates opportunities for assessment because there is scope for:

- simulation of a work situation: real client (the drama teacher and director who chooses the play), real budget, real deadlines and all the real-life problems that can occur;
- there is scope for creative work based on first-hand research and study into the work of others: generally a visit or meeting with a professional in the field is arranged by the student;
- application of number and communication key skills, again in real situations. Students have to keep an account of costs, show that they can specify quantities, work to scale, etc.; they communicate ideas verbally and visually using drawings, models, etc.;
- working with a wide range of materials and technologies including audio-visual equipment and IT;
- team work: 'working as a team' is an additional key skill that can be claimed for credit at the end of the course;
- planning, evaluation and modification;
- consideration of health and safety issues, required by specifications as part of the considerations.

Enabling students to work in a real situation clearly has many intrinsic benefits. It

A R T & D E S I G N

UNIT 7 advanced

PROJECT BOOKLET: WORKING TO SET DESIGN BRIEFS

designs for the theatre

7.1	Clarify and confirm appropriate design briefs
7.2	Research brief and develop ideas
7.3	Produce quality final work
7.4	Evaluate project management and final work

Duration: Half a term   Tutor:

The full range for elements 7.1, 7.2, 7.3 and 7.4 are included in this assignment. In this pack you will find a series of tasks, the unit specifications relating to the vocational elements listed above, the key skills elements listed below and resources/information to help you get started.

KEY SKILLS

This assignment also offers opportunity for evidence in:

Communication: 3.1, 3.2, 3.3, 3.4

Application of number: 3.3 shape, space, measurement.

IT: 3.1, 3.2, 3.4

ASSESSMENT

You are required to focus all grading themes.

ASSIGNMENT TITLE: ....

....

DEADLINE for completion of this unit: ....

Student: .... Tutor: ....

Figure 7.4.1 GNVQ Unit 7: working to design briefs

A
R
T
&
D
E
S
I
G
N

UNIT 7   working to design briefs

Task 1

NB: Think about how you will record and present all of your information; this information will form the basis of your design proposals for this entire unit's work.

- PLAN the tasks carefully
- draw up a TIMEPLAN (a visual plan is easier to follow)

You have been asked to design a REAL stage set (brief 1) <u>and</u> costumes (brief 2) for the forthcoming college production of 'THE TEMPEST' by William Shakespeare and you will need to begin writing out a clear design brief based on your client's needs.

The brief must include the answers to the following:

- Identify who your client is.
- What are their requirements?
- What are the technical constraints?
- What are the production constraints?
- What are the limits on money and materials?

You will need to set up a meeting with your client to discuss the above.

- Write up the findings of this meeting in full and from this report clarify your design briefs.

This done, you should be in a position to begin initial research ready to present preliminary ideas and thoughts to the client at a mutually agreed time (decide when this will happen at the first meeting).

In planning your reseach consider the following:

- existing designs; look thoroughly at the work of professional/ student set designers for a range of theatre, TV and other performance sets;
- places you could visit for ideas and what to look for there;
- people you could talk to for ideas/advice; what questions will you need to ask them?

EVALUATE the information collected as you go along.

Figure 7.4.2 GNVQ Unit 7: task 1

A R T & D E S I G N

Produce a project booklet containing the following:

- a clarified design brief for the set design/costume design;
- planning for both tasks;
- notes from any meetings or discussions;
- all relevant research for both tasks including costings, etc.;
- notes and drawing of initial ideas and thoughts;
- any appropriate reviews/revisions of your plans.

From your initial ideas work up your drawings and visuals for **both** briefs to form a series of presentation sheets to give to your client (<u>at least</u> **2** ideas for each brief). These should be annotated with evaluations, costs, etc.

Evaluate as you go along:

- does the idea answer the brief?
- what does your client think of your idea?
- is your proposal realistic in terms of:
    - time constraints?
    - equipment/materials available?
    - costs constraints? (Is it too expensive?)
- what are the strengths/weaknesses of each idea?
- what improvements could you make?

Choose one idea to develop into 3D – justify your choice.

Figure 7.4.3 GNVQ Unit 7: task 2

A R T & D E S I G N

You should be ready to make a model stage and costume!

Design and construct a scale working model of your favoured set design and a scale toile of your favoured costume design to present to your client.

Plan your work and keep a diary during the construction phase noting difficulties encounted, changes of plan, etc.:

- what materials will you need and why?
- what skills will you need to learn/develop?
- include working drawings to show how you intend to make the models;
- what needs to be done and how will you do it? – proritise!

The best ideas will actually be constructed/produced and used for the performances taking place 4,5 and 6 December.

Task 5

EVALUATE your project management.

- did you use your time effectively?
- how did you overcome your problems?
- were your plans and working drawings helpful?
- what were the views of the other people about your work?

Figure 7.4.4 GNVQ Unit 7: tasks 3/4 and 5

inspires confidence in handling complex tasks, dealing with other people to achieve positive outcomes and a sense of achievement.

4   Several externally set and assessed tests (multiple choice and short written answers) relate to mandatory and key skills units.

5   From September 2000 there will be an externally set and internally assessed Benchmark *assignment*: practical and theoretical research to a set theme leading to the planning, development and execution of an Art & Design outcome, and key skills assignments (linked to Art & Design).

Most mandatory units follow a pattern of requiring the student to make broad initial investigations in a range of materials, processes or ideas and then select and refine them so that development is informed by experience. Optional units tend to be more focused from the outset. In devising a course year plan there are two basic choices of model:

a   SoW devised to generate some of the evidence for several units, including key skills;

b   SoW devised to generate all of the evidence for one vocational unit and part of the key skills unit (remember that where possible, key skills should be integrated into everyday and practical situations).

The first model, (a), allows you to focus on one material or theme but has the distinct disadvantage of being very complicated when tracking student progress in terms of completed units. I have adopted the second model, (b), as it has the major advantage of being discrete: it is much easier for the students, who should be encouraged to self-assess against the unit specifications before submission of the unit portfolio and to track their progress unit by unit.

If you take this second model, you might treat some units such as Units 1 and 2 of the Advanced course, '2D Visual Language' and '3D Visual Language' respectively, as 'long thin units' assessed over the whole first year because they are the foundation to the work in other units and therefore require coverage of many different materials and processes. The first term of the standard two-year Advanced programme might be used to deliver two projects which cover the bulk of the units' 'evidence indicators' with scope to generate supplementary evidence in the background of work of later projects. You could view other units such as 'Working to Design Briefs' as short fat units for which tasks aim for full coverage of the requirements in a concentrated 'one-hit approach' over a period of approximately one term.

ASSESSMENT

A GNVQ candidate must meet all of the requirements of their course to obtain the full award. Key points in GNVQ assessment:

1   portfolios of 'evidence': each unit in the course document outlines the evidence required to gain a pass;

2   demonstration of skills in both process (planning, information seeking and information handling, and evaluation) and outcome (quality of outcome);

3   work marked internally by the tutor/assessor, aside from externally-set tests;

4   internal moderation takes place regularly to maintain standards within an institution when an internal verifier samples the work for a given unit to confirm assessment decisions;

5   external moderation to ensure national standards takes place once a term when trained standards moderators work alongside internal verifiers to monitor the assessment;

6   the Benchmark assignments perform a similar function to practical exams in GCSE and A Level. However, although criteria-based they are not translated into the numerical marks used in GCSE and A Level.

TYPES OF EVIDENCE

Evidence is generated in response to instruction or stimuli: it is the response to these stimuli by the candidate that is evaluated, described or compared for the purpose of assessment. There is a wide range of evidence that you can use including:

1   Direct performance evidence: you observe the candidate's performance and practical skills in normal classroom conditions.

2   Alternative performance evidence: the candidate demonstrates knowledge or skill through products, e.g. written work, drawings, diagrams, models produced away from normal classroom activity (i.e. homework and studio-based work carried out in personal study time and self-directed study).

3   Supplementary evidence: assessment decisions are informed by comments in response to a direct question to confirm candidates' understanding of the work set.

Sources of evidence include:

1   'Natural performance' is where you directly observe a student: e.g. using a word processor, carrying out studio-based practical tasks, etc. This is an unobtrusive method of obtaining supplementary evidence putting minimum pressure on the student's performance. The student must be aware of assessment taking place and the assessment must have clear objectives. You would record the evidence using a signed check list of criteria or a witness statement by you in the role of the assessor.

2   'Simulation' is an alternative method of skills demonstration which involves role play or workplace activities and can involve teamwork.

3   'Written questioning' seeks to test a student's learning by focusing on their ability to apply practical skills and knowledge in theoretical situations.

4   'Oral questioning' is often more useful in determining comprehension than written answers as it can be directed to specific skills at a given moment. Again you could use a witness statement to record results or use video/audio tape.

5   'Prior experience' is evidence which arises from activities which the student has carried out in the past, e.g. evidence of experience of workshop practice in GCSE Design and Technology provides evidence of a certain knowledge of health and safety.

6   'Candidate and peer reports' is evidence which is assessed either by the students individually or by their peers. This could involve the use of a check list to determine work covered. Self-assessment by the student is always verified by you, the assessor.

CONCLUSION

GNVQs should not present unique problems in terms of content when designing SoW or assessing students' work. The framework outlined by the awarding body is open to flexible interpretation and classroom delivery and can easily be tailored to meet the needs of students using available resources.

GNVQs work very well as an alternative to A Level or GCSE. Every year I have encountered students whose GCSE results are so poor that you might not expect them to qualify for further study at Art college and apprenticeships: GNVQs provide such students with a second chance. Others enter the Advanced programme with results that entitle them to take more traditional routes. But students choose GNVQs because of the variety of experience they offer and the value they place on individual development. This allows them to progress to higher levels with sound vocational, intellectual, practical and personal skills.

FURTHER READING

Steers, J. (1996) 'GNVQs: the context, the rhetoric and the reality', *Journal of Art & Design Education* 15 (2) 201–213.

8 Issues in Craft and Design Education

Lesley Burgess and Kate Schofield

Through the units in this chapter issues of craft and design and their positions in secondary Art & Design education are discussed. You are encouraged to question traditional definitions, attitudes and values, consider the roles of 'practical thinking' and 'intelligent making' and explore your understanding of these concepts and their pedagogical application in forming an inclusive curriculum.

Unit 8.1 focuses on craft and looks briefly at its history in relation to teaching and learning in secondary schools. It challenges the limited definition perpetuated in many Art & Design departments. The role of intelligent making is investigated and different models of delivery are explored. Examples of work by contemporary crafts-people are used to show how crafts are not static but are constantly being updated and redefined. The important role played by the Crafts Council in developing an awareness of the above is acknowledged. Reference to their research projects is highlighted. Gender, class and ecological issues are referred to and artefacts and products from different cultures are briefly considered in order to raise awareness.

Unit 8.2 focuses on design, and subjects it to a similar analysis, this time using contemporary designers to define new directions and their potential application in the secondary classroom. Initiatives developed by the Design Museum and Design Council provide examples of good practice.

8.1 CRAFT

OBJECTIVES

By the end of this unit you should be able to:

- challenge preconceptions and question existing orthodoxies both in school and in the wider community;
- formulate your own working definition of craft;
- understand different ways of promoting craft in the curriculum.

WHAT IS CRAFT?

Craft is a fluid, technological activity that cannot easily be categorised into one set of attitudes or lifestyles.

(La Trobe–Bateman 1997: i)

. . . an object designed and made by one person, possibly with the use of assistants . . . craft is the workmanship of risk.

(Crafts Council 1998)

Craft is like a four-letter word; it has such grungy connotations.

(Blackburn 1998a: 27)

. . . craft is a shaky business.

(Perry 1998)

. . . the decorative and applied arts

(*Crafts Magazine*)

As you can see there are many conflicting and partially formed, even ill-formed, definitions of 'craft'. This can create confusion and ambiguity.

Craft does not appear as common currency in relation to the arts until the mid-nineteenth century with the beginning of the Arts and Crafts movements. Prior to this there were many partially formed definitions of craft. Greenhalgh (1997) charts the evolving meaning of the term. In the eighteenth century the term was used to describe political acumen and shrewdness: 'to be crafty'. The term also had connotations of power and secret knowledge, e.g. Freemasons, witchcraft. Greenhalgh asserts that craft was gradually divorced from design in the 1920s when it was, 'intellectually isolated from both the pursuits of beauty (art) and purpose (design)' (Greenhalgh 1997: 40).

Practical skills have been taught under a variety of labels and categories. Handicrafts were later separated into light crafts and heavy crafts. As terminology changed previous labels became unpopular and even disparaging.

> The term 'handicraft' has developed connotations implying elementary practical activities which require a minimum of skill and thought and little knowledge to achieve success.
>
> (Glenister 1968: 6)

Berkeley (1987) suggests that an old ethos exists where practical and creative school subjects remain on the fringe. Art and Music are exceptions as they are considered 'culturally credible'.

BRIEF HISTORY OF CRAFT IN SCHOOLS: THE SPLIT BETWEEN FINE ART AND CRAFT

As Dormer pointed out, 'A hundred years ago the link between craft knowledge and fine arts was taken for granted' (Dormer 1994: 7). Today the situation is different.

In the 1970s and early 1980s Art & Design teachers perceived craft as outside their remit. It was firmly located in departments of Craft, Design and Technology (CDT). In the main the work undertaken by CDT teachers was seen as belonging to a separate discipline as their training route was often radically different from that of Art & Design specialists. Although CDT encompassed a wide variety of activities and teaching styles it was felt that a pupil's level of understanding and overall intellectual grasp could only be comprehensively manifested through designing and making. This was managed in a linear fashion, in response to a perceived need and with a prescribed range of materials and tools. In contrast teachers of Art & Design emphasised a less prescriptive, more experimental and open-ended approach. Importance was placed upon process and individual responses rather than problem solving and working towards a predetermined outcome. This concurs with Dormer's view that, 'an emphasis upon individuality in art has aided the decline of craft' (*ibid.*: 26).

During the 1980s with the growing importance of critical and contextual studies in Art & Design education, the division between art and craft became increasingly pronounced. Art was seen to promote the creative and aesthetic dimension. In contrast, craft still emphasised skills within a given tradition, the efficient use of different materials, tools and techniques. This concept of craft was not limited to schools but was reinforced by the general public.

Dormer sums up this position:

> There is a view that craft knowledge, because it is communal (it has been created by many people), conflicts with originality. The prejudice against craft tends to be crude: craft, it is thought, is bound by rules, and it is assumed that rules necessarily conflict with freedom of thought, imagination and expression.
>
> (*ibid.*: 7)

At the beginning of the twentieth century craft and art were easy to tell apart. They were defined by materials. Art used materials, such as paint, canvas, marble, bronze whereas crafts used materials such as fibres, clay, wood, paper and glass. Art had 'serious' content and no utility function while craft objects were made to be used. This suggests that applied art is not serious art, an entirely prejudiced view. Craft has evolved and old distinctions have begun to dissolve.

CRAFT AND THE NATIONAL CURRICULUM ART

The NC Art Working Group (DES 1991b) were clearly of the opinion that it was, 'unhelpful to dwell on the divisions between art and craft that seem to exist at a professional level' (p. 13, 3.28) suggesting that art and craft are so intimately linked that they are best treated as one. However, as a student teacher it is important that you have an informed understanding of these issues. It could be argued that in order to understand how they relate to each other you must first understand their individual and particular qualities. Indeed, they have different histories which overlap and interweave.

The NC Art 5-14 (DES 1991a) insisted that, 'a craft approach to materials and tools should not be subordinated to an art directed one' (p. 13). However, many schools took this as an opportunity to neglect craft, craftspeople, or craft history, as a valid component of the art curriculum. They assumed that through the realisation of ideas and intentions Art & Design provides insight into the properties of materials and their sensitive handling, thus craft is subsumed. The Crafts Council has been keen to dispel this idea. Teachers have been criticised for adopting an elitist position in which craft is secondary to fine art sculpture: they have failed to afford it a place within the curriculum. The NC 2000 states: 'Art and design includes craft.'

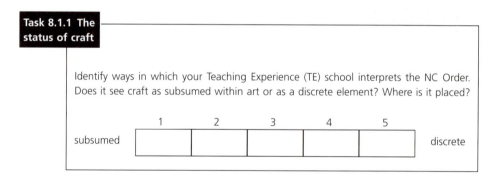

Task 8.1.1 The status of craft

Identify ways in which your Teaching Experience (TE) school interprets the NC Order. Does it see craft as subsumed within art or as a discrete element? Where is it placed?

	1	2	3	4	5	
subsumed						discrete

THE CRAFTS COUNCIL

An important contributor to the debate is the Crafts Council who have played an important role in raising the profile, status and understanding of the crafts. Following the implementation of NC Art they identified the need to revisit, redefine and raise the profile of the crafts in education. This was a timely debate as OFSTED (1995a) identified that many schools give insufficient attention to this area of the curriculum. The Crafts Council has worked closely with centres of initial teacher education initiating a dialogue to question preconceptions and to investigate issues and concerns in craft education, particularly contemporary craft practice. They have given a high priority to the support of critical and contextual resources in school.

The findings from the survey 'Pupils as Makers' (*Crafts Council Survey Part 1*: 1995; *Part 2*: 1998) reinforces concerns about the inward looking nature of school practice and highlights the need for teachers to support pupils' practice through contact with professional makers and their work. It is important that you question preconceived ideas and question the boundaries through your school-based schemes of work (SoW). With the support of Crafts Council publications you can move away from the definition of the crafts as merely traditional, decorative and unchanging and recognise that contemporary crafts have a sense of dynamism and diversity akin to the work of contemporary fine art. You need to consider not only the qualities that differentiate the disciplines but the blurred boundaries where some of the most exciting practice exists.

The term craft brings with it connotations which Art & Design teachers have found problematic. At a conference of the Crafts Council in October 1997 writer and critic Pam Johnson suggested that most people are limited in their definition of the term. She insists that it is important that educators establish a shared understanding of ways of defining intelligent making (her definition of crafts) and the role and the importance of practical, creative learning. Contributing to this debate can increase your knowledge and understanding of the scope and range of professional practice in the crafts and consider the relationship between developing practical skills and knowledge and understanding of craft's historical, technological and cultural contexts. Too many teachers still define the crafts in terms of traditional practice, e.g. corn dollies, thatching or a representation of craft through reductive commercialised kits such as tapestry packs, model aeroplanes, kites.

Task 8.1.2 Defining craft

a   In no more than 12 words give your own definition of craft:
    Discuss your definitions in tutor groups.
b   Identify from the following those terms which fit your definition of craft:

   aesthetics, system, hand-thrown, manufactured, concepts, product, process, mass-produced, personal, unique, pattern, decoration, traditional, contemporary, precious, three-dimensional, artefact, object, model, icon, symbol, pictures, two-dimensional, realistic abstract, private, multimedia, ICT, cottage, primitive, industrial, sculpture, imagination, identity, self-expression, ideas, creative, materials, skills, heritage, non-European, reproduction, restoration, applied, hand-built

---

Johnson (1997a) suggests that craft can fall into four different cultural positions. These are:

- Utilitarian
  Made for use, distinctive, batch production, mindful of unit cost, profitable: the sort of objects and artefacts found on sale in Habitat or at a crafts fair.

- Decorative
  Embellishment of a place or person, celebratory, sensuous, sublime, lyrical, affirming rather than disturbing.
- Interface
  Collaboration with manufacturers, architects and designers, part of a whole, emphasis on skill and knowledge of materials.
- Expressive
  Production of meaning, transmission of experience, narrative, challenging perceptions, defamiliarised, exploration of cultural values, associative potential of materials and forms, one-off, posing of questions.

However, Johnson acknowledges that there is a danger in fixing these categories which she claims are constantly shifting and open to reinterpretation. She recognises the crafts as a field with uncertain boundaries which encourages debates. This reflects the statement by Buck:

> Borders are never comfortable places to be but they are often where the most interesting things happen. Nowhere are the boundaries more volatile and complex than those that lie between art and craft. It is a particular feature of Western culture that such a distinction between these two branches of artistic practice exists at all; and even the most cursory examination of the art–craft dialectic throws up some disconcerting insights into our current cultural climate.
>
> (Buck 1993: 8)

Because many contemporary craftspeople are beginning to imbue their work with art concepts, they are demanding more recognition; and because many artists are combining craft materials and processes into their work, the boundaries are being blurred and constantly redefined. As student teachers you need to establish working definitions and communicate them to your pupils, recognising that you will need to redefine them as they in turn become old categories and ideas.

Task 8.1.3 The crafts and visibility

In Chapter 5 you were asked to name 6 artists. Can you now name 6 craftspeople?
Now name 6 contemporary craftspeople.
Discuss in tutor groups:
To what extent are craftspeople studied in your placement schools?
How are they used and referenced within SoW?
In what ways have you included craftspeople in your own SoW?

1 *The semiotics of materials*

2 *The semiotics of materials*

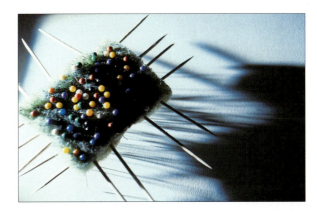

3 *The semiotics of materials*

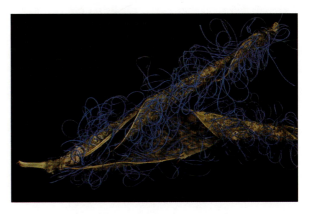

4 *The semiotics of materials*

1, 2, 3, 4: PGCE students (Institute of Education, University of London)

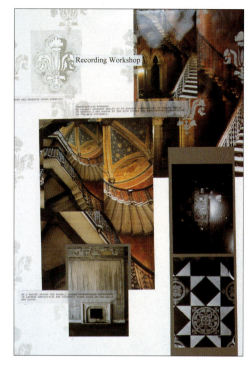

5 *Recording a 'Sense of Place': St Pancras Chambers*

6 *Recording a 'Sense of Place': St Pancras Chambers*

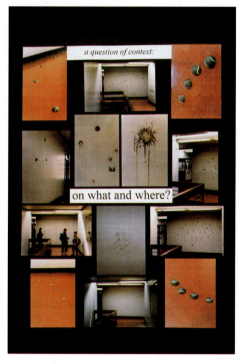

7 *A Question of Context*

8 *Cabinet of Curiosities: Sketchbook?*

5, 6, 7, 8: PGCE students (Institute of Education, University of London)

9 *Playing with Fire* (1998), Lucy Casson

10 *Spice Boxes* (1998), Natasha Hobson

11 *Gazebo* (1995), Thomas Heatherwick

12 *Heart Development Hat* (1998), Helen Storey

13 *Between Words and Materials*

14 *Untitled* (1998), Meera Chauda

15 *Seat of Learning*

16 *'I can't draw either'*

13, 15, 16: PGCE students (Institute of Education, University of London)

17 *Robot Bodies* (1998), Keith Piper

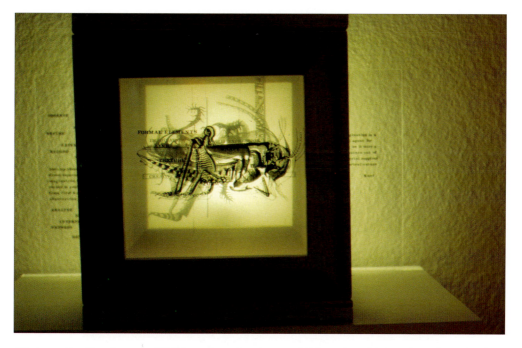

18 *Investigating Insects*, PGCE student (Institute of Education, University of London)

19 *Walking in Space* (1998), Michael Brennand-Wood

20 *Steppenwolf* (1997), Caroline Broadhead

21 *'Evidently Not'* (1998), Cathy de Monchaux

22 *Diasporic Pyramid: Monument to Asylum Seekers*, PGCE student (Institute of Education, University of London)

23 *Sculpture: colour workshop*

24 *Layering Cultures*

25 *Shaman's Coat*

26 *Installation: Domestic Turn*

23, 24, 25, 26: PGCE students (Institute of Education, University of London)

STATEMENTS BY CRAFTSPEOPLE

Michael Brennand-Wood (Plate 19)

Patterning is a visual language; a genealogical study of patterns reveals a rich history of association and meaning. I am interested in understanding the connections between the information I collate, the reasons I associate specific ideas together. The adventure is to discern *why* and make visual sense of the clues amassed. The examination of materials, both conceptual and physical, is an important aspect of my work. Layered shapes and patterns can act as mnemonic devices, triggering memories of places, experiences and alternative realities. Currently I am investigating the relationship between the bio-morphic and the cellular aspect of textile constructions. Fabrics are a skin, structural elements unite to form a flexible surface, a process that parallels the evolution, growth and mutation of cells within the body.

Caroline Broadhead (Plate 20)

Over the last 25 years the emphasis of my work has evolved from jewellery, and the relationship of objects to the person, through to work that expresses ideas about the person and their relationship to the world. I have worked on ideas around the changing nature of an object when worn or not; the ability to manipulate and change the identity of an object; the potential for movement and/or restriction when being worn. I have worked with the idea of garments being metaphors for people carrying information that reveals the personality. The continuing theme is the body, the way we present, package and decorate our bodies, and how that wrapping can reveal underlying tensions and contradictions. More recently I have become interested in shadows – the negative double of the body – as an indication of a presence, or absence, the exaggeration or distortion of a person, the transparency of shadows and the way they affect the surfaces on which they fall, shadows that trace the pathway of the sun. The shadow as an image with no substance, the alter ego, the dark side, the inescapable other, gives me the opportunity to focus on the non-material. By giving 'body' to a shadow I am giving more importance to the things that are not immediately apparent.

Thomas Heatherwick (Plate 11)

For me, the process of making – the hands-on experience of creating things – is the most important way of learning. Making helps us to better understand and appreciate the world in which we live. Everything that surrounds us is made, either by man or nature. We can all learn through making. It informs the link between abstract concepts and physical reality. Through making we can learn to solve problems, adopt more practical ways of thinking, understand the way things work, the nature of materials and the relevance of theories and mathematics, as well as witness the realisation of our ideas. I believe that making should be an integral part of the curriculum for all schools and colleges, not just as a means of teaching art and design, but as a source of learning and inspiration about the world around us.

Natasha Hobson (Plate 10)

My sculptural ceramics were inspired by a five-month placement at the Bezalel Academy of Art and Design in Jerusalem. The 'cocktail of nationalities' I discovered on my placement enabled me to study different cultures and ceremonies. I grew especially interested in the Jewish 'Havdallah' spice boxes. After the Sabbath, Jews light the spices contained in the boxes as a sign of renewal for the coming week. These ceramic sculptures are made from porcelain clay; thrown, hand-built and press-moulded from plaster casts of fruit and bubble wrap. They are decorated with oxides and coloured clay slips, then salt and lustre fired to 1280°C and 780°C.

Helen Storey (Plate 12)

In June 1997 I embarked on a project with my sister (a developmental biologist at the University of Oxford). Our work together was instigated by The Wellcome Trust's Sci/Art initiative – a concept which sought to bring artists and scientists together to produce bodies of work that could explore each others' worlds and capture the public's imagination at the same time. Our collection entitled 'Primitive Streak' portrayed the first 1,000 hours of human life in textiles and clothing. It has been through the experience of this project that I have come to understand, and fully appreciate, first hand the power of visual communication. This work has succeeded in a way none of my previous work has. It underlines the vital need for an education system that encourages and harnesses the skill of the eyes; not just as a means to copy or take in information but as a gateway to engage an individual's imagination and the ability to make personal sense of the world around us.

Task 8.1.4 Responding to craft practice

Consider the statements and work (Plates 10, 11, 12, 19 and 20) of the craftspeople represented in this chapter.

How would you:

1   Develop a SoW which has as its focus one of the above?
2   How would you integrate one or more of these craftspeople into a SoW?
3   Can the methodologies used for critical and contextual studies for fine art be applied directly to craft?

RECYCLING, SCRAP, REINVENTION

> Craft finds us the diamonds in the landfill.
>
> (Press 1996: 13)

The idea of recycling materials is not new. People from all ages and in all cultures have recycled as a necessity for their existence. Denise Mucci Furnish, a contemporary quilt maker, compares her work, which includes lint from electric clothes dryers, with the earliest examples of quilting found in ancient China. The Chinese believed that the wearer's spirit lived on their clothes: worn out fragments were stitched onto new fabrics and the cycle began again. Today this idea of reusing has become an aesthetic in its own right.

Recycling is currently one of the important issues in contemporary crafts. An awareness of environmental issues has led to a reassessment of materials and an investment in recycling technology to which craftspeople have responded with both interest and ingenuity. It is important to establish a positive attitude to the reuse of materials by presenting examples of craftspeople who have continued to develop a recycled aesthetic: Jane Atfield, Lucy Casson, Michael Marriott, Tejo Remy, Lois Walpole. The use of recycled materials should always be carefully considered and selections made on the basis of appropriateness to your projects. They should not be regarded as an easy cheap alternative in which financial and resourcing implications may outweigh pedagogical concerns.

In the catalogue 'Recycling', Press (1996) states that the value of the stuff people throw away draws attention, in a way only the crafts can, to the properties, uses and abuses of materials to create models of 'sustainable enterprise in an unsustainable wasteful industrial system' (Press 1996: 14). The following statements provide differing rationales for recycling:

> I use recycled and found materials for environmental and aesthetic reasons
> . . . I have always found it liberating to use these materials because it takes the
> angst out of experimenting.
>
> (Lois Walpole in *ibid.*: 52)

> I utilise found objects and do not exclusively recycle but start by making
> shapes from existing objects. They come from a wide variety of sources
> including plumbing shops and cookery stores. The biggest challenge would
> be trying to get a client to pay a commercial price for a piece of scrap.
>
> (Tom Dixon in *ibid.*: 80)

> My primary reason for using reclaimed materials is aesthetic. I like the
> quality of old wood and printed time. If it's been washed by the sea too then
> so much the better. Obviously the economic factor is a bonus and it can
> allow more experimentation without fear of expensive failure . . . working
> with something that's already had a life and that past history can sometimes
> work against you as well as with you.
>
> (Kirsty Wyatt-Smith in *ibid.*: 56)

I am sculpting with recycled tin plate, wire, wood, plastics and found objects. My way of working has been to discover as I work, evolving different methods of using the materials to obtain a certain effect, my techniques are discovered through messing about with metal. I like to use non-precious materials and simple techniques. The fact that it is a non-precious material enables a freedom that a similar new and therefore expensive material would not allow.

(Lucy Casson (Plate 9) in *ibid.*: 48)

Task 8.1.5 Using scrap materials

In tutor groups discuss the above statements and their application to the classroom. Devise a SoW based on one of the concepts about 'scrap' offered by these three crafts-people. Support your SoW with the work of others that relates to these categories.

COMMENTS: ATTITUDES AND VALUES

Consider the following statements:

Outside the insular world of contemporary crafts, the word 'craft' carries all the associations which are guaranteed to hasten its educational decline. Craft in our culture is dumb. It is associated with manual skills, domesticity, occupational therapy and our less than smart school children. In other words, crafts is working class, female, mentally unstable or just plain thick – and very usually all four. It succeeds in scoring highly on all the prejudices of a dominant culture that is anti-working class, patriarchal and believes that intelligence is only demonstrated, and thus can only be measured and valued, through the use of words and numbers – most especially when the latter is preceded by pound signs.

(Press 1997: 47)

Craft today is seen as an accessible quasi-art form. It is essentially a middle class concept. Crafts are purchased by the educated elite – they buy craft objects because they are 'one-off' or limited editions and therefore expensive – Liberty's posh!

(Johnson 1997b: 42)

Craft practice promotes a set of values that recognise 'making' as a basic human need – values that will become increasingly important as we progress through an information age. Crafts, education and practice integrates creativity, problem solving and manufacture, a process which facilitates innovation – and therefore has an important contribution to make to the needs of industry and the business world.

(Ball 1998: 4)

Task 8.1.6 Self-positioning in relation to the crafts

In tutor groups discuss the above statements as a starting point to debate your own position.

As you have discovered, personal narratives, informed and developed by personal experience and prior learning, including gender, race and class, condition your position to and within art, craft and design. It is essential that you question your own preoccupations and identify existing school orthodoxies and their current craft practice.

Gombrich (1992) insists that people's judgement of what they see is affected by the mental framework they use when looking. He suggests that if people are told that something falls into the craft category their attitudes towards it are going to be different than if they had assumed it to be 'fine art'. (p. 27). Blackburn (1998), organiser of two selling exhibitions of modern crafts for Sotheby's, admits that getting people to take crafts seriously as an art form is an 'uphill struggle'. She states: 'I've known wealthy collectors who admire something and ask the price, you say £1,500 and they almost pass out with shock. But they look at a painting for £20,000 and can't get their hands on it quickly enough' (p. 24).

It is worthy of note that the persistence of this hierarchy is a particularly British phenomenon. Michael Brennand-Wood (Plate 19) who has been at the forefront of British textiles for over 20 years, is better known abroad, particularly in Japan.

POLITICS

Attitudes and values are in fact perpetuated, albeit unintentionally, by government institutions and organisations such as the Crafts Council, the Arts Council and the Design Council, and, to an extent, the Engineering Council. They are responsible for establishing boundaries in relation to their funding. The Crafts Council is structured around materials: ceramics, glass, wood, metal, textiles, etc. The subsuming of the Crafts Council under the auspices of the Arts Council is likely to have a profound effect on attitudes. A new cultural body is called for in a paper published in 1998, '*The Comprehensive Spending Review: A New Approach to Investment in Culture*,' where it is posited that there is a need to break down artificial barriers within art forms. Johnson states:

> The categories of art, craft and design now strain to contain the diversity of hybrid objects circulating within visual culture. The hybrid textile object tests the limits of our current exhibition funding structures . . . why not a Council of Visual and Material Culture and with it a more fluid use of exhibition space?
>
> (Johnson 1998: 26)

Practitioners have for many years endured feelings of inferiority and are unable to articulate and theorise about their work, 'their craft is a wallflower at a cultural ball' (*ibid.:* 34).

SOCIAL, CLASS DIVISIONS

Class divisions are also significant. Parker and Pollock (1981) state that traditionally, History of Art constructs hierarchical classifications for art and craft separating artist from artisan. Fine art becomes the possession of the privileged classes while the applied art or craft is associated with the working class and the unpaid work of women.

> The art/craft hierarchy suggests that art made with thread and art made with paint are intrinsically unequal: that the former is artistically less significant. But the real differences between the two are in terms of *where* they are made and *who* makes them.
>
> (Parker 1984: 5)

In the middle ages women worked alongside men in embroiderers' guild workshops. By the eighteenth century such work was perceived as a female occupation and by the nineteenth century it was recognised as the pursuit of women of wealth, fine stitchery, or the skilled work of working-class women for which they were poorly paid.

During the Industrial Revolution mass-made factory products separated out skill from knowledge and understanding. The workers on the production line, unlike their predecessors, had no real understanding of the objects they were making. Each worker was restricted to the knowledge required to produce a small part of the production. The worker was no longer a craftsperson but 'an animated tool' (Frayling 1990: 163).

GENDER

It is impossible to write a unit on crafts without addressing gender. In the past craft has displayed a gender divide, which was based on historical circumstances. Right up until the 1980s schools' curricula dictated that activities such as wood and metalwork should be followed by boys, while girls were assigned needlework and cookery classes. Traces of this gender divide continue to linger in many schools today. It is your responsibility to challenge these traditional attitudes towards gender and examine any recurrence of stereotypical attitudes. It might be useful to look back at the list of craftspeople you provided earlier in the unit. (How) does your list reflect these issues? Have you provided an equal balance between men and women craftspeople?

In a short succinct paper, Dalton explains the position of craft in schools in the 1980s. She claims: 'In schooling the conditioning process continues to train young women into habits and skills that make association between women and textiles a natural one' (Dalton 1987: 31).

Attempts to include boys in textiles and to afford equal status and value to metal, wood, plastic, fabric, food and clay, have failed to change the divide which equates boys' craft activity with the workplace and girls' craft activity with domesticity. Dalton reveals how fashion and textile manufacturers further perpetuate this divide by recognising the potential to exploit a young female audience. They provide textile teachers with ready-made packs, including wall charts, visual aids and cheap patterns. Girls are taught garment and soft-toy making through a systemic use of pre-designed, precut paper patterns.

Shreeve (1998) suggests another reason for this gendered divide. She claims that characteristics of tacit knowledge, such as intuition, hunches, know how, are ascribed to the feminine and given lower value. Society in general places more emphasis on explicit knowledge and craft, for its sins, falls into a less-valued category.

Parker (1984) cites the way advertising and the media constantly reinforce a gen-dered view of the crafts. The changes in the secondary examination system in the 1980s, from GCE to GCSE, provided an opportunity for educators to reconsider stereotypical gender divisions. The NC Art proposals (DES 1991a) reiterate the need to remedy earlier divisions insisting that it is important for teachers to select method-ologies, topics and materials which attract the interest of both sexes. The document claims that it is important to present pupils with examples of the work of artists, craftworkers and designers of both sexes insisting that this way, 'both girls and boys can grow up knowing that the full spectrum of media techniques and skills is open to them' (p. 60).

You must avoid the notion that gender stereotypes in the crafts are unusual. They are not. Roles are often reversed across history, place and time, for example, in sixteenth-century England knitting was primarily a male occupation and in West Africa the Fori men of Dahomey produce woven cotton cloth which is used for appliqués. Feminism in the arts during the 1970s and 1980s protested against the distinctions between 'art' and 'craft' grounded in different materials, technical train-ing and education. The way activities such as sewing, piecing, cutting, appliqué have been assigned in history to women, despite the fact that men also practised them, can be seen, in part, as responsible for the shift in attitudes in education. Chadwick claims that during the 1970s distinctions between 'art' and 'craft' began to 'fray around the edges' (Chadwick 1990: 332). The movement known as Pattern and Decoration, which attracted both men and women, used fabric and surface elaboration partly as a reaction to the gender-biased use of the term 'decorative'. Judy Chicago's *Dinner Party* (1979) blurs distinctions. It celebrates women's historical and cultural contribu-tions in a large installation which incorporates sculpture, ceramics, china, painting and needlework (Chicago 1996). Similarly, Faith Ringgold and Miriam Schapiro explore issues of exploitation, race and gender in their work (Chadwick 1990). Jefferies (1995) insists textiles is a hybrid in the expanded field. She quotes Sarat Maharaj's essay, 'Arachne's genre: towards intercultural studies in textiles', which argues that avant-garde textiles practice, 'maps out an inside/outside place – "edginess" – it cites estab-lished genres and their edges as it cuts across them', to throw out of joint, 'handed down notions of art/practice/genre/gender' (*ibid.*: 168).

LEARNING THROUGH INTELLIGENT MAKING

Theories about learning can be called upon when discussing the crafts. Perhaps the most widely used excuse for not linking thinking with making in craft is the belief that learning in the crafts is tacit. The dictionary defines tacit as: silent, passed over in silence, unspoken, implied or indicated, but not actually expressed, arising without express contract or agreement. So, within the context of making in craft, tacit learning could be thought of as the ability to do something, to weave willow, to stitch fabric, to weld silver, as having an inherent ability to work with one's hands and eyes. It was Polanyi (1964) who insisted that we know more than we can tell, that tacit learning includes: implicit thinking, intuition, hunches, skilled performances.

Polanyi's explanation of tacit knowledge, as the kind of knowledge that cannot be fully articulated, is commonly used by makers to account for their difficulty in explaining how they have worked their materials and what skills they have employed in making. As Dormer reminds us: 'Craft relies on tacit knowledge' (Dormer 1997: 10). Tacit knowledge of craft making is acquired through experience and it is this knowledge that enables pupils to do things as distinct from their talking or writing about them. Closely allied to this theory is that of implicit learning or the inherent ability to learn from a task without rational or conscious deduction. In education, explicit knowledge is the given. The unseen and the unspoken aspects are under-valued or ignored within Art & Design. Often you find that pupils have, what appears to be, an innate ability to work with a material and realise a successful outcome without being able to articulate how they have done so.

Haptic knowledge has close links with both tacit and implicit learning. Haptic knowledge is that which you acquire through tactile and kinaesthetic physical sensa-tions. It depends on feeling and touch as a means of communication. This sense of touch can be undervalued in Art & Design. Shreeve (1998) provides a powerful poetic example of a student working on a piece of embroidery. She describes the way different parts of the body respond to the machine and production of the work:

> Without conscious effort, the eyes, hands and feet all respond and adjust minutely to changes in the total production of the piece. The body gently moves in response to the work, the performance of the machine and to the desired outcome. Fingers walk along the fabric to gently guide and obtain feedback from the embroidery process. At no point has the student needed to articulate exactly what is required; this has been processed at a level below the conscious threshold.
>
> (Shreeve 1998: 43)

This example highlights the importance of kinaesthesia, the faculty of being aware of the position and movement of parts of the body. In addition, the synaesthetic response should not be overlooked. Sounds of tacky ink sticking to the printing roller, the purring of the sewing machine or the smell when flux melts letting in the silver solder, are part of and inform intelligent making. Shreeve suggests that sometimes language has no function in craft learning; she identifies simultaneous doing and speaking as difficult.

Task 8.1.7 Tacit and haptic learning

In your experience of Art & Design what other examples can you give that rely on tacit, implicit and haptic learning?

Watch a teacher demonstrating a making activity, e.g. throwing a pot, carving a piece of wood, using a sewing machine. Identify the extent to which s/he uses language to explain the activity. Do you think that language is always necessary?

CRAFTS IN CONTEXT

> In many cultures the clear distinction between the major or fine arts (painting and sculpture) and the minor or applied arts (craft), is an irrelevance. The European Renaissance established this division by raising the fine arts to a liberal, and thus cerebral, art, fabricating its intellectual credibility through a supporting body of theory, and debasing its sibling by designating it mechanical . . .
>
> (Addison 1997: 4)

Traditionally in many cultures the hierarchy of crafts or the Western representation of a divide between art, craft and design does not exist. Crafts are not special, pigeon-holed as particular representations or with particular domains. Teachers have for many years recognised that non-Western visual and material culture provides a rich resource for teaching and learning in Art & Design. The NC Art Order (DFE 1995) states that pupils' work at KS3 selected from non-Western cultures should exemplify a range of traditions from different times and places. However, you must recognise the need to locate artefacts from these cultures in context avoiding decontexualised comparisons and Western categories of art and craft. You should realise the danger in projecting a Western aesthetic.

You need to move away from formal analysis as a way of perception and think about the following questions: how are crafts made? what materials are used? why are they made? and by whom?

CRITICISM

> They say: 'Learning a craft is easy, it's thinking that's difficult'. Yet in learning a complex craft it is the head that hurts, not the hands.
>
> (Dormer 1991a: 38)

The identification of ways in which the crafts contribute to an understanding of contemporary, visual and material culture has preoccupied makers and crafts critics throughout the last decade (e.g. Harrod 1997; Johnson 1998; Greenhalgh 1997; Dormer 1997).

It is rare to find an Art & Design department that differentiates between art and craft in their approach to interpretation and evaluation. One of the main anxieties around craft criticism is the attempt to talk about a wide range of practices as if they were one thing. Johnson (1995) asks us to consider models of craft criticism and questions whether it is appropriate to use art historical models, 'Let us hope that the critical debate will go forward by acknowledging and outlining the differences and contradictions rather than trying to resolve them prematurely or, indeed, to hide them away' (p. 44).

> **Task 8.1.8 Critical evaluation of art and craft**
>
> In tutor groups discuss:
> Do you need to adopt different critical approaches to art and craft?
> What possibilities or problems does a differentiated approach produce?
> Does such differentiation serve to reinforce or question hierarchies?

Secondary school pupils are fascinated by, and surprisingly well informed about, contemporary technology. However, they often fail to make the connection between innovative practice and the crafts they experience at school. As student teachers you need to ensure that you do not perpetuate only the traditional definition of craft. Stungo (1998) suggests that we:

> envisage a world where fabrics are not inert materials to be cut and stitched but live, intelligent substances; a world where dyeing is redundant because natural fibres can be grown in any colour, a world where fibres carry vitamins and sunscreens in their weave to protect your skin and clothing comes with built in pollution monitors, mobile phones or just any other kind of technology that you care to think of.

(p. 26)

WAYS FORWARD FOR CRAFT

In an article written in 1998 we tentatively suggest two alternative ways forward:

- to foster craft as dependent on a distinctive form of intelligence, with a unique but alternative way of knowing with its own history and traditions, secure within its own boundaries;
- to promote craft as an important and integral facet of contemporary culture; a field that is not static or fixed but is sufficiently confident in its distinctive features to reject limiting boundaries and transcend the restrictions of conventional categorisations.

(Burgess and Schofield 1998: 129)

The first promotes learning in the crafts while the second emphasises learning through and in the crafts. Our preference is for the latter: which do you choose?

THE CRAFTS COUNCIL

Resources and publications concerning the crafts are available from the Crafts Council. It has an extensive reference library, a reference desk and a photostore (*www.craftscouncil.org.uk*).

- Reference Library: contains over 4,000 books, 100 journals and 40 videos.
- Reference Desk: provides listings and contact details of craftspeople in any crafts discipline throughout the country.
- Photostore: is an interactive, user-friendly picture library system which enables visitors to search quickly for images and information by maker, object, material and technique. It contains over 35,000 images of contemporary British craft from architectural glass to toys and spanning twenty years.

FURTHER READING

Burgess, L. and Schofield, K. (1998) 'Shorting the circuit', in P. Johnson (ed.) *Ideas in the Making*, London: Crafts Council.

Crafts: The Decorative and Applied Arts Magazine.

Dormer, P. (ed.) (1997) *The Culture of Craft*, London: Crafts Council.

8.2 DESIGN

OBJECTIVES

By the end of this unit you should be able to:

- understand why design remains an area of contention and confusion;
- develop different ways of promoting design in the curriculum.

Established narratives of design, like those of craft, are being challenged and deconstructed. Myerson (1993) suggests that design as an object fixed in time and space, has been replaced by design as a process, fluid, changing, perplexing, and increasingly unable to be contained by traditional disciplines or methodologies.

Perhaps one of the most contested areas of the curriculum is design. Its position and status have shifted regularly over the last three decades.

Task 8.2.1 The status of design

In tutor groups: think back to when you were at secondary school.
  What did design mean to you then?
  Did you perceive it as an integral element of the Art (and Design) department's teaching or was it located in a different area of the curriculum?
  Was it part of Craft, Design and Technology (CDT) timetabled with Home Economics, or, an aspect of Design Technology (D&T)?
  Did it part company with Craft and if so why?
  How is design defined in secondary education now and in what ways is this definition still shifting?

You may identify a lack of consensus and begin to appreciate the problems associated with attempts to classify this elusive and changing subject. In order to understand how this situation has arisen it is important to look briefly at the recent history of design education. Even today teachers working in different departments in secondary schools use the term 'design' to mean different things; sometimes as a verb, sometimes as a noun. This confusion has no doubt arisen from the way design has been shifted from one subject area to another. The field too, is vast including: architecture, fashion, ceramics, silver, furniture, interiors, town planning, graphics, environments, systems.

In the 1970s design was, in the main, perceived as part of CDT where the work undertaken by CDT teachers was seen as a separate discipline from Art. Here, an emphasis was placed on identifying particular aspects of the design activities and this was evidenced by the teaching of specified skills, materials and outcomes; a process model. Pupils were taught the purposes of products and systems and how this knowledge might be applied to their own designs. Through this, pupils improved their manual dexterity and developed an ability to solve specific design problems. This rigid model was not wholeheartedly endorsed by Art & Design teachers.

The CDT GCSE Guide for Teachers (1986) emphasises particular aspects of the design activity through its assessment objectives which focus on the following sequential development:

- identify and analyse problems;
- generate ideas and potential solutions;
- plan the production;
- produce the selected solution;
- compare and evaluate against its specification.

(p. 1)

Integral to this is a requirement to foster economic and industrial understanding. With this in mind pupils are taught to identify needs, both in terms of the uses and purposes of products and systems, and how this knowledge can be applied to their own designs. Additionally, stress is put on discovering the efficacy of different materials, tools and techniques. Although CDT encompasses a wide variety of activities and teaching styles, it subscribes to the principle that a pupil's level of understanding can only be manifested through designing and making in response to a perceived need and with a prescribed range of materials and tools.

The late 1980s saw the growth of large faculties in some secondary schools which incorporated art, craft, design technology and home economics, and in some cases the performing arts. These faculties often operated a carousel system, which some schools have retained. They function by dividing the curriculum into discrete, six- to nine-week units. Sometimes these are brought together with an overarching theme or common topic, entitled, for example, 'local architecture' or 'old and new'. This arrangement gives all pupils a taster course working with different methods and materials, but it tends to remain superficial. For it to succeed, it requires rigorous coordination and time for teachers to plan, review and evaluate. Such arrangements are not without their problems. Teachers may have different degree specialisms, philosophies and ways of working. An uneasy relationship can be formed between subject areas which may have little in common. However most educators now regard carousel arrangements to be no more than an administrative convenience, guaranteed to ensure that there is little depth, coherence or progression. As Thistlewood (1990) remarks: 'Design is not easily taught; it is far simpler to teach a form of pseudo designing in which pupils draw and make pastiches of existing consumer products' (p. 6).

DESIGN AND THE NATIONAL CURRICULUM

> It [Design & Technology] is a messy story of political and ideological meddling combined with a good deal of confusion and frustration in schools.
>
> (Kimbell 1996: 29)

With the development of the NC Subject Orders following the Education Reform Act (ERA) (1988), the carousel approach was largely abandoned in favour of discrete areas. The ERA proposed that Art & Design, and Design & Technology (D&T) become independent. However, the legacy remains in some schools, for example, specialist rooms and teachers. Following many discussion documents and views fielded by the Design Council, the Engineering Council, the Crafts Council and the NSEAD about the ways in which Art & Design could contribute to the proposed NC Design and Technology Order, design became decoupled from Art & Design and embedded within a separate and often discrete new NC subject, Design & Technology. It is interesting to note that D&T was not originally intended as a subject in its own right, but more as a vehicle for cross-curricular work. This is in line with the Design Council's view that design is a whole-school issue.

The NC, D&T Order (DES 1995) has driven a specific approach to design through knowledge of engineering and technical instruction linked to industry via vocational pathways. It promotes a particular way of working, that of playing the role of 'designer as producer'. Less emphasis is placed on a critical understanding of the manufactured world.

This is not to devalue what can be taught successfully. Through group work, pupils can actively investigate a design brief and by working with materials and processes, create a 'quality' outcome which draws on their existing knowledge and experience. Such an approach develops an understanding of research as well as fostering inter-personal skills. This mirrors the way a real life design team works, where each designer cooperates and collaborates by 'chipping in' something from their own experience and imagination to add to the final outcome. Pupils can be empowered if they are given control over the process of designing through the teacher fostering an environment for creative questioning, intelligent making and pertinent evaluation.

All too often design in schools leans towards the proliferation of limited and pre-scriptive design work reflecting the need for quantitative and comparative assessment outcomes rather than process-led open briefs. If a simulated 'neo' design for public view is to be made by pupils it needs to be costed, trialled, evaluated and made to work, activities which have social and educational benefits.

THE PROBLEM-SOLVING APPROACH

> 'Problem solving' has, over the years, had a very full treatment from psychologists, educationalists and philosophers, seeking to describe and explain the processes of thought. Thompson (1959) outlines the characteristics of problem solving behaviour; Kubie (1962) relates it to a cybernetics theory of learning; Vygotsky (1962) links it to the development of language and particularly to symbolic representation; Rogoff (1968) relates it to schools and education, 'a difficulty is an indispensable stimulus to thinking'.
>
> (Kimbell *et al.*, 1996: 30)

Since the 1980s a particular approach to design teaching and learning has become a recognised model of good practice in secondary school. Green (1982), Baynes (1982, 1990), Kimbell *et al.* (1996) and Eggleston (1996, 1998) have all promoted variations of the problem-solving approach. It has been successfully adopted by both Art & Design, and Technology departments.

Problem solving can provide a framework for experimentation and learning, where pupils are active participants. It is the teacher's role to provide a design brief, which replicates, within the classroom, the idea of the client who requires an identified need to be resolved. This may be, for example, a logo design for a brand of sports shoe, a local council requiring its bus shelters to be re-designed, or a wall hanging for a local hospital. Pupils at this stage begin by brain-storming and concept mapping using both verbal and visual means. This can be done individually, or in pairs, and can be discussed later with the whole group in the classroom. Pupils need to map their ideas keeping in mind the possibilities and limitations of suitable processes and materials along with historical precedents, ergonomics and social, economic and environmental concerns. After considerable discussion, interviews, research, reading and drawing, pupils can reflect upon their findings in order to propose a number of tentative solutions to the given problem. They may find it useful to present these to small groups within a whole-class situation. Reflecting upon their findings and the views of others can lead them to select materials and processes and produce a prototype.

This model of design learning combines the active with the reflective. As the problem-solving task becomes clearer the constraints on the solution are subject to critical analysis. Kimbell *et al.* (1996) point out that there is an astonishing similarity between the design process and more general thought processes. They suggest that the introduction of a problem disrupts routine and forces the learner to '. . . stop drifting and think about what they are going to do . . . this creates the cognitive conflict that is essential for a subsequent reformulation of a new, more comprehensive schema' (p. 30).

This process of reflection on action is described by Schon (1987) (see Chapter 2). Often the problem-solving process is perceived as a linear model which starts with the identification of the problem and ends when the solution has been evaluated. Figure 8.2.1 illustrates this diagramatically:

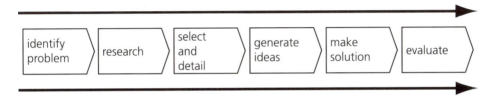

Figure 8.2.1 Problem-solving process (after Schools Council Design and Craft Project 1972)

The strength of this way of working is that thought and decision-making processes are made concrete. Kimbell (1996) explains how, by looking through pupils' design folders, their decision-making processes are revealed. He suggests this facilitates the following:

- You can go back over it with the pupil to examine where critical decisions were made.
- You can look to see the basis of evidence for that decision.
- You can examine points at which alternatives would have been possible.
- You can use these as jumping-off points into new lines of development.

(p. 31)

We consider that all the above points should be recognised as a two-way process or dialogue between pupils and teachers, extended through group evaluation and paired discussion between pupils.

However this activity has its limits. Pupils are often given a design brief without their having knowledge of the socio-economic and cultural values through which real designers operate. You need to ensure that problem-solving encourages pupils to stand outside themselves and consider the needs of others, recognising differing tastes and values.

ART & DESIGN TECHNOLOGY: THEIR POSITIONS IN SCHOOLS

Unless you advocate the role of design in Art & Design most headteachers are unlikely to recognise your practice as a key contribution to technology in the curriculum. If Art & Design is to be adequately resourced to maintain its present wide range of activities and experiences it is necessary to argue this case by providing evidence of good design practice within Art & Design departments. Teachers of Art & Design are keen to identify the contribution that they can make to D&T in particular, teaching pupils to formulate design proposals, to apply aesthetic judgements, to evaluate products and to work with materials and ICT. D&T can be seen as an important component in the Art & Design curriculum and teachers have sought to establish the relationship between these subject areas and ways that one might complement the other. However, others have suggested it is not which 'bridge to build but whether there should be a bridge at all' (Buchanan 1993). Buchanan suggests that the D&T curriculum is, 'too dull, too mechanistic, too linear and not enough about design' (*ibid.*).

However the D&T Working Group Interim Report (DES 1988) identifies the role art could play within this foundation subject:

a   . . . a greater emphasis on creativity, curiosity, imagination and the aesthetic dimension.

b   a recognition . . . that the way of thinking which governs the route leading from ideas to tangible designs often involves a considerable amount of intuition. Many of the most innovative ideas are initially intuitive and subsequently are rationalised to meet a variety of constraints.

c   an ability to analyse the relationship between design and environmental and social concerns.

(*ibid.*)

Steers (1994b) points out that in D&T the creative and aesthetic dimensions are still neglected. He claims: 'It is doubtful whether the [Technology] proposals, constrained as they are by specified skills, materials and outcomes, would inspire adequately pupils to experiment, challenge assumptions and take conceptual risks' (pp. 3–4).

Art & Design newly qualified teachers (NQTs) often find themselves working in art, craft, design, technology faculties as specialists in either, creative textiles, 3D or graphic design. Therefore it is essential that you are familiar with the D&T Orders and examination syllabuses.

Task 8.2.2 Design in schools

In your placement schools look for evidence of design practice:

- in the Art & Design curriculum;
- within school documents;
- on public display;
- in the technology department.

Task 8.2.3 Questioning the status of design

Consider the following statements and use them as a starting point to help you devise a questionnaire for both the Head of Art & Design and the Head of Technology.

- Design is future-orientated.
- To design is always to develop some form, structure, pattern or arrangement for a proposed thing, system or event.
- Design is problem solving.
- Design education must address values. It should be informed by environmental, social and moral issues: not which bridge to build but whether there should be a bridge at all.
- Design is real-life activity.
- Design is as much about aesthetics as function.
- Design is designer labels.
- Design requires an understanding of skills and techniques.
- Design makes the boring beautiful.
- Design is central to all our lives.

This questionnaire should enable you to identify the position on design that has been established in your TE schools.

How is design education perceived (high or low status, from differing perceptions and philosophies)?

Where is it located in the curriculum (as a discrete subject or cross-curricular)?

Share your findings with other members of your tutor group. Is there any consensus of opinion?

What are your views of design and how it should be taught? Find an opportunity to discuss these issues with your subject co-tutor (SCT) or other colleagues.

Baynes (1990) claims: 'Art & Design are not synonymous. They are not identical. Art & Design is only one pair. Technology and design is another of equal importance' (p. 51).

Rigid boundaries between Art & Design do not exist, nor would we want to draw strong demarcations between them in this chapter. The 1980s was the time when design filtered into every aspect of our culture and an obsession with labels continues to pervade the classroom. It is important that you ensure that you do not become 'prisoners' of the present, unable to foresee alternatives or recognise possibilities of choice. You need to ensure that pupils are introduced to design histories including their wider social and cultural implications.

Task 8.2.4 Craft and design links

> In discussion with other student teachers reflect on what you have read about craft earlier in this unit to consider the links between craft and design.
> How might you contrast and compare these terms which are often used indiscriminately in schools? Both terms can be used as nouns or verbs.
> How have you used these terms in your planning and teaching?
> As a noun, verb, interchangeably, independently, separately, why?

DRAWING AND DESIGNING

Perhaps one easily identifiable difference between craft and design is that designers are more likely to start with drawing. Drawing, it is generally agreed, is a fundamental element in the design process, and as Kimbell *et al.* (1996) so aptly remind us: 'it is no coincidence that designers frequently talk of themselves as "thinking with a pencil" ' (p. 30). This is in direct contrast to the craftsperson who often starts by working directly with materials. Design drawing takes the form of exploration and communication. It enables the designer to both organise thoughts and express ideas to others. Garner (1990) points out that the drawing strategies employed by the designer aim to: 'explore problems, manipulate information and visualise responses', and that these have no clear divisions between them; he states:

> Whilst a designer may be exploiting drawing creatively and personally, these same drawings may communicate form, detail, scale, or other information quite readily. Similarly much sketching activity is used to simultaneously clarify conceptual development, facilitate evaluation and provoke the further generation of ideas.
>
> (p. 51)

Although new software packages (CAD) enable designers to draw and visualise electronically, the designer's sketchbook is still one of the most important tools used for generating ideas.

Task 8.2.5 Drawing and design

> Design a SoW in which a need is resolved through a process of sequential drawing.

SOCIALLY RESPONSIBLE DESIGN

This is an age of mass production in which everything is planned and designed. Design has become very powerful in that it demands a high social, moral and ethical responsibility from the designer. Design must become more responsible, more creative and more responsive to people's real needs, and not merely respond to the demand for designer artefacts, labels or more throw-away objects. Now that many things have become possible through technology, you as a teacher need to alert your pupils to the ethical and moral issues surrounding design and its marketing power.

The most public champion of green design is without doubt Papanek. His seminal text *Design for the Real World* (1971) calls for a responsible design movement. Papanek is concerned about the mismatch between the power and influence of design and the lack of social and moral responsibility shown by many professional designers. He bemoans the fact that design is dominated by consumerism and profit. He proposes an agenda for design which comprises the following six priorities:

- design for the Third World;
- design of teaching and training devices for disabled people;
- design for medicine, surgery, dentistry and hospital equipment;
- systems design for sustaining human life under marginal conditions e.g. polar icecaps, underwater, deserts;
- design for breakthrough concepts, a radical rethinking of approaches to design, rather than the more conservative approach of continual refinement of existing products.

Papanek has been described as unnecessarily reactionary and sensationalist by some design critics who dismissed him as 'a cult figure' while ecology was fashionable during the early 1970s. Others have heeded Papanek's call for social responsibility. These include the Unit for the Development of Alternative Products (UDAP) based in Coventry, Sheffield Centre for Product Development and Technological Resources (SCEPTRE), London Innovation Limited (LIT) and the Centre for Alternative Technology, Wales.

Task 8.2.6 Designing for social responsibility

Working in small groups develop a visual resource pack for use in school which promotes socially responsible design. For example, you might include products or systems designed for the physically disabled, promotional literature for a specific local environmental concern, products which rely on natural power to function.

Using this resource as starting point devise a SoW for year 10 students.

CRITICAL, HISTORICAL AND CONTEXTUAL DESIGN STUDIES

Consider Attfield and Kirkham's (1995) distinction between art and design:

> Designed objects surround us and form part of our culture in a much more intimate way than art – we interact differently with artefacts like clothes, cars, buildings, streets and food packaging directly in our everyday lives. But

because we encounter these in a much more mundane way – we wear shoes, walk on carpets and drink coffee from mugs – design tends to become invisible and absorbed into the less thinking part of our daily existence. Art, on the other hand, has to be sought in special places such as galleries where it is set apart for contemplation and appreciation. While the fashion and style aspect of design is so self-conscious and visible, depending as it does for its very existence on ephemerality, the fascinating depths of meaning it embodies are disguised and hidden beneath its apparent triviality.

(p. 1)

Design understanding and decisions are often taken by pupils in school in a vacuum. The need for a design education that links the study of formal and functional elements with training in visual language, communication and contextual studies has been recognised for some time. Baynes (1982) advocated this when he said: 'I want to look towards social and cultural aspects of design education and away from the debate about the content of the curriculum. I am disenchanted with it, it seems to me that it has rapidly become sterile' (p. 106).

Buchanan (1990) reiterates the importance of an extended notion of design education when, nearly a decade later, he bemoans the fact that, 'Little has been done in most schools to introduce pupils to the work of designers . . . to explore the range of influences on their work or to judge or value directly, and in a critical way the outcomes and effects of their work' (p. 9). Design history is an extensive field but it is only since the late 1980s that it has become a subject in its own right at degree level. It is not surprising therefore that teachers have not felt comfortable introducing and integrating this into their SoW. Pupils are often asked to design without the benefit of understanding either the history of design or the socio-economic and political context in which designers operate. It is widely believed that as Art & Design educators it is our responsibility to encourage pupils to view their internal (domestic) and external environment and receive messages critically rather than passively. The concept of design as described in this unit and promoted by the NC Art & Design Order is not universal. You should be aware that the attempt to demarcate definitions between art, craft and design is a particularly Western preoccupation. It would be wrong to assume that Western definitions have been accepted globally. Design needs always to be contextualised in time and place.

Design history is concerned with analysis of the forces that influence visual and material culture. Design has for the most part escaped critical analysis especially within the Art & Design school curriculum. Designed artefacts surround us in our everyday lives but it is far rarer that teachers and pupils choose to investigate these in the classroom in a critical way. Although such objects are ubiquitous in Art & Design departments as part of still-life groups, few teachers make use of them as resources for pupils to learn about their values and significance. As Conran (1993) states: 'The history of the world can be documented by the design of objects and the study of those objects gives a clear message about the changes that were taking place in society' (foreword). If Conran's word 'by' is substituted with the word 'through', then there is ample room for critical enquiry. It should be noted that contexts do affect the perception of artefacts. An artefact takes on different connotations depending on its location;

imagine the same object displayed in a museum, shop or classroom. Designed objects of whatever type, chairs, shoes, computers, cars, telephones, may be examined for their significance and values. It may be more relevant for teachers to provide such objects for handling and discussion, or ask pupils to bring in, objects which are central to their lives. Alternatively, artefacts collected by the teacher or borrowed from museums alert pupils to different and unfamiliar types of design. These act as triggers for the investigation of function. Pupils' mobile phones, bags and drink containers might be objects to start the ball rolling. Appropriate questions should be asked so that pupils can analyse and make 'use' of objects and specialist language can guide pupils towards asking effective questions about their artefacts. Formulating the 'right' question is as important as its answer.

Schofield (1995) has revealed that by using a carefully constructed framework, pupils' perception and understanding of an object will change over time. She suggests that five domains are appropriate, within which questions and ideas about objects can

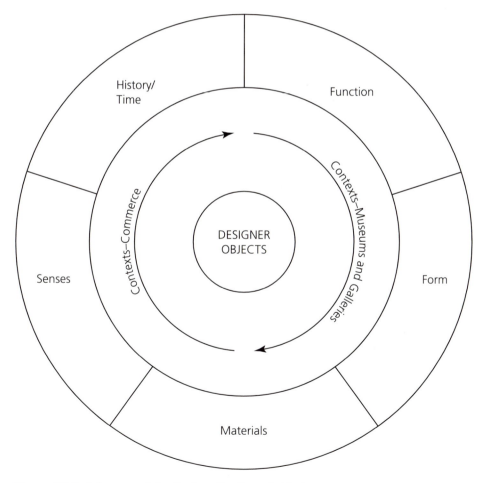

Figure 8.2.2 A framework for the investigation of objects

be located. These are: history and time, function, form and style, materials and senses. (See Figure 8.2.2).

Schofield (1995) has shown that these domains are not exclusive and that no one domain should be seen in isolation when discussing or investigating a designed arte-fact. Emotive, aesthetic and sensory responses to designed artefacts can be manifested in both verbal and visual ways. Pupils may record and document their reactions and perceptions by communicating their feelings and knowledge of the object in question.

Task 8.2.7 Reading design

Collect a selection of designed artefacts chosen by yourself or your pupils.
Produce a SoW which makes use of the domains suggested by the framework.
What other domains would you like to include?
How would you link critical/contextual work with practical making in this SoW?

THE POSITION OF THE DESIGN COUNCIL

It is no coincidence that the Design Council and the Design Museum have concen-trated their attention towards NC Design & Technology rather than NC Art where the traditional Western fine art canon continues to dominate.

In an attempt to clarify the place and role of design in mainstream education the Design Council (1994) convened a consultation forum. It invited representatives from a wide range of interest groups such as HE and FE. The resulting paper identifies four key issues which the Design Council believed form the core of the current range of design opportunities and problems in education. They are:

- the status of design in schools;
- the relationship of design to other subject areas;
- the values which should inform learning and teaching in design;
- working towards a total design policy for schools.

However, rather than locate design firmly in one area of the school curriculum the Design Council calls for an expanded definition:

> We recognise that 'design' as a word is problematic within schools. But where we use it, 'design' should be read as covering both subjects of the National Curriculum (Design and Technology and Art & Design) and *indeed a much wider remit* [our italics].

> (Design Council 1994: 4)

They go on to suggest that what is missing in design education is an element of risk taking. It called for design-related activities that deal, 'with subjective and interpretative issues without the safety net of a fixed knowledge base. Such activities offer pupils a challenge and an opportunity, but teachers must allow pupils to take supported risks in their own learning' (*ibid.*). This raises the pertinent question whether these approaches can be encouraged in initial teacher education and in in-service training. The lack of a fixed, static definition for design can be recognised as a strength as well as a cause for confusion.

THE POSITION OF THE DESIGN MUSEUM (LONDON)

The mission of the Design Museum education department is to encourage and enhance the study of design using its collection and temporary exhibition programme as a primary teaching resource. It aims to broaden public perceptions of design includ-ing both the products that are used and the environment in which people live. The museum believes that design can be defined as the purposeful use of knowledge, skills and physical resources to create products that meet a perceived need or opportunity. The Design Museum's programme provides publications, resources, workshops, lec-tures, seminars and professional development courses for all levels of education. It suggests that pupils can contribute effectively in a technological society by being able to:

- create practical solutions to problem solving;
- communicate effectively and work in teams;
- develop visual awareness by appreciating the importance and value of design;
- appreciate the relationship between design and environmental issues;
- understand the relevance of design to industry.

Task 8.2.8 Visiting the Design Museum

How would you prepare pupils for a museum showing contemporary design?
What issues would you ask them to investigate?
Design a work sheet for pupils to complete during their visit.
How would you follow up this visit in the classroom?

FUTURE: THE TWENTY-FIRST CENTURY

As you can see a rather two-sided view of design in education has emerged:

- the vocational training of a 'neo'-designer; the creation within the classroom of mini workshops, 'pseudo' factories or design studios where cardboard mock-ups act as substitutes for real prototypes;
- design activities within a broader cultural context that promote a critical understanding of the visual and made world.

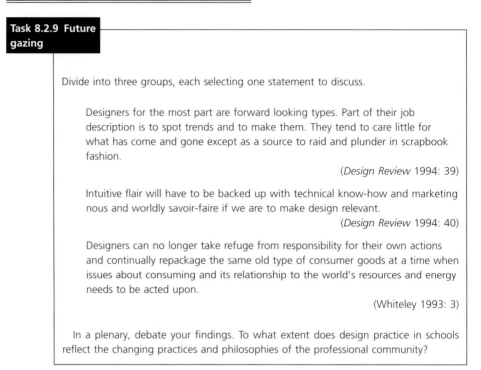

Task 8.2.9 Future gazing

Divide into three groups, each selecting one statement to discuss.

Designers for the most part are forward looking types. Part of their job description is to spot trends and to make them. They tend to care little for what has come and gone except as a source to raid and plunder in scrapbook fashion.

(*Design Review* 1994: 39)

Intuitive flair will have to be backed up with technical know-how and marketing nous and worldly savoir-faire if we are to make design relevant.

(*Design Review* 1994: 40)

Designers can no longer take refuge from responsibility for their own actions and continually repackage the same old type of consumer goods at a time when issues about consuming and its relationship to the world's resources and energy needs to be acted upon.

(Whiteley 1993: 3)

In a plenary, debate your findings. To what extent does design practice in schools reflect the changing practices and philosophies of the professional community?

FURTHER READING

Many of the books mentioned in Unit 8.1 on craft do not provide a clear line of demarcation between craft and design and should be recognised as relevant for both areas of study. Likewise many of the authors who have written about craft also write about design. Such is the overlap between these two closely related, interrelated spheres.

Attfield, I. and Kirkham, P. (eds) (1995) *A View from the Interior: Women and Design*, London: Women's Press.

Dormer, P. (1991b) *The Meanings of Modern Design*, London: Thames and Hudson. Dormer suggests that contemporary design is more restricted than contemporary art; it is always wary of its audience. If design moves too far ahead of what people understand then it fails them as consumers and they stop consuming.

Papanek, V. (1995) *The Green Imperative*, London: Thames and Hudson.

Sparke, P. (1995) *As Long as it's Pink: The Sexual Politics of Taste*, London: Pandora.

Walker, J. (1992) *A History of Design and Design History*, London: Pluto.

9 Attitudes to Making

Andy Ash, James Hall, Pam Meecham and Rose Montgomery-Whicher

9.1 SKETCHBOOKS AND ARTISTS' BOOKS

James Hall

A sketchbook is . . . a personal visual memory bank that can be used as a resource for collecting and developing ideas . . . a repository of ideas which come faster than they can be realised.

(Robinson 1995: 14) (see Plates 8 and 15)

INTRODUCTION

The sketchbook is regarded as the visual notebook and key research tool of the artist, craftworker or designer. Sketchbooks can offer insights into the personal vision, source material, ideas and working processes of the artist. Equally, the educational value of sketchbooks has long been recognised. Now, the National Curriculum for Art, GCSE and GCE A Level syllabuses all emphasise the development of pupils' personal responses, their recording and observation skills and their abilities to investigate and analyse ideas. Sketchbooks provide evidence of pupils' realisation of these aims. As well as supporting pupils' Art & Design education, the use of a sketchbook can help pupils to develop self-awareness and skills as independent learners and critical observers.

OFSTED inspections have shown that poor standards in art are evident in schools 'where pupils have not been taught how to draw or use sketchbooks to record their visual research' (AAIAD 1994). Taylor and Taylor (1990) found that it was possible for a pupil to reach the sixth form without ever having kept a sketchbook. For pupils to be taught and encouraged to use and keep a sketchbook, it is necessary for teachers to have a clear understanding of the purpose of sketchbooks in art education.

Sketchbooks support all aspects of the National Curriculum for Art and children should be encouraged to keep sketchbooks so as to function, from an early age, as researchers (Robinson 1995: 95). Pupils can be taught to develop their sketchbooks in increasingly diverse and personalised ways as they mature and their confidence and skills grow. Furthermore, the development of critical studies in art education has increased the amount of written work which pupils produce in their art courses and submit for examinations. There are examples of some highly original 'personal study' submissions for A Level examinations, which can challenge any division between so-called 'academic' and 'practical' work and cross boundaries between text and image, essay and artefact, into a synthesis of written and visual work. Such personal studies and sketchbooks often come closer to the art form known as 'artists' books', which will be discussed in the concluding section of this unit.

OBJECTIVES

By the end of this unit you should have:

- reflected upon and understood the educational value of sketchbooks;
- developed awareness and understanding of the sketchbook as an invaluable research tool for pupils;
- investigated ways in which sketchbooks can be used in a variety of ways to realise aims and learning objectives in art education;
- planned, taught and assessed some work for pupils' sketchbooks.

THE ROLE OF SKETCHBOOKS

Why should sketchbooks be used by pupils in a secondary-school art department? One of the key functions of art in the secondary-school curriculum is to give pupils the opportunity and the skills to respond to their personal experience of the world in a visual way. There are other areas of the curriculum in which the learning content is prescribed. However, it is in art where pupils' personal responses and ideas are highly valued. In many departments, pupils work increasingly to their own personal agenda within the framework of study offered by the teacher and department. By the time pupils reach their GCSE or A Level years, they are encouraged and expected to generate their own focus for their studies. Robinson (1993), in her research into pupils' use of sketchbooks, saw them as a means for pupils 'to sustain and celebrate their individuality'.

The sketchbook is a creative tool: its use should encourage information-gathering, experimentation and risk taking in the search for a creative solution to a self-generated idea or problem. Six key functions of the sketchbook are to encourage pupils' development of:

1   A personal response
     Pupils are encouraged to develop and express their own unique and individual vision and experience.
2   Investigating and making skills
     Pupils develop their ability to record their responses, as keen and critical observers. They build their confidence in handling and controlling a variety of media in a spirit of experimentation. They should also develop their visual vocabulary through using the elements of art to communicate their ideas and intentions.
3   Critical and analytical skills
     Pupils are encouraged to develop a critical and inquiring approach to their art work. They develop their ability to select and reject information and to evaluate their responses. This leads to informed choices of approach and media, which pupils can justify as their intentions become clearer.
4   Self-awareness as independent learners
     Through these processes, pupils are developing knowledge about themselves and confidence in their own ideas, interests and abilities as a self-directed and independent learner.
5   An active and creative approach to learning
     Pupils are encouraged to develop an active and creative approach to learning, in which thinking and planning is conducted visually and actively within the investigative process and does not precede the action.
6   Documentation skills
     Pupils develop the skills and habits of collecting and storing data to provide source material for work in the classroom, to suggest starting-points for ideas. The sketchbook can be a valuable aide-mémoire, a repository of ideas in which to record complex events or phenomena simply and quickly for future reference.

In relation to each of the above functions, pupils should compare their own work with that of artists.

INVESTIGATING THE ROLE OF THE SKETCHBOOK

Task 9.1.1 Your own use of a sketchbook

- How do the functions of a sketchbook outlined above relate to your own experience?
- Do you value or use your sketchbook in other ways?
  Discuss and compare your use of sketchbooks with those of other teachers or fellow students.

Task 9.1.2 Your placement schools' approach to sketchbooks

Do the functions of a sketchbook described above fit with:

*   your observations of how sketchbooks are used by pupils?
*   the department's written policy for sketchbooks?
*   how teachers see the functions of sketchbooks?
*   how pupils see the function of sketchbooks?

*   Do pupils of different ages use sketchbooks for different purposes?
*   Are pupils taught ways in which they might use their sketchbooks?
*   What is the nature of the tasks given for sketchbook work?
*   Would you say the tasks are largely pupil directed or teacher directed?
*   What might you expect to see in a 'good' sketchbook?

Task 9.1.3 A broader picture of the role of sketchbooks

*   Read reports from OFSTED inspections for their references to the use of sketchbooks. You could read the report for your particular school or a summary of national findings (OFSTED 1995b; AAIAD 1996).
*   Read Clement (1993) chapter on drawing, Robinson (1993) or AAIAD (1994) and compare their ideas with your findings.

THE FORM OF SKETCHBOOKS

The term 'sketchbook' may suggest it is only used to 'make sketches' e.g. 'a preliminary, rough, slight, merely outlined or unfinished drawing or painting, often as an experiment to assist in the making of a regular picture' (Concise Oxford Dictionary 1985). This definition implies that a sketch is a preliminary drawing done in advance of and in preparation for 'the real thing' or 'finished piece'. Is this the only way in which a sketchbook can or should be used? It implies that the creative process is a linear development of ideas in which the artist moves logically and sequentially from preparation, to execution and conclusion. The ways in which artists research and create artworks is highly personal, diverse, complex and idiosyncratic. Indeed, artists may not be conscious of the processes they engage in. Certainly, it is clear that artists make drawings, in books or otherwise, in their own right and during or after the creation of the 'final' or 'finished' piece. Does it not demean drawing for it to be seen merely as a preparatory process?

It is well known that GCSE pupils often produce their preparatory studies after they have completed their final piece. This practice is supported by the notion of the creative process as a cyclical rather than a linear process. Ideas are informed by the process of making and handling clay, paint, print, plaster or whatever the chosen medium. Sketches can be made in the chosen medium, not simply as preparation for it. Pupils could be encouraged to use their sketchbooks not only to prepare for, but to document the evolution of their idea from conception to conclusion.

Task 9.1.4 What's in a sketchbook?

Look at some sketchbooks belonging to pupils to discover whether they contain 'sketches', done in preparation for a 'finished' piece, or whether they are used for other, more varied purposes. The sketchbook can be a unique artefact. For example, is the sketchbook used:

- as a visual diary – a daily record of observations, thoughts or feelings;
- as a travel journal – a record of a journey;
- as a personal museum – a scrapbook, an album of memorabilia;
- as a visual notebook – a record of visual data, noted and stored.

Task 9.1.5 Where do pupils use their sketchbooks?

Survey the sketchbooks of a whole class to discover where they have used them, e.g. in the classroom, at home, in their locality, on holiday, in a museum or gallery.

Feelings of self-consciousness or preciousness can inhibit the use of sketchbooks in an open and experimental way. Most of you, no doubt, will have experienced similar emotions leaving you bashful about drawing in a public place or worried about what teachers or other pupils might think about your sketchbook. One can feel the compulsion to 'edit' out works which do not come up to a predetermined standard. These are natural inhibitions so how can teachers help pupils to overcome them?

Task 9.1.6 Overcoming inhibitions

Think of tasks or strategies which might help pupils to get over these inhibitions and adopt a truly experimental and creative approach.

EXPERIMENTING WITH MEDIA

If it is a function of sketchbooks for pupils to explore and experiment with a variety of media, how will you encourage and teach them how to do this? If pupils are working on the theme of the built environment, they could be asked to investigate an aspect of their own home or locality and to record their observations. As the teacher, you could help the pupils to select from the following list:

- drawings from direct observation, memory or imagination in a variety of media – pencil, pen, charcoal, crayon, pastel, felt-tip, ballpoint, or freehand drawing using computer software;
- collecting images using photography or a digital camera, to be manipulated using computer software;
- collage – using scrap or found materials, photographs, cuttings, perhaps worked into with drawing or other media;
- printmaking – crayon rubbings, monoprinting, block-printing and stamp-printing techniques can all be employed directly in sketchbooks;
- writing – drawings can be annotated with notes, poems, diary entries, diagrams, etc.

USING A SKETCHBOOK TO DEVELOP RESEARCH SKILLS

One of the main purposes of the sketchbook is to encourage pupils to develop research skills. Robinson (1993: 104) suggests the four steps in the scientific and mathematical process identified by Poincaré can also be applied to the artistic process:

- Preparation: investigating the problem, gathering relevant data. Sketchbooks develop skills of data-gathering and action planning, and provide an arena for the concrete working out of ideas.
- Incubation: consciously getting away from the problem and waiting. Sketchbooks develop skills of assimilation.
- Illumination: the sudden insight or breakthrough when the solution comes. Sketchbooks offer opportunities for reflection and moments of illumination. Drawing in sketchbooks is a means of discovery.
- Verification: evaluating and testing the solution before applying it. Sketchbooks foster skills of self-criticism and self-evaluation.

Task 9.1.7 Research skills

Do you recognise these stages of the artistic process in your own art work or in pupils' work? Could you use this four-stage model to help plan some work for pupils?

SKETCHBOOKS AND ASSESSMENT

Sketchbooks can be key documents for assessment: they inform the dialogue between teachers and pupils which is at the heart of the assessment process in art education. Sketchbooks can provide evidence of the pupil's progress towards the learning objectives of a project or course and can reveal their developing ideas, skills and ability to work independently. Used in this way, as part of a process of continuous assessment, sketchbooks can help teachers to teach more effectively through understanding pupils' ideas, intentions and abilities and planning for individual needs accordingly.

How pupils' sketchbooks should be marked is more of a vexed question and will depend upon the department's aims, policies and ethos. Indeed, teachers' views differ as to whether sketchbooks should be marked at all. On the one hand it is argued that if teachers want to encourage pupils to value their own thinking then sketchbooks should not be marked and no adverse comments should be written in them by teachers. On the other hand, there is the view that pupils need and enjoy receiving marks or grades as an indication of the standard they are reaching and as an incentive to improve and develop. Many schools have developed systems to track or monitor pupils' progress and achievement and this is often done using quantifiable data such as grades or marks. Target setting involves teachers setting or negotiating targets for pupils' improvement, such as higher grades or marks. Thus, art departments may be required to follow whole-school policies on assessment.

In fact, both views serve to highlight the nature of assessment in art education: that negotiated and formative assessment are at the heart of the dialogue or conversations between teachers of art and their pupils. Assessment in art is a continuous process, which is not only a flow of views and evaluations between the teacher and pupil, but an internal conversation of the pupil with her or himself. In this process, the sketchbook can serve as a valuable self-evaluative or self-critical tool.

There follows the question of when pupils' sketchbooks should be assessed or marked. If sketchbooks are used continuously for work undertaken both in school and elsewhere, they are likely to be looked at by teachers every time they teach a pupil. Teachers may look at and mark sketchbooks during the lesson, as they move around the class talking to pupils. Alternatively, work could be collected for teachers to assess in their own time, which has the obvious disadvantage of depriving pupils of their sketchbooks.

Task 9.1.8 Sketchbook assessment

- How frequently do teachers look at pupils' sketchbooks in your department?
- What form does the assessment of sketchbooks take? e.g. does the teacher discuss them with pupils individually, in groups, or as a whole class?
- How are pupils' sketchbooks assessed and/or marked in your department?
- If marks and/or grades are awarded, how frequently is this done and are criteria for the marks or grades published and known to pupils?
- Are teachers' comments written in sketchbooks or only given orally?
- Do pupils evaluate their own work in their sketchbook?

MONITORING PUPIL PROGRESS THROUGH SKETCHBOOKS

Task 9.1.9 Monitoring progress

Select a number of pupils in a class which you are observing or teaching. Monitor their progress through their use of sketchbooks, asking yourself the following questions:

- How are the pupils responding to the sketchbook tasks set by the class teacher or by you? For example, how well do they understand and interpret the task?
- Are they resalising the objectives of the project or task?
- Do they use work done outside the lesson to support their classwork?
- How could the pupils be helped to use their sketchbook to make more progress in the project?

Sketchbooks can reveal progress or development over longer periods of time than a project or scheme of work. Sketchbooks produced in year 7 or even in primary school can be compared with those produced during a GCSE or GCE A Level course. Michael Rothenstein looked back over many decades to the sketchbooks he produced when aged 4 to 9 years old and found that the imagery reflected his concerns and obsessions at the end of his professional career in his seventies and eighties (Rothenstein 1986).

ARTISTS' USE OF SKETCHBOOKS

In simple, physical terms, a sketchbook will usually comprise a collection of sheets or pages of drawing paper bound between covers. However, the ways in which the sketchbook can be used are infinite and there is no single or orthodox method for using them. The ways in which an artist or art student uses a sketchbook can be as diverse, creative, imaginative and original as the ways in which artworks themselves are realised. One major difference is that while the finished artwork represents the artist's ideas realised and made public, the content of the sketchbook often represents the artist's thinking, musings, seeds of ideas and observations – the raw material and data which may be used by the artist to work towards more complete statements.

Turner's sketchbooks provided him with a means of painting watercolour sketches wherever he travelled. Constable kept sketchbooks to record his beloved Stour Valley:

> During the years 1813 and 1814, he carried two minute sketchbooks around with him which contain the seeds of most of his mature paintings, sometimes with the whole idea there to be barely altered in the paintings of ten and twenty years later.
>
> (Taylor and Taylor 1990: 59)

For Picasso, the sketchbook appeared to have been a means of capturing fleeting ideas on paper. Picasso filled innumerable notebooks, often composed of lined writing paper, with drawings, notes, collages, shopping lists (Glimcher 1986).

Smith, writing about the drawings of Joseph Beuys, offers insights into the value of sketchbooks:

> It is in his still active drawings which may be found the most poignant and accurate insights. These works on and of paper were torn from squared jotters, used notebooks and diaries, or were simply scraps. They were often combined, sometimes later, to produce new images . . . in these works we can experience the faint scratching-around the thought-made sculpture . . . the drawings are in pencil, watercolour and unorthodox media such as blood and stains, and were used to search out, to depict and to document. They are often annotated.
>
> (Smith 1995: 181–182)

Whilst there may be art departments in which what Robinson (1995: 26) calls the 'sketchbook habit' is not fully developed, there are others in which teachers have encouraged pupils to develop highly personal and idiosyncratic approaches. Many schools and pupils use hard-backed varieties of sketchbook which can be 'doctored' and personalised. For example, the cover can be collaged or faced with materials such as papier-mâché, leather, fabric, found materials, even wood or metal. The pages of hard-backed sketchbooks are often made of medium- to heavy-weight cartridge paper which can cope with the range of media and the transformation of surface to which the pages are often subjected. These sketchbooks, when displayed in GCSE or GCE A Level exhibitions, are often sculptural in quality, tactile, object-like and highly individual and personal in nature.

Within the typical A Level syllabus pupils are required to make a personal study of an artist or art movement. This provides students with an opportunity to link critical and contextual study with their practical course work. Students learn about an artist's work through researching the literature, in some cases interviewing the artist, critically responding to original artworks, and through practical and theoretical investigation. These investigations can result in ingenious ways of integrating different modes of investigation, combining text with image or object, and theory with practice. However, the personal study can also be narrowly interpreted as a 3,000–5,000 word illustrated essay.

ARTISTS' BOOKS

So diverse, individual and creative have some pupils' responses become to the notion of the 'personal study' and 'sketchbook' that they relate to that growing field of the visual arts known as artists' books. It is difficult to provide a reliable definition of the artist's book as the contemporary genre traverses traditional boundaries between image and text, book and art object. However, some useful descriptions are:

Artists' books refer not to literature about artists nor sculptures constructed from books but to works by visual artists that assume book form. Courtney (1995), has described artists' books as 'meeting spaces', a particular form of collaboration between a story or idea and a succession of images unavailable in more traditional art forms such as individual paintings, prints or sculptures. Some of the following characteristics of artists' books may have implications for developing the concept of the sketchbook in art education beyond the obvious. If children acquire the sketchbook habit early in their school education and are taught ways in which they can develop increasingly individual and diverse approaches, their natural curiosity, enthusiasm and need for personal realisation and expression could find a focus in the sketchbook.

Task 9.1.10 Artists' books

To what extent do the following elements of artists' books relate to the characteristics of pupils' sketchbooks:

- The viewing or reading of an artist's book requires the interaction of the viewer or reader.
- The experience of the work is kinetic and temporal in that the work or cover must be opened and the pages turned.
- The work offers a tactile and spatial experience: the book may fold or open out in unexpected ways; the paper, binding and materials have their own smell and sound as the pages are turned.
- Only one person can experience the work at a time: it is an intimate and private experience; the scale of the work is small, hand-held.
- The element of surprises is an important characteristic: our expectations may be confounded or confirmed; the notion of chance which was so important to the Dadaists and the collaging and recycling of materials which so informed the work of Schwitters are significant concepts and processes for book art.

Task 9.1.11 Alternative forms

Think of ways of extending the given definitions and uses of sketchbooks and artists' books taking into account other cultural forms e.g. scrolls, new technologies.

FURTHER READING

On art education

Association of Advisers and Inspectors in Art and Design (AAIAD), Midland Group (1994) *Sketchbooks*, Bromsgrove: AAIAD. This is one of a series of pamphlets on Art, Craft and Design in the National Curriculum produced by the Art Advisers' Association, offering a useful summary of the principles of good sketchbook practice in primary and secondary schools.

Association of Advisers and Inspectors in Art and Design (AAIAD) (1996) *Art: A Review of Inspection Findings 1995–96*, AAIAD.

Clement, R. (1993) *The Art Teacher's Handbook*, (second edition) Cheltenham: Stanley Thornes. The chapter on drawing offers useful implications for the role of sketchbooks.

OFSTED (1995) *Art: A Review of Inspection Findings 1993–94*, London: HMSO.

Robinson, G. (1995) *Sketchbooks: Explore and Store*, London: Hodder and Stoughton. This book focuses on the primary age range but is also relevant for secondary teachers, particularly at key stage 3 of the National Curriculum. It sets examples of sound practice within a philosophical framework informed by Robinson's research into pupils' use of sketchbooks, which is also described in: Robinson, G. (1993) 'Tuition or intuition? Making and using sketchbooks with a group of ten-year-old children', in *Journal of Art and Design Education* 12 (1) 73–84.

Taylor, R. and Taylor, D. (1990) *Approaching Art and Design: A Guide for Students*, London: Longman. This is a practical guide and resource written for students, particularly at A Level. Section 4 is entitled 'Personalised use of the sketchbook'.

On artists' use of sketchbooks

Cézanne, P. (1985) *A Cézanne Sketchbook*, London: Dover Publications.

Glimcher, A. and M. (eds) (1986) *Je Suis le Cahier: The Sketchbooks of Picasso*, New York: The Pace Gallery.

Moore, H. (1972) *Henry Moore's Sheep Sketchbook,* London: Thames and Hudson.

Rothenstein, J. (ed.) (1986) *Michael Rothenstein: Drawings and Paintings Aged 4–9, 1912–1917*, London: Redstone Press.

Smith, M. (1995) 'Joseph Beuys: life as drawing', in D. Thistlewood (ed.) *Joseph Beuys: Diverging Critiques*, Liverpool: Liverpool University Press and Tate Gallery Liverpool.

Turner, J. M. W. (1987) *The 'Ideas of Folkestone' Sketchbook, 1845*, London: The Tate Gallery.

Van Der Volk, J. (1987) *The Seven Sketchbooks of Vincent Van Gogh*, London: Thames and Hudson.

Victoria and Albert Museum, London (1985) *John Constable: Sketchbooks 1813–14*, London: Victoria and Albert Museum.

On artists' books

Courtney, C. (1995) 'Private views and other containers: artists' books', reviewed by Cathy Courtney for *Art Monthly 1983–1995*, London: Estamp Editions.

Lyons, J. (ed.) (1985) *Artists' Books: A Critical Anthology and Sourcebook*, New York: Visual Studies Workshop Press.

Phillips, T. (1980) *A Humument*, London: Thames and Hudson.

Turner, S., Tyson, I. and Courtney, C. (eds) (1993) *Facing the Page: British Artists' Books, A Survey, 1983–1993,* London: Estamp Editions.

The Artists' Book Fair is held in London every November.

The Hardware Gallery, 162, Archway Road, Highgate, London N6 5BB, specialises in artists' books.

9.2 THE THINKING HAND: RECONSIDERING THE ROLE OF MIMESIS IN TEACHING DRAWING

Rose Montgomery-Whicher

As a teacher of Art & Design, you will inevitably teach drawing. Historically, drawing is thought to be the primary means by which artists and designers initially depict three-dimensional form, conduct inquiry and conceptualise their works. Consequently, drawing is considered the most intellectual of the visual arts disciplines and the foundational skill essential for the conceptualisation of finished works in other media as diverse as painting and computing animation. At the secondary level, proficiency in drawing provides pupils with an important means of visual analysis and invention. While drawing is central to a wide range of art, craft and design practices, this unit will focus on drawing mainly in the context of the European tradition of painting.

Although most Art & Design educators agree on the importance of learning to draw, theories about how drawing is learned, and how drawing should be taught have been vigorously debated and contested. Different theories of learning and teaching drawing reflect different orientations to art and art teaching.

Many teachers still adhere to the expressive orientation to art teaching which has dominated art education for much of the twentieth century, believing that copying is at cross purposes with what they see as the aim of art education, namely 'self expression'. Interestingly, injunctions against copying and the promotion of master drawings or paintings as a source for student work are both founded on the premise that drawing is a mimetic activity: left to their own devices, pupils will not draw from observation, memory or imagination, but will inevitably imitate already existing graphic forms – an inclination which some educators want to suppress, and which others actively encourage.

The next section offers an overview of the expressive and mimetic orientations to learning and teaching drawing, showing how they are aligned with different theoretical perspectives in aesthetics and psychology, and how they imply different ways of teaching. My description of these 'orientations' is derived from Efland's (1979) article outlining four orientations to art teaching: expressive, mimetic, pragmatic and objective. This article offers a sound conceptual framework for considering the theoretical roots of different approaches to art teaching, although today we might add important contemporary 'orientations' to teaching such as feminist or multicultural orientations.

Awareness of the theoretical underpinnings of differing orientations to learning and teaching drawing will offer you a sound basis for questioning current teaching practices and developing your own.

DRAWING FROM THE HEART

Whether people draw from imagination, memory, or the visible world around them, the act of drawing is necessarily filtered through their feelings. This is the premise on which expressive theories of learning to draw are built. Drawing, in this view, is understood as primarily subjective expression and response. In one art educator's terms, this is 'drawing upon the heart' (Anderson 1992: 47–49). Expressive drawings are often inventive, energetic and flamboyant. Exaggerated or distorted forms and arbitrary colour are not considered errors in perception or representation, but means of expression, whether intentional, or unintentional (Anderson 1992: 48; Efland 1979: 23). Some examples of expressive drawings could be found in the work of Käthe Kollwitz, Emily Carr, and Vincent Van Gogh.

An emphasis on the artist and on the visible rendering of feeling places this orientation to teaching drawing in line with expressive theories of aesthetics (Efland 1979: 21). In this view, it is not the subject matter in itself which is important, but the maker's feelings about the subject matter, or in the case of abstract work, the maker's feelings as embodied in expressive gesture.

Lowenfeld (later Lowenfeld and Brittain 1987) has been the most well known proponent of an expressive approach to art teaching. He defined self-expression as 'giving vent in constructive forms to feelings, emotions, and thoughts at one's own level of development' (p. 18). Lowenfeld based his theories on the Freudian premise that art is an expression of deep feeling in a 'sublimated symbolic form' (Efland 1979: 22). To put it differently, drawings make feelings visible. For this reason, as well as a general emphasis on self-understanding and self-expression, an expressive orientation to teaching drawing is closely aligned with psychoanalytic psychology.

Approaches to teaching drawing within an expressive orientation encourage playfulness and experimentation, and involve little or no intervention on the part of the teacher who nevertheless must create a 'nurturing and sheltering environment' (Efland 1979: 24). Teaching is a matter of facilitating personal expression while preventing the influence of others. In this orientation, a drawing is generally considered successful to the extent that it expresses the feelings of the maker. However, evaluation of success in terms of expression is more difficult than evaluation in terms of accurate representation or faithfulness to a model. This difficulty can lead to an unfortunate 'anything goes' attitude on the part of a teacher which is frustrating for confident young people who are attempting to develop skills of representation. However, a conscientious educator's purpose for promoting an expressive approach to drawing would be to foster personal expression as well as the pupil's confidence in her or his own unique manner of mark-making.

DRAWING FROM THE HAND

Mimetic theories about learning to draw are based on the premise that people learn to draw mainly from other drawings. According to this view, we draw what we see – not in the world, but in graphic images, or our memory of them. Art imitates art. In the words of the art historian Wölfflin, 'all pictures owe more to other pictures than to nature' (Pariser 1984: 145). Some examples are Picasso's drawings based on Velasquez's *Las Meninas*, and drawings by Hockney which have been consciously influenced by Picasso.

This orientation has a strong affinity with mimetic theories of aesthetics, in which 'the quality of the work is judged by its faithfulness to the model' (Efland 1979:23). This approach to learning to draw is also called 'conventional' as it is based on the premise that we learn from graphic conventions (Duncum 1984: 93; Pariser 1984: 143).

A mimetic orientation to teaching drawing coincides with a behaviourist orientation to psychology, which asserts that 'learning is acquired by imitation' (Efland 1979: 22). Accordingly, we learn to draw by imitating the graphic marks of others, and we are motivated to continue to draw by the reinforcement that we receive from others for successful copies (Efland 1979: 23). A strictly behaviourist orientation however, would focus on imitating and mastering the form of graphic marks, rather than attending to the ways of seeing and thinking which shape particular graphic marks. Whether copying drawings is conceived of as imitation, or as a conversational relation, a mimetic approach to drawing gives importance to the maker's relation with others: those who have made the drawings copied, and those who comment positively on copies. This focus on others contrasts sharply with an expressive orientation which focuses on the self.

Teaching with a mimetic orientation involves providing models to copy: of finished works or of procedures and processes (Efland 1979: 23). Learning involves practice and repetition until the visible mastery of certain predetermined graphic marks or conventions has been achieved. This may include conventions of European art such as linear perspective and the proportions of the human body, or artistic conventions of non-European cultures and pre-Renaissance art. Pariser suggests that acquiring a repertoire of graphic conventions can, like learning the established cultural forms of music or poetry, offer pupils 'categories for ordering personal experience' (Pariser 1984: 143). An important consideration is what drawings pupils copy. Hurwitz, Wilson and Wilson (1987) propose that children, like young adult art students, should copy the work of established artists. In this way, learning to draw is conceived of as apprenticeship to master draughtspersons.

Task 9.2.1 The experience of drawing

Think back to your own education in Art & Design. With a fellow student, recall a specific instance in which drawing was taught. What was this lesson about? What were the activities? Discuss how the teacher responded to your work. Was there any discussion, and if so, what was it about? And perhaps most importantly, what was your experience of drawing like? What did you learn? Can you identify elements of either an expressive orientation or a mimetic orientation in this drawing lesson? Would you repeat this lesson with your own pupils? Why or why not?

Now that you have a sense of the different theoretical bases of expressive and mimetic orientations to teaching drawing, and through your own reflections, how might these be played out in classroom settings and pupils' experiences? I want to invite you to reconsider the pedagogical value of a mimetic orientation to the teaching of drawing.

THE THINKING HAND

Drawing is widely believed to be the basis of education in design and the visual arts. This belief has its origins in the association of drawing with the conception of ideas, a notion which can be traced to the French and Italian words for drawing, dessin and disegno (Dinham 1989: 324–325). This connection of drawing with thinking forms the basis for persuasive arguments for teaching drawing. For example, Hockney states: 'I've come to the conclusion that drawing should be taught very seriously everywhere, in all schools, not just in art schools . . . Drawing helps you to put your thoughts in order. It can make you think in different ways' (Camp 1981: 6) (see Plate 18).

How is it that drawing orders thought and helps us to discover new ways of thinking? Consider how, when you draw, your thinking is mediated through the actions of your hand. 'The hand,' writes Heidegger, 'is a peculiar thing' (Heidegger 1954: 16). He goes on to point out that while we usually think of the hand as a bodily organ of grasping, the 'essence' of the hand is the human capacity to think.

> Apes, too, have organs that can grasp, but they do not have hands. The hand is infinitely different from all grasping organs – paws, claws, or fangs – different by an abyss of essence. Only a being who can speak, that is, think, can have hands and can be handy in achieving works of handicraft.
>
> (*ibid.*: 16)

The ways in which drawing is described suggest that the hand has a significant role to play in the experience of drawing. Art historians study artists' 'handwriting' in order to attribute drawings to one artist or another. Teachers assign exercises in 'gesture' drawing. Art critics comment on the quality of gesture in drawings or paintings. Thus experiences of drawing are often described in terms of feeling, gesture, touch and grip. And yet, the drawing hand does more than simply manipulate. Heidegger writes:

> The craft of the hand is richer than we commonly imagine. The hand does not only grasp and catch, or push and pull. The hand reaches and extends, receives and welcomes – and not just things: the hand extends itself, and receives its own welcome in the hands of others. The hand holds. The hand carries. The hand designs and signs, presumably because man is a sign. Two hands fold into one, a gesture meant to carry man into the great oneness. The hand is all this, and this is the true handicraft.
>
> (*ibid.*)

When you draw, you think through your hands; this is not verbal, thinking, but a non-verbal 'know-how.' Thus, Elderfield states that 'drawing, within the visual arts, seems to me to hold the position of being closest to pure thought' (Edwards 1986: 50). Just as speaking and thinking are uniquely human, so is drawing. Only human beings have hands; only human beings think; only human beings draw.

WALKING IN THE HAND'S FOOTSTEPS

Although copying is believed by some art educators to promote inhibition, dependency, and false security (Lowenfeld and Brittain 1987: 69, 176, 179–180) as well as social conformity (Read 1958), it is a time honoured way of learning the graphic conventions of a tradition. In the fourteenth century, Cennini advised aspiring artists to 'take pains and pleasure in constantly copying the best things you can find done by the hand of great masters' (Hill 1966: 108). By the nineteenth century, this had become an accepted part of artists' academic training; twentieth-century artists have also used copying as a means of learning. In a study of how artists in the past learned to draw, Duncum (1984) points out that children who later became successful artists, including Eugene Delacroix, Paul Klee, and Diego Rivera, taught themselves how to draw by copying the work of other artists as well as graphic images from their contemporary popular culture. 'I've been fifty thousand times to the Louvre' said Giacometti, 'I have copied everything in drawing, trying to understand' (Audette 1993: 29). Although perhaps Giacometti exaggerates, his point is clear: master drawings and paintings, whether historical or contemporary, are rich resources to which we may return again and again to learn (see Plate 16).

Barthes (1991) writes about an experience with American artist, Cy Twombly's drawings which further illuminates how it is possible to learn from copying:

> This morning, a fruitful – in any case, an agreeable – occupation: I very slowly look through a book of TW's [Twombly's] reproductions, and I frequently stop in order to attempt, quite quickly, on slips of paper, to make certain scribbles; I am not directly imitating TW (what would be the use of that?), I am imitating his gesture, which I, if not unconsciously, at least dreamily, infer from my reading; I am not copying the product, but the producing, I am putting myself, so to speak, in the hand's footsteps.
>
> (p. 171)

Here, Barthes alludes to why working from artists' drawings is so instructive: you are retracing 'the hand's footsteps' as if taken by the hand of a more experienced artist to feel for yourself the movement, direction, speed and weight of his or her gestures. And, because, as Heidegger suggests, the gestures of the hand are so intimately connected with thinking, when you thoughtfully retrace 'the hand's footsteps' you discover, wordlessly, intuitively, perhaps 'dreamily,' something of the maker's thinking.

A QUESTION OF INFLUENCE

'All of us, children included,' declare the Wilsons, 'draw mainly through imitation and influence' (Wilson and Wilson 1977: 5). Mimesis describes the way in which much childhood learning happens. 'This is the way after all' writes the curriculum theorist, Grumet, 'we found the world in the first place, as we mimicked our mother's song, followed her glance, and then the sweep and destination of her gesture' (Grumet 1990: 102). Extreme forms of an expressive orientation to teaching drawing however, deny the potential value of learning through imitation and attempt to minimise influence. Cizek was the first well known educator to establish an expressive orientation to teaching in his art classes for children in Vienna at the end of the nineteenth century (Viola 1936: 13). While encouraging his pupils to express freely themselves, he forbade copying, as well as corrections of children's work by adults, believing that a child's art will develop at its own rate (Viola 1936). 'When children do only what they wish' he said, 'there is the danger that they may copy or imitate or may be influenced by tradition' (p. 18). Such a view romanticises 'child art' and assumes that all adult influence hinders the natural development of a child's unique manner of expression. Perhaps those who subscribe to such a view do not trust teachers' judgement in selecting and mediating influences upon pupils' work.

Seen from a different perspective, the very notion of influence can be understood as central to pedagogical practice. The curriculum theorist, van Manen, uses the term 'pedagogical' to refer to the normative influence in inter-personal relations, situations, and activities that leads towards learning, maturity, and identity development – a process which takes place in childhood and throughout life (van Manen 1991: 13–38). Broadly interpreted, 'influence connotes the openness of a human being to the presence of another' (ibid: 16). In learning to draw, Hockney insists that 'one shouldn't be afraid of being influenced' (Camp 1981: 7). Nor should one be afraid of influencing. As a teacher, you will inevitably influence your pupils: your speech, attitudes, and responses to situations all constitute subtle forms of influence. Thus, you may notice pupils 'trying on' your gestures or tone of voice. Whether intentionally or not, you extend to your pupils an example which they may imitate, of a way of being in the world; you teach through your relationship to the world and to art, craft and design. 'To adopt the stance of the artist' writes Grumet, 'is to perform one's relation to the world for one's students' (Grumet 1990: 102). Of course, you will influence your pupils through your curriculum choices. When you make an informed selection of works for your pupils to draw from, you are demonstrating a 'strong relationship with the sources of influence' (van Manen 1991: 16). When you facilitate critical thinking about the wide range of visual culture they encounter and draw from outside the classroom, you can mediate these influences, moving between the world and the pupils in your charge. In doing so, you are leading your pupils to find their own way in a world in which the ability to read and interpret visual information is increasingly vital.

The term pedagogy holds still more significance for the teaching of drawing. It originated in ancient Greece, where an adult guardian accompanied the child to school. Thus, implied in the word pedagogy, is a gesture of taking by the hand, by which the child is led into the world, including its cultural heritages and practices. In a mimetic orientation to the teaching of drawing, the meanings of pedagogy as formative influence and as leading into the world are gathered in the gestures of the hand. The teacher's hand points to works. Then as pupils begin to draw from these, they find that the maker's hand is silently extended to their own, guiding their effort to retrace, to mime, not only the path of a line on paper, but a line of thought.

FURTHER READING

Efland, A. (1979) 'Conceptions of teaching in art education', *Art Education* 32 (4): 21–33.

Hurwitz, A., Wilson, B. and Wilson, M. (1987) *Teaching Drawing From Art*, Worcester, MA: Davis Publications, Inc.

van Manen, M. (1991) *The Tact of Teaching: The Meaning of Pedagogical Thoughtfulness*, Albany, NY: SUNY.

9.3 SCULPTURE IN SECONDARY SCHOOLS: A NEGLECTED DISCIPLINE

Andy Ash

INTRODUCTION

> If you know exactly what you're going to do, what's the point in doing it?
> Since you know, there's no interest in it . . . It's better to do something else.
> (Picasso 1974: 71)

An interesting insight into one artist's way of *making*. Yet, Picasso's humble materials and modest proportions revolutionised the formal assumptions that had dominated Western sculpture for a millennium, even when deviating from classical norms.

To a degree this is what I set out to do in this chapter and reconsider the formal assumptions which dominate secondary-school education. Sculpture has been neglected due to assumptions held by Art & Design teachers about what it is, what it entails and how to teach it: too much bother, too much to learn, too much time taken! In this chapter I form a picture of the daily reality of sculpture in schools and the problems it faces. I realise that not everybody is sure of the value of sculpture and there are those who need convincing of its place in the Art & Design curriculum. I devote some time to highlighting the educational benefits of teaching sculpture.

OBJECTIVES

By the end of this unit you should be able to:

- consider the possibilities for sculpture in the curriculum;
- examine why sculpture is neglected;
- identify ways forward for sculpture in schools.

SCULPTURE

Sculpture is a developing, growing thing. Any definition needs to be organic. I use an artist's perspective, for artists speak intimately on the subject. However, you should not assume they all agree!

> Over the last twenty years, sculpture has increasingly become art's dustbin . . . Anything and everything that moves away from or no longer fits into the concept of its parent *language*, has a *sculpture* label slapped on it . . . Sculpture's nest is heavy with cuckoos.
>
> (Williams 1994: 2)

If you think of sculpture as a living and growing entity rather than fixed and static, you can appreciate that it changes with the addition of each new 'thing': this is why definitions are confounded by experience. 'The key I had discovered was this: a sculpture, like any work of art, is a thing, a made thing, and more of a thing than any other thing' (Tucker 1974: 22). Tucker believes that attempting a definition is not futile. By saying what sculpture is or is not, in setting out its limits, you are inviting others to break the law. It is important that pupils understand that orthodoxies can and should be challenged. Therefore it is necessary to include developments in sculpture, areas such as: construction, assemblage, found objects, installation, happenings, performance, time-based and environmental sculpture.

> With the rapid changes that sculpture has passed through since 1945, there is a particular interest in what other arts, or disciplines outside the arts, sculpture butts up against, what attitude it takes to subjects as diverse as history, memory, landscape, theatre, architecture, the museum, the art market, the manufactured object . . . Sculpture in this period borrows its terms of reference from many other areas; its resonances and inflections come from countless sources. It is a peculiarly open discipline.
>
> (Causey 1998: 7)

WHY DO WE NEED SCULPTURE IN SCHOOLS?

I am a multi-disciplined practising artist who trained in Fine Art Sculpture at Trent Polytechnic. This was a very important and influential time for me. I was extremely fortunate to be learning in a department which valued and nurtured the individual's right to expression and exploration. Art was seen as a process, a way in which an individual organised experiences, not just simply making an artefact.

> A way of broadening personal consciousness . . . a sense of situation and awareness of self and the world around, whether that purpose is conceived of as a large abstract (broadening personal consciousness), a personal search (everything came from what I wanted to do), a particular need (getting particular energies out of your system).
>
> (Gentle 1990: 269)

I slowly began to realise that the word 'art' should be interpreted as meaning 'to do'. I chose to specialise in sculpture at a very early stage, after experiencing next to nothing of it in school. Why? After all appreciation of sculpture depends upon the difficult task of responding to form in three-dimensions:

> That is, perhaps, why sculpture has been described as the most difficult of all arts, certainly it is more difficult than the arts which involve appreciation of flat forms, shape in only two dimensions. Many more people are 'form-blind' than colour-blind. The child learning to see first distinguishes only two-dimensional shape; it cannot judge distances, depths. Later, for its personal safety and practical needs, it has to develop (partly by means of touch) the ability to judge roughly three-dimensional distances.
>
> (Moore 1952: 7)

I think I was seduced by sculpture, and perhaps I still am. I remember the releasing feeling of being able to look at, and appreciate art which existed in the same space as myself. I could walk around it, look at it from 360 degrees, view it close up, or from great distances, touch it and smell it. It was an opportunity to interact with it, to have a relationship. Sculpture demands active participation on the part of the viewer, rather than passive enjoyment. I can stand next to it, and measure myself against it, literally and metaphorically.

Art educators establish values and foster opportunities for learners. I want pupils, in sculpture, to actively challenge beliefs, partly through knowledge and appreciation, and partly by developing personal responses through making. I hope the result is that the learners make some sense of their physical world. 'It is this process of discovery and realisation which is at the base of art in education' (Barrett 1982: 14).

Golomb (1989) realises that sculpture has a special and important role to play in the educational process:

> The tactile contact with the material fosters a more intimate involvement in the making process and a special appreciation for volumes and surfaces. Modelling (and sculpture generally) also requires more persistence and a re-thinking about the object to be represented and it may well lead to a deeper understanding of the object and of the self.
>
> (Golomb 1989: 117)

There is a need for a broad and balanced Art & Design curriculum acknowledging both two and three dimensions. Pupils come into contact with: volume, space, plane, form, movement, materials and tools, in all aspects of their daily lives. Sculpture is vital because it helps decipher complex sensory and intellectual experiences in a tangible and physical way. Individual experience depends upon the kinds of qualities the sensory system picks up. The kind of meanings an individual produces are determined by experiential factors: the character and distribution of qualities in an environment and the particular focus an individual brings to that environment. When two-dimensional representation is emphasised, only one particular sensory system is util-ised. The kinds of meaning this type of representation expresses is limited. If the skills necessary for engaging with three-dimensional form are not taught, the individual is denied access to the full range of meaning. As an educator providing opportunities for these kinds of experiences, it is important to remember that the relationship between the pupil and the environment is an interactive one:

> It is simply not the case that the qualities themselves determine what will be selected; nor is it the case that the individual will entirely project his [sic] internal conditions on the environment. There is a give-and-take process. Each factor makes its own contribution, and out of the interaction experience is born. This point is particularly germane for designers of curriculum.
>
> (Eisner 1982: 55)

By preparing pupils with alternative ways to understand experience, you equip them with tools to decipher the concepts of the world and their behaviour in it (by tools I do not mean literally an implement for working upon something, but a conceptual framework).

From a very early age children develop and learn naturally through sensory exploration. They desire to experience the three-dimensional world around them. A child's sensory exploration automatically leads them to touch, grasp and even place in the mouth objects which they investigate. As children develop into young adults they are guided away from direct, spontaneous and egocentric responses. They are constantly told, 'Do not touch' and 'Keep off the grass', yet there remains a primitive need within, which, if starved, hinders the development of the whole person, cognitively and artistically.

> Although as adults we grow away from the need to experience directly the sensation of things that are part of our environment, it is apparent that our need for physical, sensory experience is very strong, even though we may not equate it directly with our capacity to learn and earn a living.
>
> (Gentle 1988: 3)

Sculpting affords the 'pupil/artist' the opportunity to consider the totality of objects and to represent their major aspects. Both the pupil and the viewer are free to vary their position *vis-à-vis* a sculpture which, unlike a drawing, can be viewed from diverse angles (see Plate 26). A sculpture can offer a more comprehensive statement about the nature of the object, and such working through of the possibilities of this unique vehicle for representation has significant implications for pupils' understandings of objects and of the symbolic activity of creating equivalences.

> Such 'working through' may affect not only the child's cognitive and artistic development but, perhaps, also his [sic] personal sense of self. To see oneself as a true 'maker' in the basic sense of creating shapes, involves the mind, the heart and the body. Because of the revisability of the medium, it offers the child new opportunities.
>
> (Golomb 1989: 117)

In a culture and a society that provides much visual information in the passive mode of watching television and playing computer games, sculpting encourages an active and constructive approach and a different experience of the dimension of time. However, more activity, if not directed towards some end, may result in developing muscular strength, but it can have little effect on mental development. Pupils need activities which reproduce the conditions of real life. This is true whether they are studying things that happened hundreds of years ago, solving problems in mathematics, or learning the properties of clay.

> When a pupil learns by doing he [sic] is reliving both mentally and physically some experience which has proved important to the human race; he goes through the same mental processes as those who originally did these things, because he has done them he knows the value of the result, that is the fact . . . They [practical activities] are necessary if the pupil is to understand the facts which the teacher wishes him to learn; if his knowledge is to be real, not verbal; if his education is to furnish standards of judgement and comparison.
>
> (Read 1943: 244)

Building sculpture relies on tools and machines, and an understanding of the physical properties of structural materials such as wood, clay, plastics, metal, paper. However, this should not be interpreted as the need to teach basic manual skills, rather like the traditional teacher of woodwork who insists on teaching pupils all the ways of working in wood. It is about allowing pupils to apply visual skills immediately, relating skills to a *real task*.

> The abstract nature of much basic design teaching, and the assumption that 11/12 year olds can make that intellectual jump from theory to practice, can all too often only lead to rather low level pattern making in materials.
>
> (Clement 1986: 201)

The quality of the material used for expression should be closely related to the subject matter. The subject matter expressed and the materials used to portray this expression should have unity. This becomes apparent if you look at the function and design of one object. Lowenfield uses the example of a single spoon:

> Such questions as 'why isn't a spoon made out of cloth, or paper, or glass?' can begin to stimulate thinking about the relationship between material and expression. 'Would the addition of holes in the handle of the spoon, or bumpy roses embossed upon the spoon bowl, help or hinder its function?' The development of aesthetic growth should not be minimised, and an increased awareness of tangible examples of both good and bad design should be explored.
>
> (Lowenfield and Brittain 1966: 249)

MAKING SCULPTURE IN SCHOOL

Task 9.3.1 Sculpture in schools

Examine your placement schools' PoS and identify the kind of sculpture produced. List the materials, processes, related concepts and type of learning which takes place.

Establish how much time is spent on sculpture. Does this change at any time: if so why?

'Well over half the schools do no 3-D work apart from ceramics' (DES 1991b). They highlight the absence of sufficient provision. OFSTED's more recent examination of schools across the whole country identifies a picture of neglect. The inspections of schools in 93/94 revealed: '3-D work tends to be weaker than other aspects of art in all key stages. Schools need to review the balance of work to make sure work in 3-D is adequately covered' (OFSTED 1995b). My research findings paint a similarly patchy picture. I was able to gain insight into the current state of sculpture teaching. Imbalances exist in many schools raising issues for consideration especially, how much sculpture is being taught and by whom?

1   Most Art & Design teachers identify with painting and drawing. Attracting sculptors into schools appears to be a factor, although not all teachers who teach sculpture need to be specialists.
2   Time constraints are a deterrent.
3   Materials are usually limited to: card and paper, scrap and junk, clay (roughly in that order), more traditional materials: stone, wood and metal, are rarely used. This is because of:

   a   Cost: card and paper are cheap, general purpose materials which can be used in a variety of projects.
   b   Familiarity: knowing a material's properties and limitations makes individuals feel 'safe'.
   c   Equipment: basic, or no tools required.
   d   Preparation or technical assistance: very little needed.

Task 9.3.2 Benefits and deficits

What benefits and difficulties have you experienced in teaching sculpture? Discuss in your tutor group the various issues and share solutions.

Teachers' knowledge base is a considerable issue. What they know and who they know produces some shocking results. Some teachers when interviewed had difficulty in naming contemporary sculptors. Examples offered to pupils in classrooms are Dead White European Males (DWEMs). By perpetuating the sole use of DWEMs and refusing to consider contemporary, female, and non-Western sculpture, teachers

restrict possibilities. All of these factors generate an image of a restricted and unbal-
anced diet of sculpture education, a particular and peculiar aspect of 'School Art'.

SCHOOL ART: SCHOOL SCULPTURE

The term 'School Art' can be traced to an article by Efland (1976), which differentiates between pupil production and the art made by artists. Taylor (1986), quotes Rose:
'What does it [school art] relate to? Only itself. I hear the argument "But you have to
teach the basics." Basics to what? School Art bears little relation to the Art of the past
and none to the Art of the present' (p. 91). School pupils' sculpture, 'school sculpture'
is no different now from these observations made two decades ago, it is obsessed in a
superficial way with 'basics'.

WAYS FORWARD

Contemporary sculpture bears little reference to dictionary definitions and school
sculpture. Hughes (1999) finds this problematic and a concern for the future of Art &
Design education: 'Much of the work taught and examined as art is too often
depressingly trivial and bears scant resemblance to the vital, radical and often anarchic
activity to which this name is attached by the rest of society' (p. 129). Looking to the
world of contemporary practice is not a new suggestion and Hughes points out that
Efland (1970), Rosenberg (1967) and Bruner (1960) all advocated similar ways
forward.

If we take the work of Cathy de Monchaux (short-listed artist for the 1998 Turner
Prize) for instance, we see an interest in exploring elements of the human condition.
Her sculpture draws upon a range of stimuli: religious paraphernalia, fantasy and
reality, sexuality, human anxieties, the beautiful and the grotesque; all these elements
are explored through her treatment of both concepts and materials. Her work draws
on a wide range of cultural references which she peronalises:

> In a way I see all my work as a form of cultural plundering in an attempt to
> evoke a culture of my own. I suppose it's a kind of world – not like a world
> view or a world truth, but just a world that has its own internal logic. Each
> piece . . . has a language which runs through it that's very much my
> language.
>
> (de Monchaux 1997: 5)

While de Monchaux's sculptures create a personal world, she invites the viewer to
extract their own imaginative interpretations (Plate 21). It is here that pupils can
engage and ask questions, fundamental questions perhaps, questions about the very
nature of sculpture and to wonder how we can define the unique experience of this
art form.

Unlike the SCAA (1996) publication which provides disappointing examples of
sculpture which perpetuate safe orthodoxies, only one project in the whole

document relates to sculpture. It is predictable, of questionable quality, yet set up as a national bench mark. Consider the table below as a starting point in your planning.

Table 9.3.1 Twelve elements of sculpture

• form	– space
• material	– process
• colour	– surface
• weight (balance)	– volume
• rhythm (composition)	– gravity
• scale	– environment

These twelve elements are by no means definitive. They serve as a starting point.

Task 9.3.3 Elements of sculpture

Consider Table 9.3.1: what elements are missing?
How might you relate contemporary practice to these elements?

In learning about these elements it is not necessary to produce finished outcomes. Some might be investigated more meaningfully through problem solving or play. For instance: stacking blocks, rearranging and re-siting objects, balancing weights, pushing materials to their limits, placing things under tension and so on. Importantly the will to discover is retained; the will to invent, try out and experiment avoids the stifling effect pure instruction has upon learning. By moving away from the mass production of skills-oriented artefacts to speculating and taking risks, a closer relationship to contemporary practice and learning in sculpture can take place.

Table 9.3.2 Sculpture vocabulary

MEDIA	PROCESSES	TOOLS AND EQUIPMENT	CORE ELEMENTS	KEY WORDS/ AREAS OF EXPLORATION/ THEMES/ISSUES/ CONCEPTS
found objects	assemblage	clamps	colour (light, tone)	anthropomorphic
newspaper	piercing	goggles	environment (context, site, time)	bust (portrait)
wax	weaving	protective gloves	gravity	kinetic

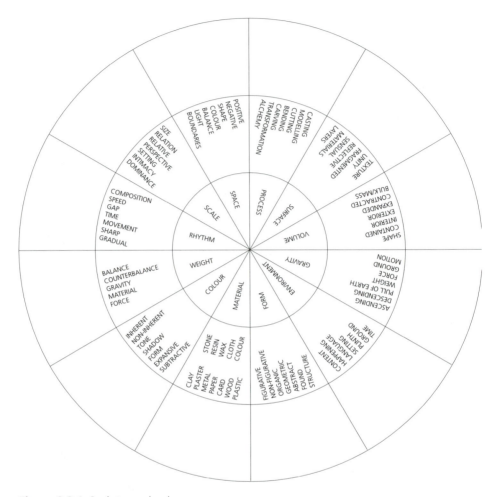

Figure 9.3.1 Sculpture wheel

Examine the sculpture wheel. Try to research and discover as many sculptors as possible to fit into the outer circle whose works may relate or are concerned with the particular issues in that segment.

Once you have completed this, revisit and establish whether you have represented the following: contemporary, non-Western, female sculptors.

The real task is to see sculpture as a process of problem-solving. Here, work is determined by problems set or defined. These can range from experimenting with materials, exploring general principles, to working with clear-cut problems which can be solved through designing, experimenting, thinking and making an artefact.

The curriculum should provide opportunities for pupils to become conscious of the world and themselves through exercises which simultaneously develop the intellectual and emotional spheres.

Task 9.3.5 Talking sculpture

Examine the incomplete sculpture vocabulary table on p. 217. Add terms you feel are appropriate to your schools' context. Add words that relate to historical, multicultural and contemporary practice. Would you want to make separate lists for the different stages, e.g. KS3 in comparison to 'A' Level? (See Plate 13.)

Look through the NC for Art and underline all words relating to sculpture.

Repeat the exercise and focus on how the Order can be delivered through sculpture using a less literal interpretation. For instance, set yourself the task of considering different aspects one at a time, and debate how they might be addressed through sculpture.

How can colour be taught through sculpture? (See Plate 23.)

CONCLUSION

There is a need for local and national forums to debate the significance of sculpture in the Art & Design curriculum. It is important to stress that you are the agents of change; you can develop and push forward sculpture in schools.

> It could be argued that art teachers need to behave more like real artists and less like bureaucrats. School art, at its worst is the art of the bureaucrat: neat, safe, predictable, orthodox and amenable to MOT type testing. School art adds up: the real thing rarely does.
>
> (Ross 1993: 161–162)

Politely, but relentlessly, confront and advocate the place of sculpture in schools. Its current situation will continue and, I suspect, proliferate, as long as you tolerate it. Address the imbalance and neglect. Challenge the orthodoxy, and produce meaningful making and discourse for pupils. In the words of Joseph Beuys: 'Thinking is sculpting'.

FURTHER READING

Golomb, C. (1989) 'Sculpture: the development of representational concepts in a three-dimensional medium', in D. H. Hargreaves (ed.) *Children of the Arts*, Milton Keynes: Open University Press.

Tucker, W. (1974) *The Language of Sculpture*, London: Thames and Hudson.

9.4 ART AND TECHNOLOGY

Pam Meecham

No art course it seems could be complete without the shade of Walter Benjamin and his essay 'Art in the age of mechanical reproduction' – its promiscuity has spawned its own age of reproductive texts that rework his title as a marker of serious endeavour. Written in 1936, the very ubiquity of the text and the reverence with which it is referred to by academics, would indicate a measure of authority that is hard to dismiss. In brief, the essay argues that the sacred 'aura' surrounding the original work of art would not survive the impact of photography, the processes of which undermine the notion of a unique creative act constructed under an autographic process. Benjamin argued: 'One might generalise by saying: the technique of reproduction detaches the reproduced objects from the domain of tradition. By making many reproductions it substitutes a plurality of copies for a unique existence' (Frascina and Harris 1992: 299).

Benjamin located the auratic art work's origins in 'ritual' which latterly is associated with the religious rite. He identified the possibility, for the first time in history, of an art in the age of mechanical reproduction which was emancipated 'from its parasitical dependence on ritual' (*ibid.*: 301). You might think that to demythologise the 'aura' surrounding the art work is a masochistic enterprise for the art teacher; perhaps it is. Nonetheless reproductive technologies do seem to hold out the possibility of demystifying the notion of the artist and their work as uniquely significant and so are an important area for you to engage with in the classroom. The incorporation of Benjamin's title into 'The Work of Art in the Age of Electronic Reproduction: Interviews with Paul Virilio, Jean Baudrillard and Stuart Hall' (1988) *Block 14* is evidence of the essay's continuing relevance but also an indication of shifting terrain. It is the continuities and changes registered in the latter-day title that are the subject of this chapter. Art produced under the aegis of an *Age of Electronic Reproduction* would somewhat belatedly seem an 'inevitable' fulfilment of Benjamin's prophetic essay.

ANALOGUE TO DIGITAL

Much has been written of the transition from mechanical to electronic reproduction, analogue to digital, with its utopian promise of a democratisation of the arts. The issue of reproduction has been with us at least as long as the earliest printing techniques. Benjamin was writing about mechanical reproduction in the 1930s and could not have foreseen the impact that digital technology would have on our understanding of knowledge, truth and perceptions. Analogue literally means similar or parallel word or thing. In art terms this indicates work that carries with it actual marks, in any medium, paint, ink or, in the case of photography, the silver salts of the photographic print. The works are also analogous in that they bear a physical relation to that which is represented: there is a transcription of some sort taking place which is medium specific. Digital works are constituted differently. They function at the level of conversion: that is, images are converted into a numerical, binary code. As this process involves the production of units that are detached from a material source, the possibilities for convergence and conversions between different forms of electronics are opened up. Digital works can therefore quickly enter into global communications networks where they are electronically transmitted and reproduced. This revolution, has not been without its casualties. 'Traditional' systems of quality control, storage, archival auditing and so on can be bypassed. The issue is more than one of control however, although this is no doubt part of the institutional angst caused by students referencing the fleeting and ephemeral works on the Net (most information only has a life of seven years) especially when they are not always easily accommodated within the Harvard referencing system. What is archived, how and when works are accessed, utilised, consumed and what system of ownership and control is in operation are all current critical issues. Electronic communication technologies have extended our capacities to create virtual worlds where information has taken the place of conventional knowledge. This has altered people's sense of what constitutes reality and has led to a process of construction and reconstruction.

Task 9.4.1 Digitalising democracy?

Discuss with other student teachers how digital technologies could open up the arts to more democratic practices and what implications they might have for the art curriculum.

Technology has always had its Cassandras, and other books that play with the title of Benjamin's work, like Birkert's *The Gutenberg Elegies: The Fate of Reading in an Electronic Age* (1994), offer a dystopic view of technology. The profit and loss of an 'electronic postmodernity' are played out across the ledger lines of the past with a sometimes fearful eye on the future. Birkert's account juxtaposes this dystopia with a broader understanding of what global networks can bring about. He proposes some benefits: the ability to operate on a multitechnical level and the ease with which new ideas are taken on, motored by relativism and tolerated. He characterises the dystopic downside by such indicators as the fragmentation of time (which can result in loss of attention span), shattered faith in institutions and explanatory narratives, a loss of history, physical estrangement, and an erosion of any strong sense of a personal or collective vision (Birkert 1994: 27). These paraphrased contrasts are acknowledged generalisations. However, they do sum up some of the difficulties you will encounter in an electronic age. Birkert's work is elegiac, as the title suggests, a song of lamentation for the dead: in this case the paper book form. On some levels the book is deeply conservative and possesses a Canute-like quality. However, it is important for you to understand that an engagement with new technologies is not a seamless incorporation of technological tools merely extending the repertoire of skills to hand. Stemming the tide of technology may be a fugitive enterprise but understanding the neutralising language that surrounds 'progressive technological inevitability' is essential for Art & Design educators.

The notion of 'technological progress' was expounded by the cultural theorist Raymond Williams. When television first entered mass culture at mid-century much was made of the 'inevitable' march of technology. Williams (1979) took up a position previously delineated by Marx, the German founder of Communism, what he defined as technological determinism. In brief, this theory attacks the naturalising language that surrounds technology. Technology is often presented as 'inevitable', 'progressive' and necessarily 'good'. However, as Williams and others have identified, technology is developed within cultural frameworks and is often linked to the research funding or the values of the market place. When economically led, this tends towards a version of the dominant order that replicates the commercial concerns of capitalism. Technology does not inevitably lead to either utopias or dystopias: it is the use to which it is applied that governs its development and reception. As Bill Viola states: 'Applications of tools are only reflections of the users – chopsticks may be a simple eating utensil or a weapon, depending on who's using them' (Sparrow 1996: 16).

If you return to the issue of 'aura' and the Utopian project of wresting the art work from the romanticism, even mysticism, that has surrounded it in Western culture during the last two hundred years or so, then electronic reproduction does hold out the possibility of works that are not wedded to notions of originality and authenticity. These hallmarks of the unique signature style, beloved of some strands of modernism, inform much of the art curriculum. Marcel Duchamp's ready-mades of the second decade of the twentieth century and Andy Warhol's assembly-line procedures, which constituted an irreverent attack on Abstract Expressionism's insistence on personal individuated self-expression, also mocked the pretensions of auratic art. Anti-auratic art practices seem to gain from a technological fix to dissemble the unique creative act through exactly reproducible digits. This possibility, however, does not take account of the durability and the necessity of the myth of individualism in Western culture and economics. Harris (1997) has pointed out the need for the identification of the creative individual even within electronic reproduction. The video works of, for example, Gary Hill and Bill Viola do not escape commodification merely by virtue of the technology invested in them. It will not have escaped your notice that the two earlier examples of anti-art, anti-creative acts, the ready-made and the production line output, did not stop an anti-hero art industry assimilating Duchamp and Warhol's unique anti-signature style into an eminently marketable and stylistically recognisable product. The critical engagement of many artists is often overlooked in the classroom when their practices could most usefully be deployed. It is the issue of the dominant order that is at stake in our use of technology in the classroom.

Task 9.4.2 To replicate or to question?

Discuss with other student teachers whether you intend to replicate or question the dominant order in your teaching?

Is your conclusion related to your experience of ICT (lack or expertise), ideological considerations or some other reason?

TECHNOLOGY IN THE ART ROOM

> In teaching us a new visual code, photographs alter and enlarge our notions
> of what is worth looking at and what we have the right to observe. They are
> the grammar and, even more importantly, an ethics of seeing.
>
> (Sontag 1977: 3)

Circular 4/98 (DfEE 4/98) is the advisory document given to subjects to ensure that
technology will rightly play an increasingly important part in the curriculum, including Art & Design. The development of a visual literacy is an important part of the art
teacher's remit. It follows therefore that an understanding of the ways in which
technology is allowed to function in our culture is part of that remit. This is not
merely a question of learning what technology can do, important though this is;
technology is only a tool that can be organised to operate within the parameters set
for it. Technology does not have to be a simple reflection of a society's dominant
order. It can also be made to determine the direction of a society. An early example of
this might be the conscription of photography as a tool in the justification of
nineteenth-century colonial expansion. Photography was invented in something like
the form known today by about 1839. It quickly came to constitute evidence. An
example of this would be the use of photography in the nineteenth-century anthro-
pometric images of colonial peoples (from the Greek *anthropos*, meaning 'man' and
ometry meaning 'measure'). Through a positivist world view, empirically based evidence could be coerced into ordering the world. Knowledge was organised and
technologically validated to exclude and control (what Foucault identifies as the
'principle of exclusion'). The development of societies becomes problematic at the
point where ideological interventions and technological inventions coalesce. Even
apparently celebratory images of other cultures are predicated on a set of power
relations that are often concealed. Gauguin's 'exotic' images of Tahitian women and
the work of other Orientalists, like the nineteenth-century academic painter Gérôme,
(who used photography extensively as a source) contributed to a colonial discourse
that rooted 'subject' peoples outside of history in an imagined romantic state. They
did not just photograph, paint or draw the 'Orient'; they created it. The European
enthusiasm for images of 'biblical' lands, South Sea islanders, of 'primitive' Africans,
increased in pace as the commercialisation of photography created images of exotic
locations and indigenous peoples for an embryonic tourist industry. Thomas Cook
began the first package tours in the nineteenth century. Photography increasingly
defined what was to be observed, measured and defined. Benjamin observed that
photography fed the almost obsessional need to bring things spatially and humanly
nearer but that this had a tendency to negate the unique or ephemeral quality of a
given event by reproducing it mechanically.

Task 9.4.3 Documentary evidence?

Discuss with other student teachers Benjamin's observations in the light of the camera's use as 'documentary evidence'.

To what extent is the photographic image an objectively truthful form of evidence?

CONTEMPORARY PRACTICE

The ordering of 'others' in photography has certainly contributed to our sense of what constitutes the self, since we can play it off against what we are not. However, in spite of knowing that images have never been an impartial, ideologically free view of the world, people still encounter them with at least a residual belief in their truth-telling capacities. But the development of digital technologies in the field of visual culture has disrupted any sense of truth telling at the level of surface representation.

The work of the Canadian artist, Jeff Wall, is apposite here. He works as an artist informed by a wide knowledge of art history and theory. He gave up art practice for a time to study art history and, while never a road to Damascus conversion, his work clearly demonstrates the changes that can be wrought in artistic practice with the cooption of a more critical and historic perspective. His work frequently appears to depict a narrative, often scenes of urban alienation, presented in large light-boxes with cibachrome images. The use of a seamless digital photography provides a sense of naturalism and propels you into a world where representation and conventional understandings of art history's past are questioned. Using the, so-called, new technology, Wall made works that referenced art history's embattled past and, crucially, his own part in unlearning the received modernist wisdom of the art history on offer to him as a student. Works like *Dead Troops Talk (a vision after an ambush of a Red Army patrol, near Mogor, Afghanistan, winter 1986)* (1991–2) a cibachrome transparency with a fluorescent light, and a display case, was constructed in a studio and digitally manipulated in sections. In it he employed the type of technical enhancement most usually deployed by film makers. In addition to utilising the ability of digital technology to erode our perceptions of reality, the composition is indebted to the grotesque nightmarish images of Hieronymous Bosch and the engravings of Callot.

FUSION?

Fusion: Art and IT in Practice (NCET 1998) (now re-formed as BECTa) is replete with examples of art using the tools of ICT. The document rightly recognises that new technologies must move away from the traditional practices of 'drawing, painting, graphic design, textiles and photography' (p. 5). It argues that the computer offers new methods and processes to replicate and develop them all. Additionally it argues that the current agenda for the raising of standards instigated by the present government will only be possible by the use of ICT in the art room. There is need for caution here. Technology does not in itself raise standards, although in the current climate of audit it may generate a gloss that traditional art forms apparently lack. The traditional areas of the art curriculum, the book continues, need not feel threatened by the use of ICT, but an 'evolution' and a 'revolution' can take place through a 'fusion' of traditional and new which will produce a stronger, broader art curriculum.

There are several points to be made here in the light of the theories outlined above. As Williams cautioned, the use of terms like 'evolution' and 'progress' in relation to technology utilises the rhetoric of 'inevitability', when, in fact, a replication of existing practices may be all that is taking place. Art under modernism has been opened up to intense scrutiny over the past two or three decades and it has often been found wanting. Technology does indeed hold out an emancipatory prospect but it may well not be through a 'fusion' with traditional practices for several reasons. Benjamin's original 'complaint' was that art had been invested with a sacred 'aura' that could only been sustained through the concept of the unique individual art work. The age of mechanical reproduction however would end this fetishising of individual work and open up art to less elitist practices. This has only in part been realised: even artists who work in new reproductive technologies are still valorised for their own unique electronic signature style. However, the field of collaboration, particularly transglobal collaboration, is made possible by digital technology. New technologies have opened up possibilities for questioning the emphasis on originality and technical skill fundamental to traditional art practice. The work of Jeff Wall is axiomatic in this respect. Technology can also be used to question what we value and why.

Digital technologies have the potential to be a tool for analysis on more than a formalist level. Ultimately, this is the difficulty that books on ICT fail to confront. The examples given are usually art works derived from the usual suspects gleaned from the received wisdom of modernist art practices, from Manet to Hockney. Reworked familiar images based on nineteenth-century French painting and the distortions through morphing of Francis Bacon can only teach a limited number of skills. Alternatively, the use of morphing by artists like Alan Schechner, can be used to question and expose structures of power; it acts as a critique of technology. For example *Bar Code to Concentration Camp Morph*, (1994) from *Taste of a Generation Series* makes visible the connection between capitalism, technology and the ideology covertly concealed behind the Holocaust theme park. It is one of many examples where artists are using technology to understand technology. In doing so however, traditional skills are often bypassed. Willie Doherty's appropriation of surveillance technology to replicate the 'banality' and repetition of surveillance in Northern Ireland is another accessible work that uses technology beyond a formalist agenda. The performance works of Coco Fusco (often shown on network TV to gain access to a wide non-gallery going audience) look at the new apartheid created by the technology industry. Fusco links the commodification of the bodies of assembly-line workers (usually black and Latino women on the US/Mexican border) with the technological goods that they produce but cannot buy. Artists like Yasumasa Morimora have used digital technology to reconfigure the self in relation to personal and national identities. Cindy Sherman has used photography and its relationship to the film industry to look at the construction of personal identities and the commodification and stereotyping of the female form. In addition to integrating technology in their work, these artists share a common theme; the function of imagery beyond fine art as a form of surveillance, control and the construction of a popular culture.

By KS3, pupils would benefit from engaging with this critical use of the medium. An involvement with modern, contemporary practice is also a way of looking at the relationship of fine art to popular culture and the ends to which technology has so far been deployed. To return to the conscription of photography in colonial discourse, the work of Keith Piper would make a more salient starting point for pupils today than the familiar South Sea paintings of Gauguin. In *Robot Bodies*, 1998 (Plate 17), Piper explores the metaphorical relationship between the image of the robot in science fiction and the history of black people. He observes that the robot *Sojourner Truth*, sent to Mars in 1997, was named after a black slave. Through the use of multimedia, Piper, at one time a member of a group called 'Digital Diaspora', makes explicit the connections between notions of cultural displacements and the new technologies.

The use of contemporary artists in the classroom is crucial to the development of young people's visual literacy skills and their understanding of the ideological bases of knowledge. The development of technical skills alone is limited unless they are applied to some understanding of the currency of the medium; this is as true for paint as it is for photography and the computer. The imperative, 'just to paint' a moment, identified in 1952 by the American critic Harold Rosenberg as 'the gesture on the canvas (that) was a gesture of liberation, from Value – political, aesthetic, moral' was, in spite of the caveats, ideologically motivated and fuelled by a complex mixture of social conditions. In some respects that is what makes art practice interesting. If the use of technology in the classroom is to be successful in enabling young people to make interesting, engaged art works that do not merely recycle early notions of expression theories using 'old French masters' as a starting point, engaging in a debate around contemporary artists' use of technology might prove fruitful. At least the newer art works are available courtsey of a Web site, CD-ROM and a Virtual Museum near you.

FURTHER READING

Crow, T. (1996) *Modern Art in Common Culture*, New York and London: Yale University Press.

Frascina, F. and Harris, J. (eds) (1992) *Art in Modern Culture*, London: Phaidon.

Wells, L. (ed.) (1997) *Photography: A Critical Introduction*, London: Routledge.

Illustrations

Plate 17: Keith Piper *Robot Bodies* Bluecoat Gallery Liverpool 3 September–10 October 1998. Commissioned by the Foundation for Art and Creative Technology (FACT).

10 Critical and Contextual Studies

Nicholas Addison

What do you understand by the term critical and contextual studies in Art & Design? Does it refer to a type of studio-based critical review, a supplementary form of art history, or does it imply a more interdisciplinary approach to the study of visual and material culture?

How can you use critical and contextual studies to extend and broaden your own and pupils' subject knowledge?

What methods of analysis and investigation can you employ to develop your own and pupils' understanding of visual and material culture?

How can you ensure that the relationship between the reception and the production of art, craft and design in your SoW is balanced and productive?

OBJECTIVES

The units in this chapter should help you to:

- understand the function of critical and contextual studies in Art & Design education (unit 10.1);
- explore different methods for investigating art, craft and design to develop visual and aesthetic literacy (unit 10.2).

10.1 THE FUNCTION OF CRITICAL AND CONTEXTUAL STUDIES

OBJECTIVES

By the end of this unit you should be able to:

- understand the purpose of critical and contextual studies in Art & Design;
- examine the range and suitability of critical approaches and methods;
- evaluate the orthodoxy of transcription;
- explore ways to integrate investigation, skills, knowledge and understanding.

No curriculum subject is an entirely discrete or autonomous phenomenon. You may have already noted that pupils' experience at secondary school can be fragmented by changes in subjects, sites, teaching styles, expectations: the way most pupils adapt to these discontinuities is remarkable. It is one task of the effective teacher to help pupils make sense of their fragmented experiences by linking the subject they teach to others in the curriculum. In Art & Design it is through critical and contextual study that a web of ideas can be constructed enabling pupils to unite the seemingly disparate approaches and dimensions of their lessons: technical, aesthetic, social, personal. This allows pupils to connect their work to the practice of others and to different areas of the curriculum.

The term critical and contextual suggests more than a knowledge of the art, craft and design of others, a sort of 'pic 'n' mix' art history of contingency or desire: this is what I have; this is what I like. What the term indicates is a reflexive process in which making and understanding (production and reception, encoding and decoding) are held in a symbiotic relationship where both are responsible for the construction of meaning. A critical and contextual approach enables pupils to investigate the possible functions of their own and others' work and to look outside the school aesthetic, beyond set exercises and the insularity of formalist practice. It ensures that the study of art seeks meanings additional and complementary to those of immediate perception or personal judgement, meanings which are culturally specific and ideologically determined. This demands a willingness to investigate art as a social and cultural practice inextricably bound to its historical contexts.

Critical and contextual studies is an important element of all Art & Design courses at secondary level, but it is worth considering how it is incorporated at KS3 as this is the only stage when all pupils follow the National Curriculum Order (NC). Because the previous Order (DES 1995) separated 'making and investigating' and 'knowledge and understanding' into Attainment Targets 1 and 2, the latter was often interpreted as art history. This was sometimes delivered in the form of a discrete and partial survey, an addition which, depending on resources: human, visual, textual and time, provided information and skills that pupils did or did not relate to their making. At worst it signalled an information–led marathon through the greats of Western art history divorced from other activities in the classroom: there is no way in which this type of survey can be said to be either critical or contextual. Art history is a vital resource and you should make every effort to develop your knowledge and skills in this discipline (Fernie 1995). However, it plays a different if complementary role to critical and contextual studies and its methods should not be your sole mode of enquiry. Thistlewood (1996) states, 'The prime criterion for admission to the critical studies canon, therefore, was (and substantially still is) the notion of what, from the realm of experience, stimulates the creativity of the representative practitioner' (p. 4).

In response to this emphasis on making and the maker, Thistlewood advocates the heresy that at some point in the future practical study might serve the needs of critical study. However, his statement reminds teachers that 'the realm of experience', the context in which making takes place, is as legitimate a focus for critical inquiry as the work of art itself. The following chart indicates something of the range of contextual evidence and provides examples:

- determining factors:

  1　environmental: climate-specific architecture;
  2　social and political: means of production and propaganda;
  3　cultural and spiritual: gendered traditions and religious practice;
  4　personal and psychological: motivations, desires and repression.

- common and relatively constant sources:
  natural or made, phenomenal or experiential; that is the originating stimulus for making: responding to light on water, an issue of personal or social significance like a bereavement, responding to the needs of others through a design brief or a commission.

- representations, literal and symbolic:

  1　visual: photographs of a city, sketchbooks, flags;
  2　textual: religious liturgy, diaries, poetry;
  3　performative: theatre, dance, ritual;
  4　artefactual: portrait busts, body adornment;
  5　simulations: these are a special and increasingly pervasive kind of representation mixing the artefactual and the performative, they are either 'fake', e.g. the First World War Trench Experience at the Imperial War Museum in London, and/or decontextualised, e.g. a reconstruction of a still life by Cézanne in a classroom.

Responding to and investigating evidence of motivations provides you with an alternative to the orthodoxy of copying or pastiching canonic exemplars. It invites pupils to explore art as something personally and socially meaningful and, at some point and at some level, requires you and your pupils to engage with the contextual and theoretical dimension of art. This sense that without context art becomes meaningless has an ancient pedigree:

> For example, a painter can paint a portrait of a shoe-maker or a carpenter or any other craftsman without knowing anything about their crafts at all; yet, if he is skilful enough, his portrait of a carpenter may, at a distance, deceive children or simple people . . . neither he nor his public know anything about shoemaking, but judge merely by colour and form.
>
> (Plato c. 380 BCa: 374 and 377)

While on your course of initial teacher education (ITE) you may find some students and teachers who believe that theory bears little relationship to what happens in practice. This division may be familiar to you from your own experience of art, craft and design education where, to put it crudely, there can be 'Those who do' and 'Those who think'. For Plato the toil of the craftsperson was honest because it did not play with the deceit of illusionism. In later centuries painting and other representational visual arts took on board this criticism and attempted to grapple with philosophical matters drawing them into a discursive and dialectical relationship with textual culture. It is important that you develop this sense of the interrelatedness and interdependence of the visual/textual, theory/practice continuums and the mediating role of language. Table 10.1.1 presents the domains of subject knowledge in Art & Design. They are categorised into separate and value laden approaches to indicate difference and complementarity. The technical and productive elements, from modelling to holography, may seem marginalised as they are reduced to one category. However, there is no reason why all approaches cannot be interpreted from the point of view of making.

Task 10.1.1 Balancing the diversity of methods and approaches

In your tutor group discuss the suitability of the approaches in Table 10.1.1.
See how you might combine them so that within the three years of KS3 you can develop SoW that introduce pupils to a diversity of approaches.
Think of omissions to this table.
Consider how your background, experience and strengths have conditioned your choices.

THE ORTHODOXY OF TRANSCRIPTION: CRITICAL STUDY OR PASTICHE?

When following a SoW which has a clear historical dimension a preferred 'contextual' homework is to ask pupils to find out all they can about a favoured artist and write it up in their sketchbooks. In this way it is hoped that knowledge and understanding can be addressed and lessons reserved for the 'fundamental purpose' of making. The result is usually a copy of the introduction to that artist's life from a biographical dictionary or general encyclopaedia and serves little educational purpose. Alternatively, critical and contextual study is delivered by a programme of copying and pastiche (Hughes 1989: 71–81). It is important that you critically

Table 10.1.1 Critical and contextual studies in Art & Design

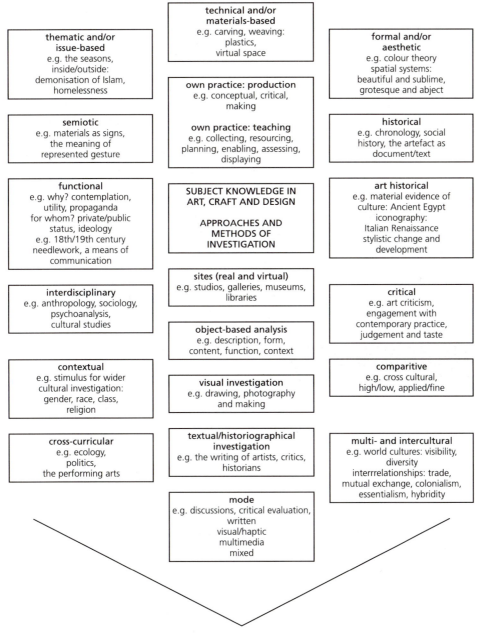

thematic and/or issue-based
e.g. the seasons, inside/outside: demonisation of Islam, homelessness

technical and/or materials-based
e.g. carving, weaving: plastics, virtual space

formal and/or aesthetic
e.g. colour theory spatial systems: beautiful and sublime, grotesque and abject

semiotic
e.g. materials as signs, the meaning of represented gesture

own practice: production
e.g. conceptual, critical, making

own practice: teaching
e.g. collecting, resourcing, planning, enabling, assessing, displaying

historical
e.g. chronology, social history, the artefact as document/text

functional
e.g. why? contemplation, utility, propaganda for whom? private/public status, ideology
e.g. 18th/19th century needlework, a means of communication

SUBJECT KNOWLEDGE IN ART, CRAFT AND DESIGN

APPROACHES AND METHODS OF INVESTIGATION

art historical
e.g. material evidence of culture: Ancient Egypt iconography: Italian Renaissance stylistic change and development

interdisciplinary
e.g. anthropology, sociology, psychoanalysis, cultural studies

sites (real and virtual)
e.g. studios, galleries, museums, libraries

critical
e.g. art criticism, engagement with contemporary practice, judgement and taste

object-based analysis
e.g. description, form, content, function, context

contextual
e.g. stimulus for wider cultural investigation: gender, race, class, religion

visual investigation
e.g. drawing, photography and making

comparitive
e.g. cross cultural, high/low, applied/fine

cross-curricular
e.g. ecology, politics, the performing arts

textual/historiographical investigation
e.g. the writing of artists, critics, historians

multi- and intercultural
e.g. world cultures: visibility, diversity interrrelationships: trade, mutual exchange, colonialism, essentialism, hybridity

mode
e.g. discussions, critical evaluation, written visual/haptic multimedia mixed

UNDERSTANDING EVALUATION INTERPRETATION

examine what is fast becoming an orthodoxy at KS3 and beyond, the 'transcription' and its companion the critical and contextual sketchbook.

For the past few years critical and contextual study in Art & Design has been managed at A Level in sketchbook form, and as a result is filtering down to GCSE and in some instances KS3 (see Unit 9.1). These books are filled with found images, an eclectic mix of the popular and the canon, a multicultural and historical potpourri. In addition pupils copy from reproductions and to a lesser extent objects from galleries and museums. This item is often separate from the 'true' sketchbook in which they draw from their environment, from life. The book also includes art historical 'investigation', usually arbitrary extracts from popular texts quoted verbatim, or personal responses which, however heartfelt, lack even the most rudimentary analytical method. They often contain impressive evidence of hours of patient copying and annotation and some pupils take the opportunity to explore, inventively, the potential of materials, but as evidence of critical investigation they are thin. What they provide is a 'look', a collection of images that act as the sign for investigation which can be replicated from pupil to pupil and from year to year.

The transcription is a singular and extended piece of work which mirrors this dependency on the 'look'. At KS3 it is often a copy from a postcard, enlarged through squaring up. It allows pupils to mix and match colour, consider the relationship between the formal elements in composition and imitate the artist's handling of paint (they are nearly always paintings). Pupils who choose their own reproductions are likely to have a very different experience of these skills if, for example, one chooses a Mondrian and another chooses a Kahlo.

Why is such practice consuming so much of pupils' time? Copying is something that most children do. It takes time and is evidence of their particular tastes. The mimetic process is innate; it is a fundamental method of learning. What teachers seem to anticipate is that the means of representation, the style, the handling of paint and so on, transfers itself to the pupils' own work. But the process of transcribing an image from a two-dimensional surface to the same is perceptually at odds with the process of representing the three-dimensional world in two dimensions; this is a process of transformation not transcription. Transcription is imitation of surface, outcome not process. It entirely decontextualises the artist's work so that its original meanings and mode of production are ignored. It also privileges painting as the dominant mode of representation, albeit in the form of photographic reproduction. Transcription in schools attempts to elevate pastiche to a critical method (for an evaluation of parody see Unit 11.3, 'Popular Culture').

INTEGRATION

When planning lessons you should avoid the temptation to tack on a critical and contextual dimension to what is essentially a techniques-based SoW. Try instead to work out how a discursive and critical approach helps pupils to develop their practical work and how, by placing their practice in a relevant context, they come to understand its social significance, for example: its environmental impact, its relationship to a history, its communicative or market potential.

The critical and contextual dimension should be an integrated part from a SoW's inception. Consider the project in Unit 3.7, 'Audience and Site-specific Problem-Solving'. Here you see that the context, the site-specific needs of under-fives, is an integral part. Had the brief been conceived from a different starting point, for example an exploration of materials and their potential in constructing open forms, continued by utilising the outcomes to construct climbing frames and then, as an extension, sought to apply these constructions to the needs of five year olds, the emphasis would be different: aesthetic experiment, design construction, application to a specific context. There is nothing intrinsically wrong with this sequence, but the later stages can easily be neglected or abandoned through lack of time. The integrated approach ensures that all dimensions are covered.

Consider for a moment how you can transform another school orthodoxy, the still life, which remains a favourite genre (Allen 1996). In order to contextualise still life, teachers often relate the displayed objects to well-known exemplars or invite pupils to employ a favoured technique or approach: Cézanne for objects, Picasso for form, Van Gogh for application of paint, Matisse for colour. This encourages pupils to develop formal and imitative processes but does little to suggest the significance of the genre beyond a mimetic exercise in painting. The exhibition, *Objects of Desire* (Hayward Gallery 1997) demonstrates how, during the twentieth century, the 'genre' was extended to encompass sculpture, photography, the ready-made and installation, approaches that provide alternatives for use in schools. But still life has a long and varied history and can also be investigated iconographically for its symbolic content. Schneider's *The Art of the Still Life* (1990) presents an extraordinary variety from the sixteenth to the eighteenth centuries in Europe. One example, Sanchez-Cotan's 'Fruit still life' (c.1602) demonstrates how a symbolic approach can encourage pupils to explore objects beyond their formal and perceptual properties.

> [It] contains only a small number of fruits so that they seem almost sanctified by virtue of their geometrical arrangement. Indeed, the festive presentation of these everyday objects was inspired by mystical notions surrounding St Teresa of Avila (1515–1582) or St John of the Cross (1542–1591), who were close to the people and emphasised the sacred nature of a simple, ascetic lifestyle, radically opposed to the wastefulness of the royal court. The parabolic arrangement of the fruit was to pay homage to neo-Platonic theories of proportion and harmony, as proclaimed in the Italian Renaissance. However, it can also be related to traditionally Christian and medieval ideas of beauty, where reference was frequently made to the Book of Wisdom: 'Thou hast ordered all things in measure and number and weight' (11:21).
>
> (Schneider 1990: 124)

After discussing this painting, pupils can construct and record still lifes which explore a particular theme, for example: necessity and indulgence, thrift and waste, popular and elite, transience and permanence. Choice of object is no longer exclusively formal as pupils set up and juxtapose objects in an attempt to construct meanings other than balance or colour contrast. This is not to suggest that formal properties should be ignored, only that they should be used in conjunction with an object's symbolic or social significance. It may be that the constructed still life turns out to be the final outcome and that it is recorded in ways other than drawing and painting: e.g. photographed in different lights, or cast in plaster.

Task 10.1.2 Contextualising still life

Select examples of still life that approach representation in different ways.
Investigate their meanings and the context of their making.
Record this information so that a pupil at KS3 can understand it.

FURTHER READING

Fernie, E. (1995) *Art History and Its Methods*, London: Phaidon.

Hughes, A. (1989) 'The copy, the parody and the pastiche: observations on practical approaches to critical studies', in D. Thistlewood (ed.) *Critical Studies in Art and Design Education*, Harlow: Longman.

Thistlewood, D. (1996) 'Critical development in critical studies', in L. Dawtrey *et al.* (eds) *Critical Studies and Modern Art*, Milton Keynes: Open University Press.

10.2 EXPLORING METHODS FOR INVESTIGATING ART & DESIGN: DEVELOPING VISUAL AND AESTHETIC LITERACY

OBJECTIVES

By the end of this unit you should be able to:

- compare and apply different critical methods to works of art;
- consider the possible role of semiotics in developing visual and aesthetic literacy;
- differentiate between the ways that images, objects and words are used to construct meaning.

A CRITICAL APPROACH TO VISUAL AND AESTHETIC LITERACY

Table 10.2.1 is a full, if not comprehensive, set of questions which mixes different methods in a somewhat liberal way. However, despite the mixture, it provides a cumulative method of investigation, parts of which can be managed empirically (directly from the object) while others need recourse to further objects and texts. The emphasis in the interpretative section is on a semiotic mode of analysis because this ensures that the interpreter considers the relation between form and content, the classic dichotomy in traditional art historical method. As you and your pupils become more familiar and expert in your observations you may be able to bypass this false separation and analyse in an integrated way from the start.

Task 10.2.1 Investigating art

Table 10.2.1, 'Ways into the object' offers a set of questions that you can use to investigate the objects of art, craft and design.

You are invited to consider whether these questions can be used to analyse all objects of visual and material culture, be it a contemporary installation, a film, a commercial textile, a musical instrument, a building, a magazine illustration.

What additions and subtractions might be necessary to accommodate different types of object?

Task 10.2.2 Comparing and applying different methods of analysis

Look at the other methods designed by art historians and educators, e.g. Pointon (1986), Feldman (1987), Taylor (1989), Kress and Leeuwen (1996). Examine and compare them. You might like to adapt them, or parts of them for use in the classroom. Using the 'Ways into the object', as it stands or with amendments, devise a whole-class activity by allocating different analytical tasks to small groups. The focus could be on one work, incorporating secondary methods of analysis. Pupils can bring their findings together in presentations, a display or in some form of multimedia activity.

How might some of these investigations be carried out through making?

ON SEMIOTICS

It is helpful, if challenging, to introduce semiotics, the study of sign systems, as a means to develop visual and aesthetic literacy. This is because the systems of cultural production in the twenty-first century, in shifting from the verbal to the visual, require pupils to understand multi-sensory forms of communication. Semiotics provides the possibility of a unified method capable of addressing the interrelationship between the visual, verbal, aural and kinaesthetic modes provided by multimedia technologies and sites. Pupils are already likely to be using semiotic methods, in practice if not by name, in English, where images are often employed as a stimulus for descriptive, analytical and creative writing. Semiotics is, in addition, a system which promotes an inclusive study of different forms of visual and material culture, from wrestling matches to fashion (Barthes 1957). It is sometimes given a bad press in art circles because it is perceived as having its roots in linguistics, particularly the pioneering work of the structuralist, Saussure (1974), and is therefore deemed antithetical to visual phenomena: this tradition can be more legitimately termed semiology (Sturrock 1993: 71–73). However, the theories of Peirce (Hoopes 1991), who formulated a

Table 10.2.1 Ways into the object: object-based analysis

0 Initial responses (optional) What is your initial response?	

1 Empirical analysis
Record your observations and cite evidence.

a *Description* (recognition and identification)	
What can you see?	• things, participants, forms
What do you think is happening?	• the relationships between the participants or the object's constituent parts/forms
Where are you positioned?	• height, angle, distance, etc.
What does this position suggest?	• e.g. subservience, dominance, privilege

Do you have to be static or can you move round the object or interact with it?
Do you need to view the object over a period of time (temporality) or can you view it, or its various parts, at any one moment (simultaneity)?

b *Form*	
Which formal elements are used?	• e.g. line, colour, texture, space
How are they used?	• e.g. is the colour local, descriptive, evocative, symbolic, arbitrary?

Is a particular perspectival system used? Is the space defined by solid or empty forms (presence or absence), mass or plane?
What are the relationships between the formal elements and how do they interact?
composition

Try using binary questions.	• e.g. simplicity/complexity, sameness/difference, unity/fragmentation

c *Materials and techniques*	
What is made from?	• e.g. bronze, fabric, light
What processes are used to make it?	• e.g. casting, embroidery, performance

Identify the material evidence for this.
Is it permanent (or intended to be so) or is it transient and ephemeral?

2 Functional analysis

a *Patronage*	
For whom was it produced?	
What did the patron determine?	• e.g. its content, form, materials

b *Purpose*	
Why was it made and what purpose did it serve?	• e.g. propaganda, decoration, ritual, habitat, self-expression
Has its function changed: is it used in ways other than intended?	• i.e. recontextualised and/or decontextualised

c *Provenance* Where was it originally? Has it moved? Where is it now?	• If so, trace its journey.

3 Contextual analysis

a *Art historical classifications*

Iconography What is the art object about, what do its images/references mean? Is it: a <u>literal</u> representation?	• e.g. the man standing with an axe through his skull represents a martyr • concerned with appearances, e.g. a portrait bust, a visual record
a <u>metonym</u>?	• a causal or associative connection, e.g. a palette and chisel suggest a traditional artist
a <u>symbol</u>	• an analogy or <u>metaphor</u>, something other than itself defining its character, e.g. the lotus blossom symbolises union
an <u>allegory</u>?	• a symbolic story, often with moral overtones, e.g. a biblical parable
Is it historical, literary, utilitarian, genre, etc.?	
Style Does it belong to an identifiable <u>tradition</u> identified with a culture, nation, period, etc? Is it composite?	• e.g. Islamic, Ming, Baroque, Finnish, Popular, Folk, International, Postmodern • intercultural, <u>hybrid</u>
Mode Does it conform to a specific way of representing/making? Does it conform to a specific aesthetic category?	• e.g. mimetic, abstract, diagrammatic, functional, documentary, narrative • e.g. beautiful, sublime, grotesque, abject
Influences What might its influences be?	• e.g. intercultural, generational

b *Contexts* How does it relate to its historical and contemporary contexts? What is its relationship to broader cultural structures, including race, gender, class?	• e.g. religious, political, economic, social • It would be useful here to consider the methods of other disciplines, e.g. anthropology, sociology, psychoanalysis (<u>interdisciplinarity</u>)

c *Interpretation: significances* How do the various elements, (forms, techniques, materials, images) work together as signs to suggest particular meanings? (<u>semiotics</u>)

<table>
<tr><td>

(c *contd*)
Are there any differences between the
explicit representations (denotation) and
the unconscious content (connotation),
in other words are there any hidden
messages? If so are they?

</td><td>

• e.g. intentional or subliminal,
celebratory, subversive?

</td></tr>
</table>

How might its changing site/function alter its meaning? (<u>framing</u>)
How has it been understood by others in its own and other times?
What is its significance(s) today?
What do you think it might mean?
What do you bring from your own experience that might condition the way you make
sense of it?

system for categorising images rather than words, may seem more accessible to you
because they are more immediately applicable to Art & Design in schools. But before
exploring his system it is worth spending time examining the relationship between
words and images from a semiotic point of view.

Words are the sense stimuli, the formal means, abstractions or signifiers which in
combination with a concept, the signified, refer the user to a thing or an idea, a *referent*
e.g. 'The letters a-p-p-l-e in combination with the concept apple, all possible apples,
refer the reader to an actual apple(s)'. A sign is produced when there is a repeatable
relationship between the signifier/signified and the referent, thus the word 'apple' is
the sign that leads us to the actual fruit(s). Diagrammatically this can be represented as:

$$\text{Sign} = \frac{\textbf{signified} \text{ (concepts)}}{\textbf{signifier} \text{ (letters, phonemes)}} \Bigg\}$$

relate to a **referent** (event, thing, idea)

to produce a **signification** (meaning(s))

Grammatical phrases and sentences enable you to combine words within a syn-
tactical structure to describe interactions in the world and to form propositions. For
example, words in combination identify more clearly the type of apple or context: e.g.
'They like to eat crunchy green apples', or allow you to speculate on why these
people like them.

Likewise, in the visual arts, the image or object is a kind of sign produced by a
relationship between signifiers and signifieds which leads the viewer to a referent,
respectively: lines, colours, volumes, textures (signifiers) in relationship with associations, representations (signifieds), refer the viewer to things seen, experiences felt,
actual events (referents). The meanings (significations) of visual forms depend on a
whole series of inherent sign systems, including their spatial placement, modality,
affectivity, material basis, etc. (Kress and Leeuwen 1996). The parallels between the
verbal and visual are not absolute (*ibid.*: 15–42, 75–78). One difference relates to the
way language tends to construct meanings in a temporal, linear fashion, syntactically
over time, whereas an image can provide information simultaneously providing complex information that can be grasped in a moment: thus the popular adage, 'A picture

tells a thousand words'. Perhaps a more fundamental difference is that the relationship between a visual sign and its referent is more direct; a drawing of an apple is more like an apple than the letters a-p-p-l-e. However, just as with language, the signified and the referent are not one and the same thing. For example, morning star and evening star are both signs that refer to the same astral body, the same referent, but their signification is very different.

Magritte teasingly invites you to consider the paradoxical relationship between signs and their referents in his painting *The Treachery (or Perfidy) of Images* (1928–29) in which a pipe is portrayed appended by the statement 'Ceci n'est pas une pipe'. The point here is that an image, whether a pipe or a pope, is only a representation not the represented object itself. This is frequently forgotten in Art & Design teaching where the perceptual model of education slavishly encourages pupils to produce the 'essential copy' (Bryson 1991), an 'unthinking' imitation of appearances suggesting that the relationship between signified and referent is absolute and universal rather than contingent and culturally specific. Gregory (1977: 161–162) examines assumptions about the nature of visual perception elucidating the difficulty of perceiving perspectival systems by peoples who live in environments in which there are no right angles. Although the research cited was undertaken in the 1960s, and today it is likely that few can avoid the right angles of industry, it proves that such systems are learned. It follows that all forms of visual description are mediated through systems of representation, and these systems are not necessarily as 'universal' as is sometimes supposed. At the same time visual representation tends to provide more semantic freedom than the verbal to make associations between the perceived image and an endless chain of other referents drawn from your own memory or fantasy. Words, of course, encourage this as well, particularly in poetry, but language is often more conditioned by intended messages than the image: this may be a cultural rather than an intrinsic phenomenon.

Task 10.2.3 Comparing visual and verbal meaning making

Collect a series of images or objects and texts which have the same referents, e.g. a biblical story and a painting of the same, a written recipe instruction and its imaged counterpart, an obituary and a monument to the dead.

In pairs or small groups try to work out by what means and to what effect your examples are communicating.

What are the differences between the ways words and images communicate?

What can words do more efficiently than images and vice versa?

Plan a SoW that incorporates this exercise as a starting point and leads pupils to produce a work in which images and words function together but in complementary ways.

Semiotic approaches are often practised in ways analogous to linguistic analysis, thus a narrative picture or a magazine advertisement is decoded in terms of who is being represented, what they are doing to one another, whether the action is literal or allegorical, what type of audience is being targeted and so on. But art, craft and design is not purely representational and its other modes of communication and being must be considered. A bentwood chair designed by Thonet c.1830 represents nothing; it is a chair, ubiquitous throughout the cafés of Europe. But the materials and forms with which it is made are signifiers which in combination with the concept; i.e. the signified chair, have multiple meanings. Its material, wood, is natural, presented simply and unadorned; its silhouette is clean, unpretentious. These are qualities, signs of utility and functional elegance which are common today, but if compared to a typical mass-produced nineteenth-century household chair, replete with machine-carved ornament, are remarkably spare. After identifying signs of difference the aesthetic and economic conditions for its production can be investigated leading to any number of associated investigations: form = function, ornament is crime; the private/public sphere, the body and domesticity. Equally, you discover very different meanings if you investigate the material basis of an oil painting rather than its iconography.

ICON, INDEX, SYMBOL

Peirce's tripartite definition of the sign may prove a useful differentiating tool for you to use in the classroom. He separates the visual sign into three types:

1   The icon refers to something through likeness, a similarity of forms: most representational art is iconic, e.g. a portrait is like a sitter; a circuit diagram is like the components that together make up an electrical system.
2   The index refers to something through association, perhaps through a relationship of cause and effect, e.g. a footprint in the sand is an index of a person; a moustache is an index of masculinity.
3   The symbol refers to something through a code or rule that has to be learned, e.g. colours in a particular formation make up a flag and refer to, e.g. a nation; a cross can refer to the Christian faith but in another context, addition.

Not all signs are discretely iconic or indexical or symbolic. They may have components of all three; thus the wing/arm of Picasso's *Nude Woman in a Red Arm Chair* (1932: Tate Gallery, London) is iconic in that it is like a wing, indexical in that it indicates softness and sensuality, and symbolic in that it signifies 'female' to the polar opposite of the 'male' paw on the other arm. Identifying the occurrence and significance of these three types of sign in a work of art is a much quicker process than the 'ways in' table (Table 10.2.1). It provides a framework by which pupils can bypass judgements of quality and taste and helps them interpret a work by looking at the way in which meaning is constructed (for an example of this type of interpretation see the article on Marcus Harvey's painting of Myra Hindley, Walker 1998).

Pluralist cultures tend not to fix symbols, they educate people to understand diversity, e.g. of languages, which are symbolic systems. In time, new symbols are constructed. They have the effect of unifying the knowledgeable group but excluding others. As such, they are frequently used to construct outward signs of identity. It is very important that you take the time to investigate the symbolic systems of the work you are showing pupils. This is not something they can understand without a 'way in' to the code. However, you should remember that some of your pupils are liable to have the key to systems you may not know well: invite them to help. Symbolic systems are clearly different and their frequent use should dispel any notion that the visual arts are in any way universal.

Task 10.2.4 Applying Peirce's system

Introduce Peirce's system to a KS3 class and assess to what extent it helps them to investigate the function of art.

How might you use this system beyond a method of categorisation?

Devise a SoW which invites pupils to make these different types of sign: e.g. pupils can design iconic, indexical and symbolic signs intended to orient people around the school; they can then test which is more easily understood.

In combining or juxtaposing signs further relationships are established producing meanings which require interpretation. Barthes (1965) suggests that meanings are of two types, the denoted, the clear surface meaning, whether literal or metaphoric, and the connoted, those meanings which may be coercive (as in propaganda or advertising) subversive (as in Dada or Pop art), subliminal or unconscious (as in all things that pretend to be natural or real but are in fact rooted in ideological and or psychological values or dispositions).

Task 10.2.5 Denotations, connotations

Select a series of images from a popular genre; e.g. magazine advertising, illustrations in *Reader's Digest*, children's cartoons.

Identify their denoted elements; consider their connotations.

Discuss your findings with other students.

How might you include this process in a SoW?

ON ADDITIONAL METHODS OF INVESTIGATION: 'WAYS IN'

Table 10.2.1 suggests an analytical, incremental model that can be used in part or as a cumulative method. Dyson's comparative techniques (see Unit 3.7) are readily applicable to the classroom in that they are easy to resource and the task of investigating into categories of similarity/difference is not difficult for pupils to assimilate. Using his system as a model you can develop additional categories pertinent to your specific investigation. Targeting signs of difference is essentially a semiotic process, albeit limited by its reliance on decontextualised reproductions.

BINARY OPPOSITES

The technique of presenting difference in the form of binary opposites is frequently chosen by teachers because the resulting extremes stimulate lively responses. Binarism has a strong theoretical pedigree as one of the preferred methods of the influential anthropologist Lévi-Strauss (1963). He proposed sets of oppositions: profane/sacred, cooked/raw, celibacy/marriage, female/male, central/peripheral through which to question cultural structures. A cultural form, e.g. a myth, is analysed by using one such opposition as an interrogative point of entry. The art historian Wölfflin (Fernie 1995: 135–151) had already produced a similar model for the analysis of Renaissance art. His five pairs of formal characteristics: linear/painterly, plane/recession, closed/open, multiplicity/unity, absolute/relative have proved influential and persistent. More recently binarism has been a favoured tool of the 'new' art historians, for example Pointon (1990: 113–134) who uses nineteenth-century oppositions: male/female, culture/nature to investigate the meanings of Manet's *Déjeuner sur l'herbe* (1863). Binarism is already adopted as a strategy in thematic or issue-based SoW, usually as a stimulus for practical responses: inside/outside, natural/made, rich/poor; such contrasts are common triggers in Art & Design. It is an appropriate system for use with pupils because of its neatness and simplicity. But its critics suggest that because it is a system based on antithesis it has the effect of imprisoning investigation into a model dependent on conflict rather than cooperation.

Task 10.2.6 Using binary oppositions

Working with a partner develop a series of oppositions and decide how you can introduce them as starting points for critical discussion and investigative making.

R. Taylor's *Form, Content, Context, Mood* (1989) is a highly favoured approach. It strongly reflects previous systems (particularly Feldman 1987), but purposely rejects the judgemental character of their later stages. However, Taylor's system does not encourage pupils to consider relationships across categories, as if each term in this particular quartet represented a quality possessing an independent existence. This can reinforce the form/content dichotomy of traditional art history. Taylor's final category, 'mood', is a particularly woolly term and invites pupils to project their own associations onto the work of art in an uncritical and ahistorical way. Cunliffe's recommendations to avoid the 'formalist labyrinth' (1996) are lucidly argued. He suggests that teachers shift between historical and contextual approaches, 'world to mind', and child-centred approaches, 'mind to world', in which the perceptions of the child are taken as a starting point. He provides a framing system for looking at works of art that he calls the 'Semantic differential Technique' (*ibid.*: 316–318) which invites pupils to identify qualities along a five-point continuum: e.g. between rough and smooth, healthy and sick. Allen's (1996) promotion of montage and simultaneity requires further practical consideration before it can be considered a teaching method. At present it stands as a welcome and salutary call to arms. Burgess and

Schofield (Figure 8.2.4) consider a method explicitly devised to analyse the objects of design.

The contexts in which you might place investigation and making are then, endless. It is important to strike a balance between pupil and adult-centred contexts, between, formalist and issue-based explorations, between art or craft or design practices and between art-specific and cross-curricular investigation.

Task 10.2.7 Contextualising practices

Select a SoW that is based on a formalist approach.
Revise it so that making is related to a wider context:

1   using a world to mind approach,
2   using a mind to world approach.

Consider ways in which collaboration with another subject area enables you to explore art, craft and design in wider contexts.

FURTHER READING

Addison, N. (1999) 'Who's afraid of signs and significations', *Directions: Journal of Art and Design Education* 18 (1).

Barthes, R. (1957) *Mythologies*, trans. A. Lavers and C. Smith (1990) London: Jonathan Cape, New York: Hill and Wang.

Cunliffe, L. (1996) 'Art and world view: escaping the formalist labyrinth', *Journal of Art and Design Education* 15 (3).

Kress, G. and Leeuwen, T. (1996) *Reading Images: The Grammar of Visual Design*, London: Routledge.

11 Towards a Plural Curriculum in Art & Design

Nicholas Addison with Paul Dash

11.1 TOWARDS A MORE INCLUSIVE ART CURRICULUM: AN AFRICAN CARIBBEAN PERSPECTIVE

Paul Dash

INTRODUCTION

> African Caribbean people like others in the Diaspora, through colonial indoctrination and constant exposure to western canons and ideologies, made sense of the world largely from the perspective of a western viewpoint ... The tensions generated by the denial of subjectivity, and the values inculcated in many black people by this dominant though antithetical culture, is an enduring feature of the relationship between black and white in society today. As the underlings in this skewed relationship, it is 'beholden' to diasporic peoples to find a means of coexistence within European cultural frameworks. Indigenous western communities, as the dominant players in this relationship have not significantly adapted the central tenets of their lifestyles to accommodate the black presence; there has not been a revisiting of the Eurocentric canons around which people shape their responses and make readings of the world.
>
> (Dash 1999: 123–124)

This unit asserts that meaningful intercultural Art & Design education, must start with an interrogation of the attitudes embedded in Western traditions and world views, and not the appropriation of practices from other cultures which may be accessed periodically and treated as irrelevant 'exotica'. Far from finding solutions to the (mis)representation of black cultures, peoples and traditions, teachers and educational authorities are trapped in structures which exclude such groups. The movement out of this culture of exclusion remains one of the most challenging issues in education and in society as a whole. In seeking to support this broad argument, some areas from which minority groups are excluded in popular and high culture are identified and solutions to this schism provided.

OBJECTIVES

By the end of this unit, you should have developed:

- a heightened awareness of multicultural development in art education since the early 1980s;
- new insights on project planning with a bearing on the multicultural classroom environment;
- your appreciation of critical studies to embrace a wider understanding of heritage;
- a more questioning position in respect of the structures and conventions which determine people's sense of identity.

MULTICULTURAL ART & DESIGN EDUCATION SINCE THE 1980s: A BRIEF BACKGROUND

The push towards multicultural approaches in the 1980s was a response to the need for a curriculum which celebrated cultural diversity. Much of the impetus for change was driven by the street disturbances which occurred in the inner cities of England in the late 1970s and early 1980s. In the wake of these occurrences the Rampton Report (1981) was written, followed five years later by the Swann Report (1985) which looked into the 'Education of children from ethnic minority groups'. As Rattansi states:

> Swann provided a liberal, semi-official legitimation for tackling issues of racism (or, more accurately, prejudice and what it coyly referred to as 'cultural pluralism') in all schools, including the the so-called all white schools which hitherto had maintained the stance of 'No problem (i.e. blacks) here'.
> (Donald and Rattansi 1992: 12)

Many teachers encouraged by Swann or simply responding to a perceived need, engaged in developing pedagogies which placed different cultural traditions at the centre of learning. Much of this teaching was supported by important museum collections and well presented publications. The influential Calouste Gulbenkian Foundation publication *The Arts in Schools* (Robinson 1982) promoted this spirit of inquiry by providing a rationale for such practices. Other publications such as MacLeod's *Cross-cultural Art Booklets* (1986) which dealt with a number of practices in a range of media from different cultures, and Birmingham Polytechnic's *The Art of Play* (1985) with a focus on 'traditional playthings from Africa, Asia, America and Europe', were made available to teachers. Mason's seminal *Art Education in a Multicultural Society* (1985, new edition 1995) raised issues, stimulated debate and brought the whole discourse of multiculturalism to the forefront of Art & Design education in this country.

Concurrent with these initiatives was the development of resource centres by some LEAs which provided material from different countries for use in art classrooms. The London Borough of Tower Hamlets, which came under the auspices of the Inner London Education Authority (ILEA), maintained an outstanding resources facility including a varied range of toys from South Asia. Garrison's *African Caribbean Educational Resource* (ACER) was a model of its kind, offering a range of user-friendly antiracist resources for the classroom.

Arts Education in a Multicultural Society (AEMS), supported a number of nation-wide secondary schemes with a focus on cultural inclusivity. Centres involved piloted innovative work in a variety of contexts. Most schemes engaged students in challenging enquiry which pushed back boundaries of practice in many media- and issues-based pedagogies. Binch and Robertson (1994) give an in-depth overview of the project's history and provide an evaluation of findings. Elsewhere teachers, often with the support of local education authorities and other bodies, introduced children to new practices and traditions including Islamic design, Japanese printing and carving, Aboriginal dream walks and other rich and diverse cultural practices. Such approaches were designed to enhance knowledge and move children out of the narrow world view culled from a traditional Eurocentric perspective on other cultures.

Towards the end of the 1980s, there was a greater demand on the part of parents and pupils for the adoption of antiracist teaching approaches. They called for pedagogies which celebrated 'blackness' and challenged racism in all areas of the educational system. This push had an echo in the requirements of an Education Reform Act in which the National Curriculum called for (i) intercultural understanding and harmony, (ii) global awareness and knowledge and, (iii) equality of opportunity.

CRITICAL VOICES

By the early 1990s significant opposition was raised against multicultural teaching, with calls for pedagogies focused more closely on traditional British values. D. Pascall, the then chairman of the National Curriculum Council, in a speech delivered at the Royal Society of Arts (RSA) in November 1992, made a telling contribution to this discourse when he said:

> In thinking about the education we offer our children we need to identify the important strands from our culture which mesh together to define and enrich our present way of life. I am thinking here, to select a number of illustrations, of the Christian faith, the Greco-Roman influence, the liberal Enlightenment, romanticism, and the development of modern humanism.
>
> (Pascall 1992)

This statement from an educator in a position of power and influence, effectively threw down the gauntlet to those teachers working to restructure the curriculum to allow greater inclusivity. Multicultural approaches were downgraded as a pedagogic concern. Despite a resolve by educationalists to press on with such approaches, the political environment fostered teaching which looked to the 'delivery' of a national curriculum with a Eurocentric ethos.

NEW LABOUR

With the coming of the New Labour Government (May 1997), there is greater acceptance of inclusivity and the need for a curriculum which celebrates cultural diversity (Steers 1999: NSEAD Somerset Conference). Such a change in perception is reflected in a revised NC Order for Art & Design.

INTERROGATING THE WESTERN CANON

> . . . today writers and scholars from the formerly colonized world have imposed their diverse histories on, have mapped their local geographies in, the great canonical texts of the European centre. And from these overlapping yet discrepant interactions the new readings and knowledges are beginning to appear.
>
> (Said 1993: 62)

More recently there has been an acceptance by some educationalists and researchers in the field (Mason 1995) that a more inclusive education requires some interrogation of the Western canon and the way traditional values and thinking impact on the lives and self-image of all Britons. Such inclusivity fosters approaches which, 'may not be multicultural *per se* [but] students will probably be dealing with issues that cross many cultural boundaries' (Chalmers 1996: 45).

Chalmers, in *Celebrating Pluralism*, quotes five approaches to multicultural education in the United States: first ascribed to Zimmerman (1990) and Stuhr *et al.* (1992):

> The first approach is simply to add lessons and units with some ethnic content.
> The second focuses on cross-cultural celebrations, such as holiday art, and is intended to foster classroom goodwill and harmony.
> The third approach emphasises the art of particular groups . . . for reasons of equity and social justice.
> The fourth approach tries to reflect socio-cultural diversity in a curriculum designed to be both multiethnic and multicultural.
> The fifth approach, decision making and social action, requires teachers and students to move beyond acknowledgement of diversity and to question and challenge the dominant culture's art world canons and structures. In this approach art education becomes an agent for social reconstruction, and

students get involved in studying and using art to expose and challenge all types of oppression. Although this last approach may not be multicultural *per se*, students will probably be dealing with issues that cross many cultural boundaries.

Our readings of history imply a Western cultural continuum with roots in ancient Greece and Rome. There is little indication of influences from other traditions on that history and notion of epistemological truth. Children in South Africa, the United States, Britain or Brazil, therefore, may perceive most architecture in their environment to exist solely within a white Western tradition. The British Museum, the White House, the Sacré Coeur or any number of churches, great houses and municipal halls in the Western-dominated world, reference a past in classical antiquity which we have been tutored to regard as a uniquely European concern (Bernal 1987). Such readings distort the truth by not acknowledging the Egyptian influence on architectural development in Greece. Our teaching should develop pedagogies which acknowledge that breadth and, by that means, demonstrate the 'interconnectedness' of our cultural heritages. There are different ways in which this could be done. I would like to focus briefly on two.

Retaining the architectural theme, a comparative approach would allow all children access to different architectural models from a number of traditions. A linkage between these could be achieved by a historical, aesthetic, geographical, religious or other common denominator. By this means and with appropriate contextualisation, the Taj Mahal could be compared, say, to the Alhambra in Spain and the Coliseum in Rome.

Another approach would be to take a more inclusive reading of 'Western' architectural achievement. The temples of Luxor and Hathor in Egypt with their prominent colonnades and classical proportions, demonstrate Egypt's obvious influence on architectural development in Greece and ancient Rome (Bernal 1987). As architectural designs from classical antiquity provided the impetus for the Renaissance and later constructions in Europe and North America, the 'interconnectedness' of our multicultural histories and cultures takes on a new significance. By this means we can begin to see influences which formed traditions of making which can no longer be regarded as the sole property of any one group.

Task 11.1.1 Making connections

> In tutor groups: think of ways to make a commonplace artefact culturally relevant to a wide cross section of children, by making connections across a range of cultural practices and shared histories. Using buildings as an example, make meaningful links between Le Corbusier's church at Ronchamp, Gaudi's Sagrada Familia in Barcelona and an example of traditional South African mud architecture.
>
> How would you organise your planning?

GALLERY AND MUSEUM EDUCATION

Visits to galleries and museums can undergird notions of truth often at variance with the experience of non-white groups. The way artefacts are displayed and the manner in which different cultures are represented can alienate some groups. Much has been done by Dabydeen (1987), Boime (1990), hooks (1994b) and others to decipher the coding devices used in Western works in which there is a black presence. Such discursive texts could inform project design and extend the scope of critical and contextual studies. Exposure to critical views of this nature would be an enhancement of school students' appreciation of great works of art and not a distortion of them. Marie Guilhelmine Benoist's fine rendering of the relaxed woman in *Portrait of a Negress* (1800) is an example of a work which celebrates the presence of the black figure in a manner rarely seen in Western art of its period. As such, it is a painting which could offer a range of possibilities for development in the classroom. In Benoist's work the sitter, unlike most black subjects in paintings of the period, addresses the viewer directly with confidence and self-assurance. Her colour is rendered in subtle tones which communicate the painter's evident enjoyment in describing the woman's body: the manner of the painting is a powerful celebration of the sitter's being.

In the whole history of Western art there is no other image that so effectively claims the right of blacks to liberty and equality than Géricault's *Raft of the Medusa* (1819), alluding to slavery but freeing them from the stigma of inferiority implicit in straightforward abolitionist iconography. If Benoist's work is a celebration of the black presence, Gericault's painting could be said to be a proclamation of our shared humanity and interdependence. This work successfully challenges the coding systems commonly employed in the representation of black subjects in the art of the period. Here, there is a reversal of the conventional ordering of humanity. As if to leave us in no doubt of his intention, a black figure is placed at the apex of the compositional triangle in a prominent position. It is as if Géricault were saying we are all castaways on the mountainous seas of life with only the security of our shared humanity on which to pin hopes of survival.

A WORD ABOUT BLACK ART

While this chapter looks at traditional Western mainstream art practices and icons to demonstrate ways in which these could be used more productively in intercultural pedagogies, I would briefly like to look at the issue of black art and the impact this has had on thinking about representation and identity since the beginning of the 1980s.

The black art which flourished in England in the 1980s provided a wealth of material with a bearing on our notion of self and nation. Shanti Thomas, Eddie Chambers, Sonia Boyce, Keith Piper, Shaheen Mirali and others provided oppositional positions on race and the experience of being 'black' in postwar Britain. In recent years that group has been augmented by other black artists dedicated to making art about issues pertaining to representation and identity. Numerous publications with a bearing on such work are being made available by different organisations and researchers in the field. Eddie Chambers the painter, curator and academic, has written extensively on 'black' art and curated exhibitions under the auspices of the Institute of International Visual Art (inIVA).

The African and Asian Visual Art Archive (AAVAA) under the supervision of Sonia Boyce and David A. Bailey has a body of resources, mostly publications and slides, which could be used by teachers in project development. Through AAVAA, access can be made to the work of artists who could offer valuable resource material for students engaging in practical and critical work.

GALLERY AND MUSEUM EDUCATION AND THE CARIBBEAN

In gallery and museum collections there is generally a dearth of material with a Caribbean focus. Indeed, outside carnival and the Anancy story, there is little visual material from the Caribbean with which most teachers are familiar. In spite of this, a growing body of publications with a focus on the Caribbean can be found in libraries and on the Internet. Clearly, islands such as Cuba and Puerto Rico are in the Caribbean region but are rarely considered in classroom planning with a focus on Caribbean art. Mainland territories such as Belize and Surinam are crucial to a better understanding of the Caribbean cultural identity. The Caribbean, therefore, should be understood to mean all the islands of the region and those Central and South American territories such as Surinam and Belize which share a Caribbean culture.

In saying that, a strong case could be put for drawing on artefacts generated within the African diaspora for project development and not merely those from the geographical space defined as the Caribbean. The enslaved people from Africa had no choice in the places to which they were taken. Many families and peoples were separated by the slavers and sold to buyers from different parts of the Americas and even Europe. As a consequence, most English-speaking African/Caribbean people identify with black Americans and their culture as easily as they relate to people and cultures of Caribbean islands such as Cuba or Martinique.

Task 11.1.2 Caribbean art

How many Caribbean countries can you name? What languages do they speak? How much do you know about their visual cultures? How much do you know about Surinam? Have you heard of the Maroons of Surinam? What do you know of their art? How would you use it in the classroom?

The people of Haiti practise voodoo. They are amongst the most extraordinary visual artists in the western hemisphere. Much of their work is based on their faith. Haitians have adopted some aspects of Catholicism in their practice of voodoo.

Many artists in Haiti are renowned for personalising famous biblical scenes, interpreting the stories in a Haitian context. Could you locate a religious scene in your school or its environment in this way?

POPULAR CULTURE

In popular culture pupils can be excluded by traditions which emanate from the time of empire. Trooping the Colour is a thinly veiled reminder of British imperial dominance of countries distant and near from these shores. Former citizens from those 'subject' nations are now elders of generations in multiethnic British families. The last night of the Proms, celebrated each year at the Royal Albert Hall, with much waving of the Union Jack and the uproarious rendering of 'Rule Britannia' with all its bitter ironies, reflects the xenophobia once held by many Britons as a matter of course. The patriotic piece alienates the descendants of people whose forebears it was intended to revile.

NEW POPULAR FUSIONS

Despite the past, young people in inner-city communities are creating hybridic expressive forms. Using computers, photocopiers and spray cans in combination with conventional artists' materials (Rose 1991; Willis 1990b) a whole new dimension is being added to our understanding of the visual world by Asian, black and white youth as they forge a new language based on the dynamics of street culture. As designer Allan Parker is reported to have said:

> If you know how to read the images, if you listen to the real soundtrack, you can now travel around the world and catch what is going on. You can see a poster from Eastern Europe and know what frame of mind they're in. You can walk into an African club for an hour and understand the vibe. You can spend the night in Bombay and realise the energy there is such that in 10 years they'll eat our mainstream culture alive.
>
> (Rose 1991: 12)

At the heart of this development is the bonding force of black music:

> As young Black Britons brought their own aesthetic into dance floor culture, the style press rewarded them with media visibility. Once relegated to the two-step soul and house party scenes of Britain's Black suburbs, by the mid-'80s Black style was ruling the heart of London club culture. It affected leisure pursuits at almost every level: fashion, pop video, music and merchandising.
>
> (Rose 1991: 31)

Rich cross-cultural fusions are a reality experienced by young people at street and club level. As Art & Design teachers building on new modes of expression and new languages as they form, you have a duty to draw on these developments to enrich your teaching and make what you do more relevant to pupils as they seek new ways to talk about the world in which they live. Given the nature of these conversations, pupils are an important intercultural resource who speak eloquently of the post-modern world of shared meanings encapsulated in the *Raft*.

SUMMARY AND KEY POINTS

Multicultural teaching should foster a greater spirit of enquiry through investigation and reflection. Such enquiry must involve the interrogation of 'attitudes' and the building of confidence to intervene in the way we construct each other. To do otherwise would be to do all pupils a disservice and remain out of harmony with the fundamental need for change in our society. Young people derive enormous satisfaction from making discoveries about themselves and others through art practice. You should make it your business to employ strategies for teaching which build on this profoundly beautiful spirit of inquiry, by adopting a more integrated approach to teaching and learning.

FURTHER READING

Tomlinson, S. (1990) *Multicultural Education in White Schools*, London: Batsford.

Lavie, S. and Swedenburg, T. (1996) *Displacement, Diaspora and Geographies of Identity*, Durham, NC and London: Duke University Press.

Mintz, S. and Price, R. (1976) *The Birth of African American Culture,* Boston, MA: Beacon Press.

11.2 HISTORIES AND CANONS AS FORMS OF IDENTITY

Nicholas Addison

OBJECTIVES

By the end of this unit you should be able to:

- identify and evaluate the canons which underpin the curriculum in Art & Design;
- question the function of current orthodoxies;
- consider the resource implications for teaching a pluralist curriculum.

ON CANONS

When you use the work of others to provide stimulus, as aids in demonstration and so on, do you concentrate on the Western canon as recommended in the NC 2000, making only passing reference to the 'unfamiliar' art, or are you more inclusive? Do you privilege the 'fine' arts, particularly painting, neglecting the 'applied' arts, craft and design, and thus perpetuate the hierarchical, gender and class divisions implicit in that neglect? (Parker and Pollock 1981) (see Chapters 5 and 8).

The gender and class bias of many reference collections in Art & Design departments is also compounded by a racial and historical bias in the form of a defence of two 'European Traditions', classicism and modernism. The former is often presented as the great Western tradition having its roots in antiquity, located geographically in Greece and Rome, and now perceived as the seedbed of modern Europe (Unit 12.3). It would be less misleading to point out that historically these locations constituted the northernmost outposts of a Mediterranean-oriented world centred around Egypt, encompassing north Africa, Arabia and southern Europe (Bernal 1987). With the collapse of the Roman Empire, partly as a result of the invasion of 'barbaric' hordes from the north east (the modern ex-Soviet Union), the classical tradition is supposed to have dissipated. The ensuing period, the 'Dark Ages' just happens to coincide with the supremacy of the Eastern Church in Byzantium and the rise of Islam (AD 622 onwards). Classicism is supposedly revived in the humanist, Italian Renaissance, finding its vestigial reappearance in Romanesque and Gothic, the styles of northern Christendom. Islam as the repository and perpetuator of the classical tradition is usually ignored. In this trajectory Britain is seen as its partial heir, adopting classicism from the seventeenth century as a unifying tradition visually manifest in the neo-classical style of Empire, although at home classicism's pagan and continental roots were questioned in the patriotic and Christian neo-Gothic during the nineteenth century.

Task 11.2.1 Mapping from a centre

Devise a SoW in which pupils collectively map a culture using its 'capital' as the centre: e.g. the late classical world seen from Rome.
How might the Roman perception of Britain and other territories be visualised? Provide texts and representations to assist pupils' investigation of these perceptions.

Task 11.2.2 Architectural style: purity and hybridity

Devise an investigative SoW based on a survey of architectural styles in the local vicinity: e.g. the local high street, where pupils could differentiate and record Classical, Gothic, Orientalist and other stylistic features in the architecture, alone and in combination.
How might you develop this SoW?

The other dominant tradition, modernism, is often perceived as an attack on classicism, and, depending on your position, blamed or praised for the individualism, essentialism, autonomy and anarchy of Western art in the twentieth century.

Individualism

Teachers often exemplify practice with reference to the work of a relatively small number of individual artists from amongst the pioneers of modernism (QCA 1998a). This tends to encourage biographical readings which reinforce the myth of the isolated male genius, the story of a series of misunderstood outsiders producing new and radical forms only to be superseded by the following generation: it makes for good copy with its archetypal antihero leavened by scatological anecdote. Walker (1983: 46–48) counters the myth of Van Gogh as a mad genius (the most referenced artist at KS 1, 2 and 3) by focusing on his social ambitions. The emphasis in schools on individualism can be countered by investigating collective practice, the relationships between the arts, intercultural dimensions, and by issue or theme-based approaches.

Essentialism

An essentialist approach suggests two paradoxical paths, either a belief in discrete traditions particular to a people or class, e.g. Expressionism as an indivisible part of the German tradition, or, the possibility of a universal language which can be understood uncritically because it stems from a basic biological core, e.g. 'All people see in the same way', therefore naturalism is universal. By ensuring that you investigate modernism as an international and collective endeavour, myths of national purity can be challenged, and by exploring difference across time and culture biological determinism can be questioned.

Autonomy

The notion of modernism as an autonomous entity results in art's disassociation from all contexts and criteria other than its formal properties. This invites teachers to train pupils in the 'basic elements': line, tone, colour, pictorial space, etc., in preparation for an eventual 'mastery' which can be applied to other contexts in the real world at a later stage. The implicit limitations of this approach can be countered by exploring the metaphoric and communicative functions of art and by shifting attention from the fine to the applied arts.

Anarchy

At its margins modernism is seen as irresponsible and anarchic. B. Taylor (1989) suggests that its ideology is strictly anti-bourgeois and that its styles are strategies for undermining academic conventions (p. 103). By extension, is modernism inappropriate for fledgling consumers of a bourgeois democracy? Modernism has indeed been allied to revolutionary politics, from both the left and the right, and has thus been associated with utopian schemes to better the lot of the masses. Constructivist, Futurist, De Stijl, Bauhaus and International modernist architecture and design have radically altered the conditions in which the mass of people live in industrial cities throughout the world. Because modernism has at its centre the notion of perpetual change and the possibility of progress, both individual and collective, it is attractive to teachers who have frequently been educated from within its precepts. The anarchy of modernism is clearly not the whole story and needs to be related to the continuous self-criticism that modernism puts itself through; in many ways this is its strength. Foster (1996) calls this the 'counter-project' and believes that this continuing critique is a constructive and reflexive check against the totalitarianism which has dominated so much of the twentieth century.

The canons which are the result of classical and modernist genealogies are reflected in available resources and sustained in the reference materials constructed specifically for the classroom, e.g. packs of reproductions (Chapter 5). But you can use these to question the canons just as readily as you can to promote them. However, there is a danger that if you investigate these canons alone, even from a critical point of view, they will be reinforced (Gretton 1986). It is vital therefore, that you transform what may appear limited or pedestrian resources through your investigative approaches, and invite pupils to add to the available store by collecting and contributing examples of their own. This way they can begin to relate the content and forms of the school resources to their immediate experiences.

In schools it is usually the originators of modernism who are held up as exemplars, in particular those oppositional artists based in Paris, from the Impressionists to Cubism. Despite their transgressive credentials the later Surrealists are not exempt from this attention and many of their stylistic if not conceptual traits, have been adopted by the commercial world and are familiar in the classroom.

Task 11.2.3 Resourcing an intercultural modernism

> Collect resources to enable you to introduce pupils to a wider and intercultural definition of modernism, extending your examples to contemporary practice.

QUALITY OR INTEREST?

Having produced a resource development plan (see Task 5.2.5) one of the difficult choices you encounter is the vexed question of quality. This is considered in relation to pupils' own work in Chapter 7. However, when you investigate the work of others, referencing available canons would appear to pre-empt the question. You have already experienced difficulties in selecting exemplars for discussion and investigation from limited resources with the added pressures of what may appear as the burden of political correctness. As recent revisionist histories have demonstrated the Western canon can be, and now is, more inclusive. Said reflecting on this collective endeavour writes:

> It was never a matter of replacing one set of authorities and dogmas with another, nor of substituting one centre for another. It was always a matter of opening and participating in a central strand of intellectual and cultural effort and of showing what had always been, though discernibly, a part of it, like the work of women, or of blacks . . . but which had either been denied or derogated.
>
> (Hughes 1993: 113)

Although quality will tend to have its day, recognition can be indefinitely postponed. It would seem that the canon is not synonymous with tyranny, although there have been instances of conspiracy. Art and politics may not be one and the same thing, but they are mutually bound in a social process of advocation, reinforcement, subversion, suppression and denial.

Revisionist histories have demonstrated that greatness in art is not necessarily the only value that is significant for education. Quality is a value that can be measured only within its own terms of reference. To be aesthetically and visually literate you need to critically encounter a more comprehensive representation of the diversity of visual and material culture. Add to this the absurdity of applying aesthetic criteria to art objects which hold no direct relationship to those criteria and you have a recipe for reinforcing difference and inferiority: for an examination of the mis-attribution of European influence on the Benin bronzes see Coombes (1994). Criteria can also be determined by generation and gender: Rubin (1969), in defining Louise Bourgeois as a 'tenacious artist of uneven but highly personal accomplishment' (p. 17) assumes that consistency and universality are fundamental criteria (modernist and patriarchal?) by which any self-respecting artist should assess their work. For others, diversity and interest may take the place of consistency and quality: posing new questions may be as illuminating as answering old ones. The lesson here is that you are liable to project your own values onto all objects from whatever culture; to overcome this you must attempt through dialogue, to meet them on their own terms: 'The cultural object can never be an empty vessel waiting to be filled with meaning, but rather is a repository replete with meanings that are never immanent but always contingent' (Coombes in Preziosi 1998: 489).

Task 11.2.4 Relative criteria

Investigate the criteria by which two different cultures value works of art.
Evaluate an object from one using the criteria of the other and vice versa.
Now evaluate each object using the criteria of its own culture.
How do the results differ?
Discuss in tutor groups what implications this has for the way you present objects from different cultures to pupils?

FURTHER READING

Gretton, T. (1986) 'New lamps for old', in A. L. Rees and F. Borzello (eds) *The New Art History*, London: Camden Press.

Gretton, T. (forthcoming) 'Loaded canons', in N. Addison and E. Allen (eds.) *Art Histories in Action*, London: Institute of Education.

Parker, R. and Pollock, G. (1981) *Old Mistresses*, London: Pandora.

11.3 ISSUE-BASED APPROACHES TO CRITICAL AND CONTEXTUAL STUDIES

OBJECTIVES

By the end of this unit you should be able to:

- differentiate and evaluate the approaches of multicultural, anti-racist and intercultural educational strategies;
- question the gender and class values embodied in the objects of visual and material culture;
- examine your own place within, or relationship to, particular traditions;
- examine art as a form of communication and meaning making;
- consider the social uses of art, incorporating the issues that arise from that use into the curriculum.

ISSUE-BASED INVESTIGATION AND EQUAL OPPORTUNITIES

Multiculturalism

When examining multiculturalism you will discover that the concept of race is central. When using the term throughout this book we do not intend to collude with the essentialist notion that race is a biological given, rather that it denotes a cultural construct which frames the way in which people identify themselves or are identified by others. As such race is not a fixed or absolute entity but a constantly shifting system of identification including notions of hybridity or indeed boundlessness.

Within Britain the origins of multicultural education are inextricably linked to its colonial past and there is much visual material in its cities and ports to serve as evidence for an investigation into such historical issues as trade, empire or slavery, (Piper 1997). More immediately, postwar demographic change within the Commonwealth has secured a multicultural and plural British society in which the influence of cultures of non–European origin is no longer peripheral or mediated through translation, but central. The impact of cultures from, Africa, Asia and the Caribbean has dramatically transformed whole areas of British culture from its cuisine and music to the fashion industry. These influences are most apparent visually in commercial and popular forms for it has been particularly difficult for Africans and Asians to gain recognition in the fine and applied arts (Araaen 1989). In asking pupils to address their immediate environments, the multicultural fabric of Britain cannot be denied (see Plate 24). However, you have already made an audit of the resources available to you and may have traduced pockets of invisibility which you are attempting to fill.

The notion of visibility is the key to a multicultural approach and it tends to be pluralist and celebratory reinforcing difference but in the name of tolerance. Holt cautions against what he terms 'the presentation of a cultural kaleidoscope', (Holt 1996: 131) because spectacular, indiscriminate exposure can distort cultures by mis-representing them through stereotypes and by allowing historical artefacts to stand in for today's; as if the cultures of the native peoples of Africa, America, Asia and Australia were unchanging. Alternatively difference in cultures can be sidelined in the integrationist attempt to find an underlying human unity. This universalist approach to the arts has found many adherents within modernism; indeed one of its most notable features, that underneath the conventions all the arts, and thus all people, are essentially the same, has been held responsible for the increasing hegemony of Western culture:

> If evolutionism subordinated the primitive to western history, affinity-ism recoups it under the sign of Western Universality . . . in the celebration of human creativity the dissolution of specific cultures is carried out.
>
> (Foster 1992: 24)

Gombrich's ubiquitous bestseller *The Story of Art* (1950) is amongst the texts most used in schools and it can reinforce the notion that significant art is the preserve of European history. Honour and Fleming's *A World History of Art* (1982), although presented in encyclopaedic rather than narrative form, is much less partial in its survey. There is a vast literature on world arts, but it is often allied to the tourist industry providing a spectacle of exoticism or domesticated difference rather than historical, theoretical or critical comment. Publications recounting aesthetic theories outside Europe do exist: take advantage of your time as a student to study these. You may hear the often repeated falsehood that 'art' is a European concept and therefore only Western objects can be examined from the point of view of 'aesthetics'. A reading of Coomaraswamy's (1934) comparative analysis of medieval theories of art in India, China and Europe soon evaporates such misinformed claims. Cheng (1994) provides a semiotic analysis of the tradition of Chinese painting focusing on its pictorial structures rather than its history. In relation to more recent cultural history, Baddeley (1994), Brett (1986), Caruana (1993), Poupeye (1998), Powell (1997), acknowledge participation by artists and designers in the development of modernism outside Europe and the USA.

Task 11.3.1 Multicultural revision

Review the SoW that both you and your colleagues have devised at your placement schools. Analyse them from a multicultural perspective.

If any of them are uncritically Eurocentric how can you revise them so that a multi-cultural dimension becomes integral?

FURTHER READING

Mason, R. (1995) *Art Education and Multiculturalism*, London: Croom Helm.

Swann, M. (1985) *The Swann Report*, London: HMSO, particularly Annex D.

ANTI-RACISM

Concomitant to this dual and contradictory attempt to embrace or integrate difference is the continuing legacy of racism. An ideology of difference was essential to the British colonial experiment, (Coombes 1991) and was particularly denigrating in relation to Africans and Asians as it relied on the notion of the 'Primitive' (Hiller 1991). The relationship between the visual arts and this concept is somewhat problematic for the primitive has been embraced as an ideal by many European thinkers from Rousseau to Foucault, and the process of appropriation evident in Primitivism (Rhodes 1996) has been central to the strategies of Western artists from Gauguin to Hesse (Rubin 1984). This is further complicated by the fact that the art of children, people with learning disabilities, even the amateur, are sometimes included within this definition (Goldwater 1986). An antiracist approach acknowledges that this legacy was and is apparent in all areas of British life and the ability to deconstruct a propaganda of difference is central to its strategies. Therefore a questioning and critical stance is the preferred position so that, for example, Primitivism or Orientalism can be investigated as ideological constructs which serve the needs of colonial oppression and more recently the tourist industry. Said's *Orientalism* (1980) provides a detailed textual examination of European colonial projection masquerading as scientific scrutiny and is an invaluable historical resource for any investigation of the continuing demonisation of Islam.

It may be that you wish to bring pupils' attention to these issues using more immediate and readily available resources. Pietersie (1992) has produced a telling analysis of the image of 'Africa and Blacks' within advertising, and his methods of interpretation could be used in the classroom to interrogate contemporary adverts. Holiday brochures can provide you with an endless image supply and it is an instructive task for pupils to collect images and texts about, for example, tropical countries, categorising them into types and comparing them with alternative information gathered from geographical and anthropological texts. Pupils would then be able to define those tropes by which the 'other' or the 'exotic' become desirable to a postindustrial/ colonial society and perhaps present their findings in the form of an annotated photomontage.

A particularly pernicious legacy of colonial racism is the perpetuation of difference through stereotype. The inIVA *Portrait: Education Pack* (Malik 1995) provides an accessible resource to introduce a range of strategies employed by artists to counter these stereotypes:

Self-representation

For example, Sonia Boyce *She Ain't Holding Them Up, She's Holdin On (Some English Rose)*, (1986) 'this English rose is black and she is staking a claim to a British identity from which she has been excluded' (Malik 1995).

Subverting stereotypes

Chila Kumari Burman in *Walk Tall – Self Portrait*, for example, (1995) questions the notion of the passive Asian woman by layering multiple signs: a demure facial portrait, a martial arts' kick, graffiti, some of which may be seen as contradictory, conflicting or perhaps complementary? As well as countering stereotypes such images suggest the possibility of multiple identities, a phenomenon which has been called 'hybridity'.

An alternative subversive strategy is to negate the power of stereotypes through ironic appropriation (see Plate 14). Chris Ofili's *Captain Shit* series (1997) presents the hero (anti?) of 1970s 'blacksploitation' movies as both a figure of uncritical adoration and absurd macho preening.

Deconstructing stereotypes

Deconstruction is a more analytical process which enables you to seek the ideological structures which support systems of power, in this instance, racism. The expanding territories and methods of Art History can demonstrate how to deconstruct colonial texts as in Coombes' exacting exposé of the racist theory of degeneracy applied to the material culture of Benin by nineteenth-century anthropologists and scholars (Coombes 1994: 43–62). Whole areas of history can be ignored to reinforce a racist ideology in which the fine arts are written up as the preserve of a dominant people who define their power in terms of race; the exclusion of Afro-Caribbeans from the development of art in the USA is a case in point (Boime 1990). Exposing these structures allows you to recognise the methods and devices used to construct differ-ence. One way in which these structures of difference can be theorised is as a form of projection, in which the secret or repressed desires of whites are played out in their representation of others (Said 1980). The endless portrayals of harem scenes in nineteenth-century painting are a case in point and the stereotypes of the 'lascivious and indolent oriental' and the 'insatiable black' are repeated in the cinema, literature and advertising to this day.

Rewriting histories

Keith Piper's *Trade Winds* (1992) for example, now included within a CD-ROM (Piper 1997), is a multimedia installation, comprising a dozen makeshift crates con-taining monitors showing images of imperialism and slavery. It explores the concept of measurement as a system and tool of control. Through image manipulation Piper is able to intervene, juxtapose, refashion and transform the representations of the past in such a way that their fictions are exposed and the viewer is invited to 'rewrite history'.

Blacks representing whites

Representations of black History should include 'white interventions', although, and perhaps because, whites find this prospect threatening:

> Black folks have, from slavery on, shared with one another in conversations 'special knowledge' of whiteness gleaned from close scrutiny of white people. Deemed special because it was not a way of knowing that has been recorded fully in written material, its purpose was to help black folks survive in a white supremacist society.
>
> (hooks 1992: 338)

> To name that whiteness in the black imagination is often a representation of terror: one must make a palimpsest of written histories that erase and deny, that reinvent the past to make the present vision of racial harmony and pluralism more plausible.
>
> (*ibid.*)

Task 11.3.2 Countering stereotypes

Devise a SoW that draws on some or all of the strategies outlined above.
How can you introduce these strategies so as to avoid the polarity between guilt and victimisation which could be the unintentional result of confrontational methods?

FURTHER READING

Gillborn, D. (1995) *Racism and Anti-Racism in Real Schools*, Buckingham: Open University Press.

Powell, R. (1997) *Black Art and Culture in the 20th Century,* London: Thames and Hudson.

INTERCULTURALISM AND HYBRIDITY

Increasingly, the difficulties faced by teachers in navigating such sensitive territory and the sometimes inhibiting effect of political correctness have encouraged commentators to develop a new approach, which, while acknowledging difference, seeks to focus on the interaction of cultures. For Maharaj (forthcoming) 'interculturalism is the scene of translation and transmission': it constitutes the perpetual condition of diaspora that has produced and transformed cultures (see Plate 22). In historical terms such inquiry would have to focus on migrations and conquests, trade routes and monopolies, and, sure enough, investigations into the transference and transformation of cultural forms through, acquisition, appropriation, imposition and diasporic cross-fertilization do exist. For example Riegl examines the so-called 'barbarisation' of the Roman Empire and the development of its decoration by Byzantine and Islamic artists in *Problems of Style* (1893) mentioned in his *Late Roman Antiquity* (1901) (in Fernie 1995: 120–126). However, it is perhaps the contemporary situation of post-colonialism and the transcultural phenomenon of hybridity that should engage your attention here.

Hybridity is the term most often used to define a condition of 'inbetweeness' (Maharaj: forthcoming), the coexistence and cohabitation, not the integration, of difference – a paradoxical unity in diversity. The construction of hybrid forms has often been used as a subversive or transgressive strategy by artists wishing to challenge aesthetic and political authority. Unlikely juxtapositions, graftings and confrontations are evident in the work of Western artists from Dada and Surrealism, Ernst's defamiliarisation of the familiar, through to the present day, Orlan's reconstruction of her own body. However, hybridity has become one of the central tenets of postmodern pluralism and has infiltrated most areas of cultural discourse. For example the ethnographic museum has become the site for cultural exchange in such exhibitions as *Lost Magic Kingdoms and Six Paper Moons from Nahuatl* (1986):

> Crucially, the deliberate focus on transculturated or 'hybrid' material culture has also been promoted as the sign of a mutually productive culture contact – an exchange on equal terms between the western centres and those groups on the so called 'periphery' . . . hybridity has often been an important cultural strategy for the political project of decolonisation. Additionally the many manifestations of creative transculturation by those assigned to the margins do potentially provide productive interruptions to the West's complacent assurance of the universality of its own cultural values. And certainly, the celebration of hybridity also implies an acknowledgement of the ways in which western culture has been, and continues to be, enriched by the heterogeneous experience of living in a multi-ethnic society.
>
> (Coombes 1994: 217)

Often the site of this hybridity has been identified as the Western, cosmopolitan city, the modernist 'centre' and its migrations: Paris, New York, etc. This has been complicated into the notion of centres: Tokyo, Mexico, Berlin and so on, but these revisions merely multiply the centres of power rather than question the city as the determinant of modernity. A new separation is ensured: only the citizens of this global and international community can produce art that is 'cutting edge', that pushes the boundaries. This still leaves peripheral territories whose inhabitants cannot be expected to participate in international discourse except as outsider guests:

> While the European artist is allowed to investigate other cultures and enrich their own work and perspective, it is expected that the artist from another culture only works in the background and with the artistic traditions

connected to his or her place of origin . . . If the foreign artist does not conform to this separation, he is considered inauthentic, westernised, and an imitator copyist of 'what we do'. The universal is 'ours, the local is yours'.

(Garcia Canclini in Preziosi 1998: 506)

In schools this can manifest itself in expectations about the behaviour and aesthetic practices of pupils from ethnic minorities who may be encouraged to conform to stereotypical, essentialist notions of their family's culture. British and USA artists have explored this issue, notably Sonia Boyce, Issac Julien, Steve McQueen, Chris Ofili, Adrian Piper and Yinka Shonibare whose work questions (celebrates ironically?) the media's representations of constructs of black identity.

> **Task 11.3.3 Pupils:**
> **a multicultural resource**
>
> Discuss in tutor groups: how to encourage pupils to investigate, value and represent the diversity of their own cultural backgrounds; how to include pupils' knowledge of additional languages and traditions in the Art & Design curriculum.

FURTHER READING

Cohen, P. (1998) 'Tricks of the trade: on teaching arts and 'race' in the classroom', in D. Buckingham (ed.) *Teaching Popular Culture*, London: UCL Press.

Hiller, S. (1991) *Myths of Primitivism*, London: Routledge.

Araeen, R. (1989) *The Other Story*, London: Hayward Gallery.

Third Text is a quarterly journal presenting perspectives on contemporary art from the developing world.

ON GENDER

Critically investigating art, craft and design in schools and, more broadly, the social and cultural contexts within which they are produced inevitably raise not only questions of race but those of gender. The same concern with visibility and the rewriting of histories became central to feminist discussion of the arts in the 1970s. Between them, both Nochlin's *Women, Art and Power* (1991) and Chadwick's *Women, Art and Society* (1990) provide you with an examination of the barriers inhibiting female participation in Western fine arts. They also celebrate those women who managed to contribute to this history despite the dominance of patriarchal discourses and their institutions. The very concept of the fine arts as the most significant visual manifestation of a given culture is questioned in *Old Mistresses* (1981) in which Parker and Pollock argue that certain forms of craft activity constitute a more representative example of women's cultural production and the exchange of meaning. By questioning the universal connotations of a hierarchical, male aesthetic dependent on the notion of genius they also add ammunition to anthropologists' and Marxist theorists already proposing more inclusive definitions of the arts in an attempt to undermine exclusively Western and bourgeois perspectives (Clifford and Marcus 1986; Hauser 1951, new edition 1999). However, because dominant practices are seen as determined by male ambition, the notion of negotiated practice has been espoused as a pragmatic measure. Pollock (1996c) quotes Susan Hiller:

> The idea was that women were creative and their creativity came out in unrecognised ways. And that this was the way of the future. This should be encouraged and one should withdraw from male notions of exhibitions, careers, vast projects and goal oriented work. Any of these sorts of things were seen as wrong. In the United States the women who made it are still seen as having sold out. Of all the political movements feminism, above all other, demands that we live our ideal feminism in a society that is not feminist in a way that socialists are not required to live their socialism in a non-socialist society, to the same extent. Yet on the other hand we agree that our actions now must presage future social relations. Now we face the situation of having to live with the struggles and make some peace with them.
>
> (p. 57)

One way of avoiding compromise is through forms of critical practice, or what Pollock (1988) has termed feminist 'interventions', an approach that is methodologically eclectic using the critical tools of complementary disciplines, in particular, sociology, semiotics, anthropology and psychoanalysis. This emphasis on interdisciplinary alliances has had the effect of forming what constitutes a new subject, namely, *Visual Culture* (Bird *et al.* 1996). Such alliances have not however, until recently, characterised critical and contextual studies in schools which, although materialising alongside these developments, tended to use a language rooted in more conventional and supposedly neutral discourses (Eisner 1972; R. Taylor 1989).

Task 11.3.4 Patterns of gender

In tutor groups discuss the gender make-up of both the students and the teaching staff when you were at college and professionals you have witnessed in 'industry'.

Can you discern any patterns that suggest gendered divisions of practice? If so, what are the reasons for these divisions?

How can you question these divisions in the way you present the work of others to pupils?

Review some of the advisory comments you have written for pupils. Can you see any differences in the type of advice you write for girls as opposed to boys?

FURTHER READING

Chadwick, W. (1990) *Women, Art and Society*, London: Thames and Hudson.

Nochlin, L. (1991) *Women, Art and Power*, London: Thames and Hudson.

Parker, R. and Pollock, G. (1981) *Old Mistresses*, London: Pandora.

ON CLASS

Art in Britain is frequently perceived as a special and privileged domain, with its objects made for and by a privileged minority. In order to question this perception:

Task 11.3.5 Grounded aesthetics, grounded taste

Ask each pupil to identify and define art, craft and design suggesting they investigate the origins of their beliefs by seeking the perceptions of their family and friends and by collecting evidence of how art, craft and design is reported, received, consumed, and produced in their home and local community.

In small groups ask pupils to present, discuss and record their findings followed by a whole-class debate in which their evidence is examined for consensus and difference.

To ensure pupils consider their perceptions in relation to broader contexts you should provide them with social and historical information to help them locate the conditions which determine or make possible such perceptions.

Additionally you can ask them to research the interplay of supposedly antithetical class-based phenomena, for example by resourcing an investigation into the appropriation of popular forms by high culture and vice versa: commercial imagery in Pop art, fine art in advertising.

Popular culture

Many teachers enter the profession in their twenties leaving little distance between their own age and that of the older pupils. Some of you may still be actively engaged in the production and reception of 'youth' culture, the aspect of popular culture with which your pupils are likely to be most familiar. Others may feel a need to familiarise themselves with it in order to get to know a significant area of pupils' lives. Whatever the motivation, it is a good idea to develop ways of allowing pupils to reference popular culture because for them it may be extraordinarily potent, a world with which they can readily identify. Hebdidge (1988) records changes in youth styles and develops modes of writing that, while remaining critical, draw on autobiography and popular culture. Willis (1990b) demonstrates how exclusive attention to 'high' culture can have an alienating effect on pupils and suggests ways in which the presentation of popular culture can aid self-esteem and allow pupils to engage in the same learning skills as they would if the subject were more canonical. It is vital that at some point you consider your pupils' own social contexts, not just in relation to how you teach them but in what you teach as well. In considering 'where they are coming from' you can pursue so-called 'mind to world' strategies (Cunliffe 1996: 315–318) but compare these to the 'world to mind' strategies (Cunliffe 1996: 318–325).

Many teachers despair when pupils turn to stereotypes and popular logos. In the last unit you have seen how such forms can be questioned, but is it necessarily a negative phenomenon when pupils have recourse to familiar and dominant signs? What if these signs become the object of parody, a process akin to the ironic appropriation mentioned in the previous section:

> Of course, parody necessarily entails imitation, although imitation does not have to be parodic. Yet when it comes to discussions of popular culture, this distinction is also highly perjorative. Imitation is seen here as an essentially unthinking process, in which particular behaviours, values and ideologies are simply reproduced – and therefore reinforced. Parody on the other hand, is generally seen to be a matter of conscious deliberation. While parody does not necessarily involve a rejection of that which it parodies, it must involve a form of critical distance from it – however affectionate it may be.
>
> (Buckingham 1998: 68)

Task 11.3.6 Parody as a critical strategy

> Discuss with other students: how to encourage pupils to parody popular forms rather than imitate them. What other strategies can you think of to enable pupils to use popular forms constructively and or critically?

When giving attention to popular culture it is worth considering the possible differences between 'youth', 'commercial' and 'home' cultures, all of which might be seen as belonging to its orbit. The popular and the commercial (mass media and consumer products) are sometimes presented as synonymous, but youth and home cultures often draw on activities outside the screen or the shopping and leisure centre, from decorating a personal space to caring for relatives; from energetic military training to profound religious devotion. If the classroom environment is a secure one, the things closest to your pupils are likely to prove motivating.

FURTHER READING

Buckingham, D. (ed.) (1998) *Teaching Popular Culture*, London: UCL Press.

Fiske, J. (1989) *Understanding Popular Culture*, London: Routledge.

Willis, P. (1990a) *Moving Culture*, London: Calouste Gulbenkian.

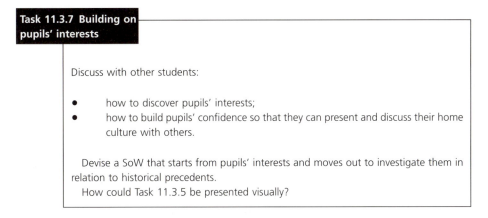

Task 11.3.7 Building on pupils' interests

Discuss with other students:

- how to discover pupils' interests;
- how to build pupils' confidence so that they can present and discuss their home culture with others.

Devise a SoW that starts from pupils' interests and moves out to investigate them in relation to historical precedents.
How could Task 11.3.5 be presented visually?

A CRITICAL APPROACH TO THE ART & DESIGN CURRICULUM

A critical approach to given bodies of knowledge is characteristic of changes in education since the 1970s. As you have seen the various canons relating to the visual arts are established ideologically and reinforced by available resources. But they can be the target of critical investigation as well as the basis of cultural indoctrination. It is important to remember therefore, that the critical procedures which question the absolute validity of such canons can themselves become the basis of new orthodoxies. The anti-aesthetic procedures of Dadaists, chance, transience, mixed or multiple modes, were partly introduced to undermine the fine arts and their relationship to capitalism, yet conservative critics would argue that they have become the established means guaranteeing access to the art market at the turn of the century. Foster (1996) has argued that the 'retroversive' power of the Right in the USA and its populist interventions into critical debate on the arts are pernicious and dangerous: to what extent is the situation in Britain different? He warns against any diminution in theoretical rigour or dilution in oppositional and transgressive practice, although he notes a move from 'grand oppositions' to 'subtle displacements'. He suggests that the Right has bestowed advanced art with such symbolic significance that it can still be an effective means of cultural and social change, and can therefore be the means to pursue critical transformations. It is important then to apply a critical approach to all phenomena, and, as would a nurse or doctor, keep abreast of current theoretical developments as they pertain to your practice. There will be periods when you need to 'rest', to consolidate your teaching, but you should be wary of formulae and outmoded practice.

The critical approach demands that nothing should be taken at face value (Price 1989). Not only does it enable pupils to review and modify their own work but it enables them to question the values embodied in the objects of visual culture, the perpetuation of canons and their own place within, or relationship to, particular traditions. Ultimately a critical approach enables pupils to understand that art is not limited to the making of beautiful, useful or imitative things alone; it is also a form of communication, an opportunity for them to find, construct and share meanings, and make sense of experiences both from within and outside the curriculum. If pupils are to negotiate their place in visual and material culture they need to understand how art is value laden and consists of a changing series of systems open to analysis and interpretation, their own no less than others.

FURTHER READING

Dawtrey, L. *et al.* (eds) (1996) *Critical Studies and Modern Art and, Investigating Modern Art*, Milton Keynes: Open University Press, is a useful dual publication, where many of the issues that can only be touched on in this chapter are examined in greater detail.

Foster, H. (1996) *Return of the Real*, Cambridge, MA: MIT Press.

12 Values in Art & Design Education: addressing the spiritual, moral, social and cultural dimensions of Art & Design education

Nicholas Addison

What and whose values are being promoted in the Art & Design curriculum?
Should the Art & Design curriculum sustain prevailing normative values?
If so, how are minority and marginal views to be represented?
How can you select and represent values other than your own, including different cultural, social, political and religious attitudes, beliefs and practices?
How can a pluralist curriculum, that attempts to represent different traditions and faiths, reconcile conflicting moral positions and beliefs?

OBJECTIVES

This chapter should help you to:

Unit 12.1 identify the values which have informed your own practice in and through different traditions of Art & Design, and their possible influence on your teaching;

Unit 12.2 identify and differentiate the intrinsic and extrinsic values embedded in Art & Design education;

Unit 12.3 consider your knowledge and understanding of your own and other value systems, and their interactions;

Unit 12.4 investigate the implications of teaching Art & Design in a pluralist society;

Unit 12.5 question and begin to resolve differences in value.

This chapter identifies some significant ethical questions arising from teaching. It helps you to investigate your own position within a pluralist society and to navigate a course for yourself and others through its complexities, its histories and possible futures. Art & Design cannot be taught in a vacuum. To neglect pupils' spiritual, moral, social and cultural values, from which they derive their identity and worth, is to limit the status and potential of the subject, the pupils and yourself.

It would be highly unusual if you were not confronted by moral paradoxes and dilemmas of conscience during a course of initial teacher education (ITE). Your expectations may not meet with the reality you find: pupils may be unmotivated by those very beliefs and issues you hold most dear; colleagues may teach in ways that conflict with your ideals. It may be that you are asked to teach Schemes of Work (SoW) that hold little interest or sense of worth for you. Personal and institutional values may conflict, compelling you to make decisions of conscience: to what extent are you prepared to negotiate, compromise or conform? The challenge here is for you to maintain your integrity despite the obstacles. Alternatively you may find yourself in complete agreement with the values inherent in the way Art & Design is taught in one placement school seeing no need to question the status quo. However, you may have to adjust your strategies radically when you move schools; a different context requires a different approach. Reflective practice is, at all times and in all contexts, a requisite of critical pedagogy.

12.1 THE VALUES OF THE AESTHETIC

OBJECTIVES

By the end of this unit you should be able to:

*   identify and reflect on the values that have informed your understanding of, and practice in, Art & Design;
*   investigate the origins of belief systems in Art & Design;
*   consider the moral and spiritual dimensions of different types of art practice.

Student teachers arrive on an Art & Design PGCE course from a range of specialisms each with its own values and attitudes. The way you value and make judgements about your own and others' practice is informed by this experience. For instance, some fine art graduates may believe that art is an autonomous endeavour and may not have considered the reception of their work by non-specialist audiences. In contrast, for some design graduates, decisions are conditioned by functional and commercial factors such as the needs of target groups and potential markets. Some students work in isolation, others collaboratively; some from within existing cultural traditions, others from emerging ones. On the one hand you may see aesthetic education as an introduction to dominant cultural norms both conceptual and technological, on the other, you may see it as a challenge to the prevailing orthodoxies; what Fuller once called the 'mega-visual tradition', the 'anaesthetic' world of the mass media and the new technologies:

> The underlying struggles . . . are between those who are basically 'collaborationist' in outlook towards the existing culture, and those who perceive that the pursuit of 'the aesthetic dimension' involves a rupture with, and refusal of, the means of production and reproduction peculiar to that culture.
>
> (Fuller 1983: 23)

For a different view you have only to turn to Williams, who, writing in the early 1960s, saw the then prevailing 'aesthetic' orthodoxy as uncritical and moribund, as totally out of touch with the realities of peoples lives:

> It is a meagre response to our cultural tradition and problems to teach, outside literature, little more than practical drawing and music with hardly any attempt to begin either the history and criticism of music and the visual art forms, or the criticism of those forms of film, televised drama, and jazz to which every child will go home . . .
>
> (Williams 1971: 172)

Fuller suggests that aesthetic practice is, in itself, a moral activity, a critique of alienating contemporary systems of communication and organisation. For him, those who advocate the virtual means of cyberspace or hyper-reality are mistaken if they think they are engaging in aesthetic practice. In relation to Williams (1979), this perception is deterministically pessimistic. He addresses a key moral question; should education provide people with the critical tools necessary for participation in democracy? In his discussion of television he implies that only through a critical approach to developing technologies can people discriminate between what does and does not answer a need. Equipped with the necessary tools they can better argue their case and effect change through the democratic processes of demonstration and negotiation.

WHY IS THE AESTHETIC DOMAIN SUCH A CONTESTED AREA?

There is a major Western tradition stemming from the Enlightenment, which posits the aesthetic as a form of knowing that is different, but in no way inferior, to logic. Until the eighteenth century, 'knowing through the senses' was considered a preliminary stage in a process leading to higher cerebral activities such as analysis and synthesis. The new philosophical study of aesthetics, or the inquiry into the nature of the beautiful and the sublime, was initially part of a moral quest to find ways of judging how things ought to be, a complement to the scientific quest to understand how things are. For some, given this moral aim, aesthetics became an alternative or surrogate religion; for others it became an end in itself. In the latter instance, all other interests, for example those of appetite, moral consequence, utility, had to be rejected so that nothing should impede the pure and disinterested act of judgement. In Kant's

Critique of Judgement (1790) beauty is no longer the outward sign of inner goodness, nor is the sublime dependent on divine revelation or the actions of great and moral people. The beautiful and the sublime are theorised as a subjective response to a phenomenon, cultural or natural, the goodness of which is judged through the faculty of taste. It is not beauty itself that is universal but the ability of people to make critical judgements: '*taste*, is the faculty of estimating an object or a mode of representation by means of a delight or aversion *apart from any interest*. The object of such delight is called *beautiful*' (Kant in Preziosi 1998: 84).

However, Kant does refer to the notion of 'general validity' which implies some sort of universality, and, for him, aesthetic and moral judgement are analogous activities. Kant's followers and apologists have frequently avoided this aspect of his philosophy and produced a tradition positing art as a transcendental and autonomous entity, the pursuit of which is, in and of itself, worthwhile. For them, the practice and appreciation of art needs no explanation, need serve no purpose, social, moral, political or otherwise. Pater wrote in the conclusion to his influential essay, 'The Renaissance' (1873):

> Of such wisdom, the poetic passion, the desire for beauty, the love of art for its own sake, has most. For art comes to you proposing frankly to give nothing but the highest quality to your moments as they pass, and simply for those moments' sake.
>
> (Golby 1986: 221)

Hegel countered Kant's speculations in his *Philosophy of Fine Art* (1835–38) by theorising art as a secondary process, the sensuous vehicle for the representation of the 'primary idea'. Art is the most concentrated and representative manifestation of the 'Zeitgeist', the character or spirit of the age as revealed in a particular culture. Seeking to understand art is therefore:

> . . . an attempt to understand the entirety of history as both a system and a *process* – a process of the unfolding of the Divine Idea in the (sensory and hence illusory) temporality of artistic change. The Divine Idea is unchanging and immutable: its changing representations over time are but confused ways in which mortal beings attempt to grasp the unchanging and singular Divine perfection.
>
> (Hegel in Preziosi 1998: 67)

Kant's speculations allow for coexistent aesthetic systems, differences in cultural outcomes that are not hierarchical, but determined by such variants as geography and climate, whereas, for Hegel, Western and Christian forms are more fully developed than any other culture or religion because they are nearer to the 'Divine Ideal'. This tradition posits art as a cultural phenomenon representing a people's collective or social spirit and thus their position on an evolutionary scale towards ultimate perfectibility, a notion which Hegel's most ardent student, Marx, was to apply to a theory of economics. These theoretical positions still manifest themselves in education: look at the *National Curriculum Order for Art* (NC) (DFE 1995) and the working document for the NC 2000 (QCA, 1999) where it states pupils should be taught skills through:

> Evaluating the work of artists, makers and designers from the community, from past and present and from familiar and unfamiliar cultures to learn about different kinds of art, the context in which it was made and how it is presented . . . Work exemplifying the Western tradition should be chosen from each of the following: i Classical and Medieval ii Renaissance and Post-Renaissance iii Modern and Post-Modern periods.
>
> (QCA 1999: 4)

Task 12.1.1 Egalitarian and/or pragmatic principles

Discuss the following questions within your tutor group.

Does an emphasis on an evolutionary and progressive Western tradition correspond in any way to Hegelian notions of cultural superiority, or is it a matter of knowing your 'familiar' culture, in all its 'diversity', more fully?

What evidence is there to suggest that recommendations to investigate 'unfamiliar' cultures relate to (1) egalitarian principles, or (2) pragmatism, responding to the post-colonial multicultural context of today's classroom.

Compare with the NC 2000.

There is then, a modern Western tradition from Kant to Greenberg (1965) which posits the aesthetic, or the contemplation of sensate experience, as transcendental, a phenomenon that through the imagination goes beyond the physiological into liminal experience, that is a disengaged state of heightened consciousness. Thus a sort of metaphysical or spiritual domain is suggested that might be considered 'religious' but without being associated directly with any specific religion. Are there any vestiges of this notion in beliefs about the way art affects people? Is this notion one with which you identify? If so, in what context does this happen to you?

For many it is in the context of religious worship or contemplation that spiritual experience is most readily manifest. It is worth evaluating the part played by the material and visual environment in which these experiences happen, whether church, mosque, synagogue, temple or other religious site. The major world religions are self-evidently the most representative vehicles for addressing spiritual needs: but in a multi-faith society it is dangerous for you to attempt to engage with the spiritual in the context of the classroom through any one system of aesthetic practices; one person's icons are another person's idols. This is an area of real difference. It does not mean, however, that you should consider religious art out of bounds: it can be viewed, discussed and provide individual pupils with stimulus for their own work. You are merely advised that whole classes should not be asked to make work in ways or with content that may be antithetical to their religious or philosophical beliefs (particularly in schools which are constitutionally dedicated to one faith). This advice also indicates some of the moral values associated with different types of educational activity: inquiry and comparative study, which are investigative processes that challenge but do not deny beliefs, and representational action, where the process is in itself a contamination of those beliefs.

> In places like universities, where everyone talks too rationally, it is necessary for a kind of enchanter to appear.

> (Beuys in Gablik 1991: 41)

Since the 1960s there has been a renewed interest in the practices of the shaman and her/his ability to contact and move through the world of the spirits. This desire to understand alternative forms of knowledge, initially driven by anthropological and psychoanalytical inquiry at the turn of the twentieth century, was pursued in Western aesthetic experimentation as an anti-materialist, anti-bourgeois strategy. Artists as diverse as Picasso and Beuys have adopted the ritual objects and performance of animistic cultures. The Surrealist quest for altered states of consciousness found in the shaman, the priestly other on which they could build their own artistic identities, the holy fool, at once foolish and wise, transgressive and sacred (Ades and Bradley 1998). Performance artists such as Fern Shaffer (Gablik 1991: 42–45) and Carolee Schneeman (Broude and Garrard 1994: 161) have also appropriated something of the ritualised activity of the shaman, incorporating gesture, dance, chanting, imaging in a multi-sensory and often hypnotic totality (see Plate 25). Susan Hiller, who before turning to art was a practising anthropologist, coordinated a group work, *Dream Mapping* (1976) in which participants gathered at a ritual site to record their dreams in visual form (Hiller 1996: 129–130).

When discussing shamanism it is important not to compound all its cultural manifestations into one homogeneous practice. Unless you approach such processes from a position of knowledge and understanding you are in danger of reinforcing myths of the primitive; 'others' as irrevocably irrational. If postmodern performance has appropriated aspects of shamanistic practice for ethical and political reasons, for example, its ephemeral nature makes it a difficult phenomenon to market and it is also open to sensationalist exploitation, the cult of spectacle, to 'shamanism'. As Hiller herself points out in response to the question:

'Post?modern?ism IS IT WORTH DISCUSSING?'

NO	YES
if it means falling back on a superficial and ideologically-constructed 'primitivism'	if it represents a real desire to break the enforced superficiality of some late-modernist/conceptual practices
if jokey historical references are used to disguise real contemporary recognition of new struggles, practices, marginalised groups, radical attitudes.	if it means we are closing the gap between 'experience' (ours) and 'reality' (theirs)

> (Hiller 1996: 141–142)

For Hiller, such practice is integral to the ongoing critique of dominant forms of Western knowledge and representation and is profoundly political in intent. Other

12.3 DIFFERENCE AND COMMONALITY: VALUES AND THEIR INTERACTIONS

OBJECTIVES

By the end of this unit you should be able to:

- question how you come to know your own and other value systems and their interactions;
- examine the benefits and dangers of using canonical exemplars to define cultures.

When attempting to identify cultures most people look first for concrete symbols to represent them. They are attracted to a process of reduction seeking out key ideas, personalities, artefacts, behaviours, etc. The resulting characteristics provide a sense of certainty and permanence standing in for what is likely to be a complex, paradoxical and diverse reality. The pressures on secondary-school teachers to educate pupils within a clearly defined and articulated curriculum invites just such a reductive process. In relation to the 'home' or 'host' culture, this usually manifests itself as a canon, that is in the form of key or exemplary 'texts'. A traditional reading of the 'West' defines it as the product of Greco/Roman culture modified by Judeo/Christian teaching passed on and developed down the ages through a succession of European heirs, and, more latterly, their colonial 'subjects'. Recent scholarship presents a less certain story acknowledging that cultures are neither strictly linear nor fixed phenomena but fluid and interactive. In the multicultural Art & Design classroom teachers tend to introduce 'world' cultures by showing a limited number of historical and emblematic reproductions: Africa = central African masks, India = court miniature painting from the sixteenth to eighteenth centuries; the past stands in for the present and contemporary practice is usually ignored. What is seen as 'familiar' or 'our' culture (i.e. Euro/British) has similar, if fuller, representation but is balanced by pupils' lived experience and the tacit knowledge they bring through living in or with a dominant culture. But its complex fabric and multiple signifying systems are distorted when reduced to canonic exemplars. A brief examination of a selection of key Western texts demonstrates just how misrepresentative it can be to represent cultures through decontextualised exemplars.

> Thou shalt not make thee *any* graven image, or any likeness of *any thing* that *is* in heaven above or that *is* in the earth beneath, or that *is* in the waters beneath the earth.
> (The Bible, King James's version, Deuteronomy: 5:8; see also Exodus: 20:4)

> The imitative art is an inferior who marries an inferior, and has inferior offspring . . . for he (the imitative poet) is like him (the painter) . . . he is the manufacturer of images and is very far removed from the truth.
> (Plato c380 BCb: 671–674)

These extracts concerned with the ethics of representation are selected from texts which are perceived as fundamental to the idea of 'Western culture'. More pertinently, as texts, they lie at the very centre of what the British claim to have achieved in the development of their culture, a synthesis of Judeo/Christian morality and Athenean civic and democratic principles (see Dash: Unit 11.1, 'Critical Voices').

Task 12.3.1 Questioning tradition

Discuss in tutor groups.
  To what extent this notion of a homogenised and linear tradition defines your identity? What aspects of your identity fall outside its parameters?
  With what criteria of exclusion do you construct this, for example by age, class, race, gender?
  In planning SoW how can you plan effectively both within and outside this tradition?

'US' AND 'OUR' CANON?

Returning to the quotations from the Bible and Plato's *Republic* you recognise that both reject mimetic forms of visual representation. Examine the following parallel texts:

> I (God) have filled him (Bezalel) with the spirit of God, with skill, ability and knowledge in all kinds of crafts to make artistic designs for work in gold, silver and bronze, to cut and get stones, to work in wood and to engage in all kinds of craftmanship . . . And all the men among the workmen made the tabernacle with ten curtains; they were made of fine twined linen and blue and purple and scarlet stuff, with cherubim skilfully worked.
> (The Bible, King James's version, Exodus; 31:3–5 and 35:8)

> 58. Myron seems to be the first sculptor to have extended the scope of realism; there was more harmony in his art than in Polyclitus', and he exercised more care with regard to proportions . . .
> 59. Pythagorous of Rhegium in Italy outclassed Myron . . . He was the first sculptor to show the sinews and veins and to give a lifelike rendering of the hair.
> (Pliny the Elder 1st century AD: 315)

Western culture is noted for its visualising tendencies, yet canonic texts provide what seems to be contradictory and conflicting advice. Which texts are indicative of western culture? Both the iconoclasts (sixth century AD) and the puritans of the Reformation (sixteenth century) relied on literal interpretations of Exodus, neglecting the context of its writing and evidence elsewhere in the Bible denoting the multiplicity of images in the Tabernacle (presumably representations of cherubim are a likeness of something in heaven). Historical 'Western' commandments and cautions show great congruence with parallel belief systems. In Hinduism, a warning that mimetic representation (likeness to appearance) is but a veil before the truth:

> That aesthetic *sadrsya* (concomitance of formal and pictorial elements) does not imply naturalism, verisimilitude, illustration, or illusion in any superficial sense is sufficiently shown by the fact that in Indian lists of factors essential to painting it is almost always mentioned with *pramana*, 'criterion of truth', here 'ideal proportion' . . . *Sadrysa* is then 'similitude', but rather such as is implied by 'simile' than by 'simulacrum'. It is in fact obvious that the likeness between anything and any representation of it cannot be a likeness of nature, but must be analogical or exemplary, or both of these. What the representation imitates is the idea or species of the thing, by which it is known intellectually, rather than the substance of the thing as it is perceived by the senses.
>
> (Coomaraswamy 1956: 12–13)

In Islam (Sura 6.74) there is an identical proscription in relation to the worship of idols: 'O Believers, wine and arrowsmithing, idols and divining arrows are an abomination, some of Satan's work; so avoid it: haply you will prosper' (in Yeomans 1993: 85) and further proscription in relation to representation of any living thing. The thirteenth-century Jurist, Nawami, stated:

> The learned authorities of our school and others hold that the painting of a picture of any living thing is strictly forbidden and is one of the great sins, because it is threatened with the above grievous punishment as mentioned in Traditions, whether it is intended for common domestic use or not. So the making is forbidden under every circumstance, *because it implies a likeness to the creative activity of God* . . . On the other hand, the painting of a tree or of camel saddles and other things that have no life are not forbidden.
>
> (*ibid.*)

These historical rules do not seem at all indicative of the visual landscape of contemporary Britain, nor indeed Pakistan or India. Clearly both the spiritual and temporal domains of Western and Eastern cultures have long come to terms with any dilemmas of conscience over the use of images, and the ubiquity of the photograph, with the distancing effect of its mechanical processes (Sontag 1977), has realised a place for the mimetic in cultures where it was once marginalised if not outlawed. In any case Coomaraswamy wrote, sometime before contemporary manifestations of religious 'fundamentalism':

> But it must be pointed out in passing that this Islamic art, which in so many ways links East with West, and yet by its aniconic character seems to stand in opposition to both, really diverges not so much in fundamental principles as in literal interpretation. For all naturalism is antipathetic to religious art of all kinds, to art of any kind, and the spirit of the traditional Islamic interdiction

of the representation of living forms is not really infringed by such ideal representations as are met with in Indian or Christian iconography, or Chinese animal painting . . . each should strictly speaking, be regarded as a kind of diagram, expressing certain ideas, and not as the likeness of anything on earth.

(Coomaraswamy 1956: 5)

You may be tempted to define cultural ancestry in terms of ideas embedded in canonical texts; this is the easy answer. If you investigate further you are likely to find them contradictory, replete with compromise, open to antithetical interpretations and embraced by cultures you might assume in all other respects to be different; this is a more demanding task, but it rewards understanding.

Task 12.3.2 To represent, or not to represent?

In tutor groups:

- discuss the possibility of constructing an Art & Design curriculum that is open to non-representational tasks;
- plan a SoW that has the choice between a non-representational and a representational outcome.

FURTHER READING

Yeomans, R. (1993) 'Islam: the abstract expressionism of spiritual values', in D. Starkings (ed.) *Religion and the Arts*, Sevenoaks: Hodder and Stoughton.

12.4 TEACHING IN A PLURALIST SOCIETY

OBJECTIVES

By the end of this unit you should be able to:

- explore the implications of teaching Art & Design in a pluralist society;
- explore how a knowledge of and respect for difference can be ensured while seeking the interrelatedness and commonality of cultures.

OTHERS WITHOUT?

Tate (1997) advocates the need to construct a cultural identity by exposing pupils to canonical exemplars. He admits that:

> Because identities are multiple, this involves a sense of how they locate themselves within a variety of cultural traditions: above all those of England, Britain and Europe, but also the traditions of those parts of the non-European world with which this country has close and long-established links.
>
> (p. 13)

Cultures typically define themselves through a system of difference and exclusion: where those differences are blurred or indistinct they tend to be written out. Tate warns against the distractions of:

> a pervasive cultural egalitarianism which refuses to recognise that cultures (especially majority ones) are very special to those who belong to them and need to be nurtured and transmitted through careful attention and special treatment. It is in part a result of the prevailing postmodernist intellectual climate with its emphasis on fictions and constructions and its sense that nothing is sufficiently substantial or objective for it to be worth passing on.
>
> (*ibid.*: 12)

However it must be one task of a critical curriculum to uncover or deconstruct the structures of difference which support a dominant culture, even if that culture feels fragile and threatened by such scrutiny. If it has anything ethically vital or cogent to offer it can only benefit from the experience. To invoke the agenda of one 'postmodernist intellectual':

> Foucault completely upsets our conventional expectations of history as something linear – a chronology of inevitable facts that tells a story which makes sense. Instead, he uncovers the underlayers of what is kept suppressed and unconscious *in* and *throughout* history – the codes and assumptions of order, the structures of exclusion that legitimate the epistemes (systems of knowledge/belief) by which societies achieve their identities.
>
> (Appignanesi and Garratt 1995: 83)

Historical acts of exclusion have left post-colonial Britain with a host of minority groups whose identity and contribution were once seen as marginal to the grand narratives of its national identity: this would include groups differentiated on such grounds as race, gender, class, sexuality, age, disability. Empowering these groups with the means to find and disseminate their voices is one of the principles of equal opportunities. But it is entirely counter-productive to lump them all together as if they were some form of homogeneous, disenfranchised 'other' who require uniform and patronising assistance. Equal opportunities does not mean treating everyone in the same way: this would deny the concept of differentiation by need which is at the heart of its strategies. For example, racism based on the grounds of skin colour is all pervasive, and teachers may be tempted to celebrate, for example, 'black arts' by showing examples of limited and usually historical artefacts. But in countering prejudice, in what ways would this benefit their pupils? What messages would it send to the class about 'blackness'? All are likely to have experienced the monolithic effects of undifferentiated prejudice, whether produced or received, but the means to counter racism cannot be achieved through a celebration of an essentialist blackness exemplified by token (mis)representation. What is needed is an understanding of the mechanisms by which colour prejudice has been constructed and perpetuated (see Chapters 5 and 11). At the same time this must be managed by avoiding strategies that engender a culture of victimisation and guilt:

> Otherization is unavoidable, and for every One, the Other is the Heart of Darkness. The West is as much the Heart of Darkness to the Rest as the latter is to the West. Invention and contemplation of the Other is a continuous process evident in all cultures and societies. But in contemplating the Other, it is necessary to exhibit modesty and admit relative handicap since the peripheral location of the contemplator precludes a complete understanding; this ineluctability is the Darkness.
>
> (Oguibe 1993: 3–8)

For many years anthropologists have been demanding that represented 'others' should be allowed to speak for themselves. The ethical way forward is to share 'textural authority with subjects themselves, autobiographical recanting as the only appropriate form for merging cultural experience with the ethnographer's own' (Clifford and Marcus 1986: 168) and museums have made concerted efforts to enable self-representation to take place. At one time the ethnographer's remit was to explain 'other' people from without their own, usually colonialist, culture, to 'us', an audience from within: increasingly they are exploring people from within their own, usually, post-colonial culture, to 'others as us' and 'us as others'.

> One of imperialism's achievements was to bring the world closer together and, although in the process the separation between Europeans and natives was an insidious and fundamentally unjust one, most of us would now regard the historical experience of empire as a common one. The task then is to describe it as pertaining to Indians *and* Britishers, Algerians *and* French, Westerners *and* Africans, Asians, Latin Americans, and Australians despite the horrors, the bloodshed, and the vengeful bitterness.
>
> (Said 1993: xxiv)

Although containing a critique, modernism's appropriation of the exotic and the other is closely bound to the colonialist enterprise – it simultaneously disseminated difference while diffusing it by appropriating it in the guise of universalism (Hiller 1991). Likewise postmodernism is part and parcel of the late capitalist world of international corporatism, exploring and embracing its technologies while deconstructing and attacking its structures. Whatever you may think of global capitalism one of its by-products is the erosion of cultural, if not economic, difference:

> Global homogenization is more credible than ever before, and though the challenge to discover and represent cultural diversity is strong, doing so in terms of spatio-temporal cultural preserves of otherness seems outmoded. Rather, the strongest forms of difference are now diffused within our own capitalist cultural realm, gender and lifestyle constructs being two prominent fields of representation for exploring cultural difference.
>
> (Clifford and Marcus 1986: 168)

Market incursions are liable both to appropriate the material cultures of indigenous peoples and gradually replace them through a system of imports either imposed or made desirable through targeted advertising. This has the effect of diluting a major basis of indigenous cultural identities. The resulting losses in difference, although welcomed by some, are perceived by many as pernicious and destructive. Postmodern pluralism is often hailed as an egalitarian triumph: all voices have their say; hierarchies are abandoned, the transcendental with the transient, the sublime with the porno-graphic; all hold hands together and dance the dance of relative values. But, 'doubt is itself in doubt, doubt should be, but isn't, tolerant of others' beliefs' (Appignanesi 1997: audio). For many in the developing world postmodernism is the bedfellow of capitalism, a cynical manoeuvre to relegate difference to marketable spectacle. For such critics postmodernism possesses no integrity or originality; it is a parasitic maw with an indiscriminate hunger to appropriate difference for mass consumption and it can do so because it has absolute control of the means of production, marketing and distribution. For people who hold fundamental beliefs the erosion of difference sig-nals a major threat and defensive positions are drawn. 'Postmodernism, with its relativ-ism, sceptism and doubt, appropriates the grand-narratives of others' certainties at its peril' (ibid.). Religious and political fundamentalisms have thus been theorised not as returns from, but as products of, postmodernism.

Hughes (1993) reminds you that all cultural and spiritual exposition, including that of marginal groups, requires critical scrutiny if its voice is to take on historical, social, and collective significance: 'Now the claims of the victim do have to be heard, because they may cast new light on history. But they have to pass the same tests as anyone else's, or debate fails and truth suffers' (p. 146). The tests and truth referred to by Hughes are not those of scientific empiricism probing divine revelation or meta-phor, but historical investigation differentiating actual and fictive events. He argues that the endeavour to empower oppressed groups within particular nation states, in this instance the USA, is doomed to failure if it fails to recognise that the myths of the oppressed are as pernicious and divisive as the myths of the dominant culture:

> Cultural separatism within this republic is more a fad than a serious proposal; it is not likely to hold, but if it did, it would be an educational disaster for those it claims to help, the young, the poor and the black. It would be a gesture not of 'empowerment' but of emasculation. Self-esteem comes from

doing things well, from discovering how to tell a truth from a lie, and from finding out what unites as well as separates us.

(*ibid.*: 150–151)

While Hughes warns against a misguided and patronising desire to accept at face value allcomers in the name of liberal tolerance, Said reminds us that indiscriminate acceptance is a recipe for chaos:

> In our wish to make ourselves heard, we tend to forget that the world is a very crowded place, and that if everyone were to insist on the radical purity or priority of one's own voice, all we would have would be the awful din of unending strife, and a bloody political mess, the true horror of which is beginning to be perceptible here and there in the reemergence of racist politics in Europe, the cacophony over political correctness and identity politics in the United States, and . . . the intolerance of religious prejudice and illusionary promises.

(Said 1993: xxiii)

He goes on to conclude:

> If the Japanese, East European, Islamic, and Western instances express anything in common, it is that a new critical consciousness is needed, and this can be achieved only by revised attitudes to education. Merely to urge students to insist on one's own identity, history, tradition, uniqueness may initially get them to name their basic requirements for democracy and the right to an assured decently human existence. But we need to go on to situate these in a geography of other identities, peoples, cultures, and then to study how, despite their differences, they have always overlapped one another, through unhierarchical influence, crossing, incorporation, recollection, deliberate forgetfulness, and, of course, conflict . . . The fact is, we are mixed in with one another in ways which most national systems of education have not dreamed of. To match knowledge in the arts and sciences with these integrative realities is, I believe, the intellectual and cultural challenge of the moment.

(*ibid.*: 401)

Task 12.4.1 Integrative realities

Discuss these questions in tutor groups.

Can you use exemplars in such a way that they will be understood as contingent (on culture and history) rather than fixed or absolute?

How can you ensure that the range of knowledge encompassed by your pupils can be shared and made productive in your lessons?

How can you match knowledge in the arts with the 'integrative realities' of the past and present?

OTHERS WITHIN?

One of the central tenets of official modernism, that 'art' has the potential to provide transcendental or liminal experiences, is a device for separating the chaff from the wheat. Thus, most applied art, craft, design and particularly decoration, is immediately placed outside the domain of 'art' and 'aesthetics' because it serves utilitarian and domestic functions (Addison 1997). Likewise popular traditions, characterised by Greenberg in 1939 as 'Kitsch', serve lesser purposes, particularly sentimentality and ephemeral fashion (Harrison and Wood 1992: 530–541). There is a powerful tradition in British education that dismisses popular forms of cultural production and upholds an exclusive and elitist notion of culture. Leavis asserted in 'Mass civilisation and minority culture' (1930):

> In any period it is upon a very small minority that the discerning appreciation of art and literature depends: it is (apart from the cases of the simple and the familiar) only a few who are capable of unprompted, first hand judgement . . . upon this minority depends our power of profiting by the finest experience of the past; they keep alive the subtlest and most perishable parts of tradition. Upon them depend the implicit standards that order the finer living of an age, the sense that this is worth more than that, this rather than that is the direction in which to go, that the centre is here rather than there.
>
> (in Easthope and McGowan 1992: 210)

This suggests that teachers, as upholders of standards and advocates of quality, should inculcate their pupils with only the very best. But if you reject the popular out of hand, as Leavis implies you should, you renounce territory with which pupils may feel kinship if not outright ownership; you reinforce the sense that they function from outside cultural codes that are valued because they identify with an inferior if majority culture. This is not to propose that you exclusively embrace the popular but that you acknowledge and approach its forms critically, seeking its function and value. Adorno's analysis of popular music (1941 in Easthope and McGowan 1992: 211–223) is scathing in its conclusions but so much more telling than Leavis's patronising certainties.

Williams (1976) reminds you that the meaning of the word 'popular' has multiple and changing meanings:

> **Popular culture** was not identified by *the people* but by others, and it still carries two older senses: inferior kinds of work (compare **popular literature, popular press** as distinguished from *quality press*); and work deliberately setting out to win favour (**popular journalism** as distinguished from *democratic journalism*, or **popular entertainment**); as well as the more modern sense of well-liked by many people, with which of course, in many cases, the earlier senses overlap. The recent sense of **popular culture** as the culture actually made by people for themselves is different from all these; it is often displaced to the past as *folk culture* but it is also an important modern emphasis.
>
> (p. 237)

It is in the context of a discussion of popular culture that you might introduce the issue of sexual orientation, in particular homosexuality, without infringing the proscriptions outlawing its promotion within LEA-run schools. In ensuring that aspects of citizenship are integrated into teaching across the curriculum the QCA (1998b) points out:

> We must recognise that teaching about citizenship necessarily involves discussing controversial issues. After all, open and informed debate is vital for a healthy democracy . . . Teachers are aware of the potential problems and are professionally trained to seek for balance, fairness and objectivity.
>
> (p. 1.9)

The investigative questions provided in *Working with Modern British Art* (Adams *et al.* 1998: 46) direct pupils' attention to a painting by Hockney, *Third Love Painting* (1960) which represents, partly covertly, aspects of homosexual desire: it was painted at a time when homosexuality itself was illegal in Britain. As with many artists of the 'Pop' generation Hockney makes reference to an eclectic and culturally diverse range of sources, from Whitman to Cliff Richard, Abstract Expressionism to graffiti, addressing the theme of same-sex love with humour and from the point of view of subjective experience. This, and similar work, is a useful vehicle for introducing an issue which is often debated from a position of bigotry and/or prurience. It can provide pupils with an avenue for discussing what is often a taboo subject amongst adolescents and at a time when they are likely to be exploring and defining their own sexual orientation. Issues surrounding gender and sexuality are central to the work of many contemporary artists. You are advised to choose examples with great care, paying attention to the wishes of parents and governors, but without being intimidated into rejecting an issue that is of vital significance in pupils' understanding of human relationships both historically and as manifest in contemporary and particularly popular culture (Mac an Ghaill 1995).

Task 12.4.2 Attitudes to a pluralism

In your tutor group discuss your attitudes to:

- forms of art other than the fine arts;
- forms of popular production;
- forms of production that acknowledge minority interests.

What implications do your attitudes have for what you choose to introduce to the Art & Design curriculum?

Task 12.4.3 Developing pluralist resources

Using the types of production outlined in this unit, form banks of reproductions which have, on their reverse, contextualising information, bibliographical references and sets of questions for further investigation e.g. such packs as *Working with Modern British Art: A Guide for Teachers* (Adams *et al.* 1998).

FURTHER READING

Appignanesi, R. and Garratt, C. (1995) *Postmodernism for Beginners*, Cambridge: Icon Books.

Said, E. (1993) *Culture and Imperialism*, London: Chatto and Windus.

Williams, R. (1976) *Keywords*, London: Fontana.

12.5 QUESTIONING AND RESOLVING DIFFERENCES IN VALUES

OBJECTIVES

By the end of this unit you should be able to:

● explore and attempt to resolve dichotomies of value in educational and other systems.

In the same way that people look for fixed points of reference when identifying cultures, you may feel it necessary to form a particular allegiance to an educational theory and pedagogical method in order to construct a clear identity as a teacher. This is almost irresistible because it seems to answer the pressures on you to resolve ideological conflicts within your own practice. However, it is advisable to remain open in your approach and not be tempted to fit one method to all circumstances.

Educational theories and traditions often pull you in directions that seem to be opposing, for example, while one suggests that you should encourage 'self-expression' and the emergence of the individual, an other expects you to provide pupils with the symbolic tools to enter a collective and social universe. The following table outlines some of these dichotomies.

Dichotomies

self-actualisation	socialisation
nature	culture
self-expression	cultural conventions
vocational	academic

child-centred	adult-centred
world-to-child	child-to-world
authenticity	normative standards
peer pressure	adult authority
feeling	cognition
intuition	logic
haptic	visual
non-conformity	conformity (behaviour)
being other	belonging
popular	elite
spontaneity	exactitude
risk	safety
certainty	doubt
action	thought

Task 12.5.1 Working with dichotomies

In your tutor group:

Add to this list any dichotomies that you have witnessed or imagine might arise in your teaching.

Consider the extent to which these dichotomous pairs or contrasting aims and values are opposed. For example the French philosopher Baudrillard says that the opposite of knowledge is not ignorance, but deceit and fraud (Appignanesi and Garratt 1995: 136).

Where you find oppositions, either in the given list or in your own additions and amendments, is it necessary to reconcile them, and if so, how?

Might they be argued as parallel or complementary?

To which of these values do you feel any allegiance? Do any patterns emerge? Where do you see a problem in applying these values to your own teaching?

AFTERWORD

In this chapter you have seen the dangers of stereotypical representation, canonical exemplification and misdirected sympathy. In suggesting that you look for contradictions where consistency is normally posed, for commonality where difference is reinforced, and by questioning 'universal' criteria used to belittle what is actually different, you help pupils to see the structures of value by which cultures choose to identify themselves. The key to effective artistic and cultural study is critical inquiry. Before you can expect pupils to adopt such a position this requires continual critical effort and demonstration by you which should not end in your student years. Without it your representations of cultures are liable to be forms of misrepresentation and mis-evaluation. In Chapter 11 it was recommended that in pursuing a multicultural curriculum you investigate the inter-cultural dimension of Art & Design. It is educationally desirable to apply its defining principle, inquiry, at the point of translation and transmission, to include and question other categories of difference: elite/popular, male/female, heterosexual/homosexual, young/old. In this way you seek out those points of contact and interaction where difference is negotiated, and transformation, appropriation, compromise and rejection are played out.

13 Further Professional Development

Andy Ash and Richard Hickman

13.1 THE PROFESSIONALS

INTRODUCTION

First, we shall discuss the notion of a 'professional' and how to build upon your Initial Teacher Education (ITE) and develop your career as an Art & Design teacher; we must try to define what we mean by 'professional' in this context. It is a word which is often used to distinguish a critical and objective approach to practice from personal preferences, although it is sometimes contrasted with 'amateur' rather than with 'personal'. In terms of the origins of the words themselves, amateur is best contrasted with connoisseur in that amateurs engage in certain activities for the love of it whereas connoisseurs engage in certain activities because they know a lot about them: with regard to art teachers, the vast majority both love the subject and know a lot about it. A more pragmatic approach to understanding the term professional is to examine its use in context. The present context is school culture, which tends to be conservative in nature and conforms in the main to middle-class values. These values may include such things as punctuality, cleanliness, tidiness, appropriateness of dress and so on. Art & Design teachers are sometimes (rightly or wrongly, probably wrongly) allowed some degree of latitude with regard to attitude – somewhat perversely, they are sometimes almost expected to not conform to the same rules as other teachers. However, it is worth remembering that Art & Design teachers are paid to teach art and to represent the school and the school's values; they are, whether they like it or not, figures of authority, not authoritarian, but authoritative, having significant knowledge and understanding of their subject and having trained and qualified as teachers. Courses of initial teacher education (ITE), such as PGCE art courses, are fundamentally concerned with 'professionalisation' – in the present case of turning art, craft and design students into Art & Design teachers. This means refocusing students' orientation so that it becomes outward rather than inward, with other people's artistic development rather than their own. It does not mean that the creativity is somehow knocked out of students, nor does it mean that your creative output is put on hold, rather the reverse. Apart from the very act of teaching being essentially a creative enterprise, an active learning environment can stimulate the art teacher into production. It is worth noting that the DfEE, in its introduction to Standards for the Award of Qualified Teacher Status states the following:

> professionalism . . . implies more than meeting a series of discrete standards. It is necessary to consider the standards as a whole to appreciate the creativity, commitment, energy and enthusiasm which teaching demands, and the intellectual and managerial skills required of the effective professional.
>
> (DfEE 4/98: 6)

It is interesting to note that one of the most scathing remarks that a person working in a profession can make to colleagues is that they are 'unprofessional'. This can mean many things, but in general it would mean not conforming to public expectations in terms of quality, character, method or conduct; the nature of public expectations will of course vary according to context. A basic requirement must be that art teachers know their subject and can teach it in a way that conforms to the conventions of the teaching institution.

Teachers' own engagement with art, as practitioners, is often sporadic and neglected. Anyone who has worked in a school for more than a few days will know of the many pressures which militate against teachers' own art production, but it remains that for many the love of, and involvement with art is a motivating factor: a source of inspiration and motivation for art teachers and central to their commitment to teaching. Prentice discusses the issue of identity and art practice, highlighting the role of 'a very strong subject allegiance and an equally strong sense of personal identity' (Prentice 1995: 11). Allison (1974) an emeritus professor of education, argued for many years for Art & Design teachers to see themselves as professional Art educators rather than as artists who happen to be paid to teach art and has developed the idea to extend to professional attitudes to research in art education (Allison 1997). The adoption of 'professional attitudes' means, in short, accepting and working within a framework of common values. We see no conflict here; the professional Art & Design teacher can also be an artist. Indeed, it is a prerequisite for good teaching that the Art & Design teacher is actively involved in some way with the subject. However, the subject is Art & Design education, not ceramics, or textiles or fine art or art history, and this should remain our central concern – the induction of young people into the world of art.

OBJECTIVES

At the end of this unit, you should:

- know about the standards required for Qualified Teacher Status (QTS);
- understand the application of the Standards in an Art & Design context;
- be able to appraise your current practice in the light of your understanding of the Standards;
- standing of the Standards;
  be able to identify target areas for your own professional development.

STANDARDS

The Standards are generic rather than subject specific and OFSTED, as a part of its inspection role, has produced guidance for Art & Design teachers to be used in conjunction with them.

Task 13.1.1 Collecting evidence for the Standards

Below are the four key areas of the Standards: consider to what extent you have expanded and progressed in them. Make a list of the type of evidence you have to support your assessments of your teaching in the following areas.

Section A: knowledge and understanding

- Do you have depth, range and adaptability in your own technical skills in art, craft and design? List your present skills/specialisms. What should you develop over the coming year?
- Do you have a broad understanding of the history of art, a variety of artistic traditions, and can you make practical connections between this and the work of pupils? Are there gaps in your knowledge, for example about design/craft history, women artists, black artists, etc.?
- Do you understand the physical, social and emotional development of children and the way this may affect an individual's approach to art? Do you understand the nature of children's artistic development and how it might impinge upon your teaching?

Section B: planning, teaching and class management

- Can you plan to achieve progression in pupils' understanding of art, and in the skills of investigating, designing and making in art and craft? How does your understanding of children's artistic development inform your planning?

- Do you demonstrate in your planning and teaching a clear understanding of how to develop pupils' creative and practical skills in art, craft and design? Continuity and progression are very important in Art & Design. How do you build upon pupils' prior learning and develop pupils' abilities say from KS2 to KS3?

- In planning and directing art, craft and design activity, do you recognise and make valid use of a diversity of ethnic and cultural traditions? How are they woven into projects without them appearing a 'bolt on'? Do your visual aids/resources represent our multicultural society?

- Do you show imagination in devising opportunities for pupils of all abilities to extend their skills and knowledge in art, craft and design? Are you constantly evaluating your projects, successes and failures? How have you changed and developed ideas, projects and resources to meet the various pupils' needs and situations?

- Can you manage a studio environment efficiently and safely and organise pupils to work effectively within it? Are you familiar with all of the materials and techniques which you are likely to find in the school environment? Do you have a sound knowledge of health and safety issues as they relate to Art & Design teaching?

Section C: monitoring, assessment, recording, reporting and accountability:

- Do you ensure that pupils can evaluate their own art work and modify it in the light of their own and others' evaluations? What opportunities have been made for discussion, reflection and re-evaluation of pupils' work and others? What ownership do pupils have in the assessment process, and how transparent is it?

Section D: other professional requirements

Section D is the most pertinent Standard for the purposes of this chapter. We asked a group of six professional tutors and ten Heads of Art & Design in secondary schools, about their views on what this aspect of the 'Standards' might include. Although their response is by no means definitive, it provided us with some insight into what is expected in schools. While the professional tutors tended to emphasise general aspects of professional behaviour such as appearance, punctuality and attendance, the art teachers, not surprisingly perhaps, emphasised the need for 'professional relationships' with pupils and a responsible attitude towards materials and equipment. Additionally, it was felt by both groups that commitment to teaching and to pupils' learning was a professional prerequisite. Commitment was exemplified in such things as participating in extra-curricular activities and willingness to put in hours on the school premises beyond the pupils' school day.

In formal terms, the Teacher Training Agency has published details of what is expected of teachers in terms of professional commitment, stating that:

> Those to be awarded Qualified Teacher Status, should, when assessed, demonstrate that they . . . have a working knowledge and understanding of teachers' professional duties . . . [and] . . . teachers' legal liabilities and responsibilities'.
>
> (TTA 1997: 11)

It would be useful to go through these responsibilities and legal liabilities from an Art & Design teacher's perspective:

- Health and safety procedures and reasonable care towards pupils
  Art & Design teachers need to know about all of the potential hazards in the art room, in particular, safe use of knives and blades (for example in lino cutting), toxicity of certain substances, such as glazes and clay dust and safe use of equipment (such as kilns). There are several publications available which deal directly with health and safety in schools, the most relevant for the art teacher is *A Guide to Safe Practice in Art & Design*, published by the DfEE (1995). It is important to be aware of the school's conventions with regard to visits to museums and galleries; this would normally cover things such as parents' permission and insurance.
- Anti-discrimination
  Most Art & Design teachers will actively promote multicultural and cross-cultural issues in their lessons; others may actively promote antiracist approaches to art and will address issues of gender, ensuring that for example, pupils are exposed to an appropriate range of work by women artists. You need to be aware of the requirements of the Race Relations Act (1976) and the Sex Discrimination Act (1975). It should be noted that Section 2 of the Local Government Act (1986) (as amended by Section 28 of the Local Government Act (1988)) prohibits local authorities from intentionally promoting homosexuality or publishing material with that intention, and from promoting the teaching in any maintained school of the 'acceptability of homosexuality as a pretended family relationship' (DFE 1994).

There are seven other aspects listed under the general heading of 'Other Professional Requirements':

Effective working relationships with colleagues
It is not unknown for Art & Design teachers to be somewhat singular in their behaviour and appearance and some might actively cultivate a maverick character.

This tends to be acceptable as long as 'personality clashes' do not hinder the effectiveness of teaching and learning. It is important to develop a positive working relationship with teaching colleagues; all have the common goal of helping to educate young people. There are several groups of non-teaching colleagues with whom the Art & Design teacher needs to work, such as Learning Support Assistants (LSAs), technicians and caretaking staff. LSAs play an important part in ensuring that those pupils who have particular needs are given appropriate help; it is up to the classroom teacher to ensure that LSAs are well briefed on the focus of each lesson. Many school Art & Design departments could not run effectively without the help of a technician. Such individuals are often well-qualified with considerable expertise and it is important that they are treated accordingly. The Art & Design teacher is sometimes the bane of the school caretaker's life; it is good practice to liaise with the caretaking staff and work out roles, duties and responsibilities. A blocked drain in the ceramics room is usually avoidable if you train your pupils in the art of disposing of plaster and the washing of clay tools. It is part of the Art & Design teacher's classroom management repertoire to enlist the help of pupils in clearing up: it is a waste of your time and that of the caretaking staff if you are left with a sink full of unwashed brushes and palettes.

Set a good example

This refers to teachers' appearance and conduct. The general rule of thumb is not to ostentatiously flout the conventions of the school. If it is a school rule that boys wear ties, then it normally follows that the male staff should do so. It is inappropriate to swear, even in the most trying of circumstances. Similarly, avoid using colloquialisms, especially the more vulgar epithets.

Ensure every pupil achieves their potential

Many Art & Design teachers when asked how they differentiate pupils' needs and work indicate they do so 'by outcome'. This means that all pupils are given the same starting point and produce work according to their ability. While this might work for most pupils most of the time, it is important to cater for those who, for one reason or another, are slow learners and also for those who might be called 'gifted and talented'. In a well-run school, each pupil with learning difficulties will have an Individual Education Plan (IEP), drawn up between the teachers for each subject and the Special Educational Needs Coordinator. It is important for the Art & Design teacher to recognise the part which the subject plays in the overall educational development of the pupil. The artistically gifted need to be stretched and this is not an easy task, but recognition of this need in itself goes some way towards addressing it. A useful publication in this area is by Clark and Zimmerman (1984).

Professional development

Professional development for Art & Design teachers can mean a whole range of things, but it is particularly important that you develop and extend your own art practice. There are usually courses available which deal directly with studio practice such as printmaking and ceramics. It is wise to be on the mailing list of galleries and museum which also run extremely useful courses on such things as working from objects and using art as a starting point to studio-based activities. Joining a professional association such as NSEAD will ensure that you are kept up to date and informed through newsletters and the professional journal (*Journal of Art and Design Education* (*JADE*)).

Responsibilities in relation to school policies

This includes pastoral responsibilities and it is important that you spend some time in familiarising yourself with school policies on such things as bullying and matters of personal safety. Art & Design teachers often find themselves acting in a pastoral capacity in addition to that expected as part of day-to-day responsibilities, perhaps because of the personal, expressive and affective nature of some work produced in art lessons. It is vital that you have an absolutely professional approach to this and do not involve yourself personally; make effective use of the structures and systems available and do not take on inappropriate, informal pastoral responsibility.

Learning inside and outside of the school context

This means ensuring that you have and maintain effective links with parents and other outside bodies, including galleries, museums and local artists. Clearly, pupils' learning, in its broadest sense, mainly takes place outside the classroom and so it is important for teachers to capitalise on this.

Governing bodies

You will be expected to have an understanding of the role and purpose of governing bodies. Their influence varies between schools and their legal position is determined by the status of the school (i.e. whether it is, for example, grant maintained). They are potential sources of funding for projects such as artists in residence; you might even find a governor who is also an artist! In well-organised governing bodies an individual governor will be assigned to each subject department. It will be important for you to establish who is responsible for the Art & Design department and encourage liaison, keeping them well informed of all events, exhibitions and achievements whilst outlining the particular and varied needs of the department.

FURTHER READING

Allison, B. (1997) 'Professional attitudes to research in art education', *Journal of Art and Design Education* 16(30): 211–215.

DfEE (1997) *Teaching: High Status, High Standards*, London: HMSO.

Moon, B. and Mayes, A. S. (eds) (1994) *Teaching and Learning in the Secondary School*, London: Open University and Routledge.

13.2 YOUR FIRST TEACHING POST

INTRODUCTION

The following sections deal with securing a teaching position and are based largely on the NSEAD publication *Finding Your First Job – A Guide for Prospective Art and Design Teachers* (Hickman 1996). Interviews for teaching positions in Art & Design are characterised by the presentation of candidates' portfolios. The preparation of the interview portfolio is dealt with in its own section, but is mentioned here to draw it to the attention of other subject specialists who might be reading this. Much can be learned from the Art & Design education student in this respect; a prospective teacher of Science, English, Geography or any other subject would benefit from presenting a well-thought through portfolio, showing appropriate material, at interview.

OBJECTIVES

By the end of this unit you should be able to:

- look for a teaching post;
- construct a CV and letter of application;
- put together a portfolio;
- prepare for interviews.

WHERE TO LOOK

The most widely used source for those seeking teaching posts is the *Times Educational Supplement* (TES) which comes out every Friday. Do not just look under 'Art & Design – CPS' (Common Pay Spine); investigate the following: sixth-form colleges and further education colleges (usually in the FE section), middle schools, expressive arts, design & technology.

Overseas opportunities can also be found in their own section. You may find working in a developing country very stimulating. Bear in mind the warning in the TES which begins: 'Advertisements are offered in good faith'. The more secure, high salaried contracts in the Middle East and East Asia are usually taken by teachers with a few years' experience, but you might be lucky and end up teaching highly motivated children in an air-conditioned classroom and enjoying a 'generous tax-free salary'.

Local Education Authorities (LEAs) publish a regular list of vacancies which are circulated to schools. You will be able to get most LEAs to include you on their mailing list; some will expect you to provide stamped self-addressed envelopes. Word of mouth is always a good source – keep in touch with your course tutors and others in the profession. The direct approach, if done professionally and with due diplomacy might pay dividends. The more you limit yourself to a particular area the less chance you have of getting a job. Obviously, if you, for family or other reasons, can only teach within, say, a radius of twenty miles from your home, then you might well benefit from writing to the headteachers of all of the schools in the area, enclosing a stamped self-addressed envelope and a curriculum vitae. It will do no harm. If you have finished your course and have not found a job by the autumn term, you might even offer to do some voluntary work in a local school; this will ensure that you are keeping your hand in. An option open to the art, craft and design specialist is working as a school's artist in residence. You should negotiate terms which are mutually beneficial.

You can register with LEAs and you can also apply to be on the books of private agencies. The latter tend to specialise in supply teaching (see section on supply teaching) – this is a challenging way to start your teaching career, and you will gain valuable if not ideal experience. Regular job seekers will note the varying thickness of the TES over the course of a year, from skeletal in high summer to positively obese in late spring. This coincides with teachers' resignation dates, which are normally by half term of each of the (currently) three school terms.

Summary of some useful places to look

Newspapers
Times Educational Supplement, Friday
Guardian, Tuesday
Independent, Thursday
Daily Telegraph, Thursday
The Teacher, published weekly
Church papers: *Catholic Herald*, *Church Times*, *Jewish Chronicle* and *Asian Times*
Local newspapers
Local Education Authority (LEA): county bulletin, or direct into a pool via teaching personnel department or district education office
Internet: *http://www.jobs.tes.co.uk*
Independent schools: the above newspapers advertise independent posts as well as the following contact agencies:

> *Independent Schools Year Book*, published by the Independent Schools Information Service, available via the university/college library or local public reference library.

> Secondary: I.S.I.S. (Independent Schools Information) 56 Buckingham Gate London SW1E 6AG (Tel: 0171-630 8793)

> The association cited below, although concerned with Primary posts, also recruits for independent secondary schools and for overseas posts:
> Primary – I.A.P.S. (Incorporated Association of Preparatory Schools) Gabbitas Educational Consultants Recruitment Department Carrington House 126–130 Regent Street London W1R 6EE (Tel: 0171-439 2071)

SUPPLY TEACHING

Supply is a temporary measure of appointing staff to cover for short-term absences/ vacancies. Increasingly it has been used to cover medium- and long-term needs in schools. The demand for supply teachers varies according to the time of year/term and your geographical area. It was once considered inappropriate for NQTs to start their career with supply work, however an increasingly problematic job market has meant a change in attitudes to what is now an important resource for schools. For those Art & Design teachers who may not want a full-time permanent appointment it enables flexibility to juggle industrial/studio work with teaching.

Check with your local LEA for their procedures, whether you have to register with them centrally or contact schools directly. In some areas LEAs and schools use agencies, which usually specialise in supply teaching. You can find these advertised in the TES and other newspapers. Your LEA may advise or see your local careers centre for locally-based companies. You will need to examine carefully the different terms and conditions and particularly the rate of pay! The agencies are generally good at providing work and most are highly reputable. They will however deduct a fee from your pay as a proportion of your daily rate.

WHAT IF I STILL HAVE NOT GOT A TEACHING POST?

Your HEI may still be notified by local schools and partnership contacts of vacancies available, or those in the offing, so keep in contact with your tutors. Let them know you are still looking and, if it is possible, keep in contact with your teaching-practice schools.

Task 13.2.1 Finding the right job

Look through the educational press for jobs making sure you examine all sections.

- How many can you find?
- Can you find a range of different types of school situations i.e. mainstream, grant maintained, voluntary aided, independent etc.? Start to get a feel for the field and the types of adverts.
- Using a highlighter pen go through each advert and pick out the key subject-specific aspects, i.e. examination boards, skills, processes, etc.
- Now make a judgement with regards to your personal strengths/needs and organise the jobs in an order which seems to be the most appropriate for you. Why have you made these decisions?
- Start to think about how you will want to respond to this advert and move on to the next section 'Applying for a job.'

APPLYING FOR A JOB

So, you have found a post advertised which you want to go for, what next? If the advert requests that you write for an application form, enclosing a SAE, then do just that. At this point do not send in a letter of application but simply write (or phone/fax if the number is included in the advert) requesting the form. Upon receipt:

- photocopy the application form;
- read it through carefully;
- write a draft using the photocopy;
- check it through yourself carefully and, if possible, ask somebody to look it over for you. Transfer this to the original in pencil;
- check it through again for accuracy of information and spelling
- fill it in with ink/or type and tidy it up.

You should include a covering letter of application – this is important. You can also enclose your own curriculum vitae (see the section 'Your CV'), unless you are specifically asked not to do so.

Remember – you are competing with many other applicants who are similarly qualified; your letter of application will either ensure that you are short-listed, or not. Spelling, grammar and syntax are important. It is not unknown for letters of application to be thrown away after the first spelling error is noticed. About one thousand prospective Art & Design teachers on PGCE courses are trained every year in the UK and all will have similar qualifications – it is up to you, in your letter of application, to make yours stand out, but do not draw attention to your letter by using a fluorescent magenta envelope and avoid enclosing jack-in-the-box puppets or other such devices. Headteachers and governors are often rather conservative; do not frighten them at this stage. If in doubt, err on the side of formality; use black ink, not green. Nevertheless, you must make yours stand out by emphasising your particular qualities, your interests and what you can offer the school. A professionally presented, interesting letter which outlines concisely your philosophy of teaching, and perhaps indicates how your previous experience has given you insights into learning, may help you to be short-listed; see Hickman (1996) for examples. Headteachers are looking for value for money, so if you can offer expertise in, for example, Information and Communications Technology, or a particular sport, or if you have experience in doing voluntary work, then put it in your letter.

You need to be honest but it is possible to be too honest. You do not have to spell out your weaknesses and natural human failings. You will, however, have to disclose a criminal record if you have one. You could be dismissed after appointment if you do not disclose a criminal record at or before interview. Under the Rehabilitation of Offenders Act 1974 (Exceptions) (Amendment) Order 1986, applicants for teaching positions (and other employment which involves children and young people) are required to disclose all criminal convictions, including bind-over orders and cautions. Circular 9/93 from the DfEE gives information on the circumstances under which a 'police check' can be made; this does not normally include student teachers but does include all employed teachers.

Your letter of application should contain enough information for the selectors to get a good picture of you. It is unlikely that one sheet of double-spaced A4 will be enough, more than three sheets is probably too much. Use your letter to highlight aspects of your skills and experience which you think will benefit the school. If you are able and willing to contribute to the life of the school outside of the Art & Design department, then specify what you can offer.

Check list

- Does the title of the post match that advertised?
- Is the address correct?
- Is the spelling accurate (do not just rely on a spell-checker; you might end up with 'discreet' when you mean 'discrete', or even 'poetry' when you mean 'pottery')?
- Does your letter say what you want it to say?
- When is the deadline?

Only use handwriting if you have a particularly elegant (and legible) style, or if you are requested to, otherwise word-process your letter of application.

Get at least one other literate person to check it for you (it may even be advantageous for a non-Art & Design person to do this). Do not get over-anxious about it and agonise over every word; try to be fluent and lively. You need to present yourself in a clear, polite and professional way.

YOUR CV

CV stands for curriculum vitae – the course of your life. It should state facts about your education (where and when you went to school, college, university), the qualifications obtained (what, where and when) plus your work experience, especially anything relevant to teaching, or to art, craft and design. You could normally start with personal details, such as date of birth and your contact address.

Do not hand-write your CV but use word processing. If your CV is on a disk then it is easy to amend and add to as necessary. Newly qualified teachers are unlikely to have a CV which is more than two sides of A4. It must be lucid, concise and easy to read; stick to basic facts and do not clutter it with irrelevant information. Put in things like vacation work if you think it is relevant, especially if you have done little else. Do not leave any gaps. Arrange your CV in sections, each in chronological order. Suggested section headings are:

Personal
- your name
- your address, telephone number
- term address, telephone number
- date of birth

- DfEE reference number (this will be given to you towards the end of your course)

You might also wish to put your citizenship or marital status.

Education
- schools attended, dates
- further and higher education institutions attended with subjects and dates

Qualifications obtained
- degree(s), subject, class, date(s)
- certificates, subject, level, date(s)
- A Levels/GNVQs, subjects, grades, date(s)
- GCSEs/O Levels, subjects, grades, date(s)
- other qualifications such as safety in workshops certificate

Teaching experience
- teaching-practice experience
- names of schools, subjects taught
- age range and levels taught, date(s)

Other experience
- work experience, especially anything concerned with art, craft and design, museums, galleries, working with young people; give all dates

Other information
- exhibitions
- give details of skills (e.g. musical ability, ICT capability)
- give brief details of interests
- give details of any positions of responsibility you have held

Referees
- These should normally be your course tutor or college principal and your school mentor or the head of department or headteacher from your school experience placements. Some schools have a policy of the headteacher signing all references, in which case you should put the Head's name as a referee, although it will be written largely by whoever has seen you teach most.
- You must ask all referees for their permission before citing their names.

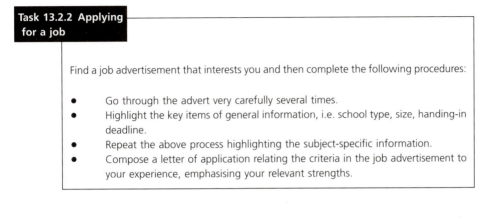

Task 13.2.2 Applying for a job

Find a job advertisement that interests you and then complete the following procedures:

- Go through the advert very carefully several times.
- Highlight the key items of general information, i.e. school type, size, handing-in deadline.
- Repeat the above process highlighting the subject-specific information.
- Compose a letter of application relating the criteria in the job advertisement to your experience, emphasising your relevant strengths.

YOUR PORTFOLIO

Together with your letter of application, your portfolio reveals your personal skills, interests, experiences and level of commitment. It is likely to form the basis for questions at interview so it is important that you select work which you can talk about articulately and enthusiastically. For general guidance you should devote no more than one third of the total examples of work to your own, with the remaining two thirds being pupils' work. It should show a high level of consideration and presentation; all work should be clearly (but not elaborately) mounted, and labelled (see the section on general layout guidelines). Avoid taking cumbersome or delicate work to interview. Large work should be photographed and mounted. It is usually acceptable to use good-quality coloured photocopies. (Slides should supplement originals; although it is unlikely that your interviewers will have facilities for, or even want, a slide show. Make sure you enquire whether relevant facilities are available.) It is essential that you include examples of your lesson planning. You should indicate what you were hoping to achieve in your lessons – your learning objectives. Remember that for the purposes of your interview, the pupils' work is evidence of their learning, not examples of art objects. Be selective, and take care to make sure each sheet has a focus/purpose.

It is important to plan the portfolio carefully; think of it as a whole as well as sheet by sheet. Consider all of the aspects of your ITE and practical teaching and how you can talk an audience through the portfolio. Make sure you are able to include for discussion a broad range of aspects such as differentiation, galleries and museums, assessment, critical and contextual studies, workshops, the NC, etc., rather than just 'Year 8 did this, year 9 that . . .'. Having a sense of audience is crucial.

> **Task 13.2.3 Portfolio: a sense of audience**
>
> Discuss in your tutor group the following:
> 'What is the portfolio's purpose?', 'Who will be looking at the portfolio?', 'What are they looking for?' 'What do I want to say?'
> Ask yourself these questions not just at the planning stage but throughout your portfolio's development and growth.

YOUR OWN WORK

Select a representative range of your best work – remember that you will probably be interviewed by people without an art background, so choose work which shows simply the skills, knowledge, concepts and techniques which you have acquired, avoiding work which expects its audience to have a thorough understanding of aesthetic theory. Art historians and others whose background is not studio-based should include visual displays of their areas of knowledge and interests, together with some indication of their use in an educational context.

PUPILS' WORK

Seek permission from pupils and teachers before borrowing pupils' work, especially examination course work. Make sure you return originals promptly if requested. Mount it and label it. It would enhance the work's impact and viewers' understanding if you include succinct notes on the overall scheme of work and its aims and on the learning objectives for lessons. You might include key terms and words, plus an indication of stimulus materials used. Include work which shows a range of responses and outcomes, and be prepared to talk about their validity, achievement and your method of evaluation. Do not just use end outcomes; it may be important to use examples from the preliminary studies and research stages. Quite often these aspects reflect more of the spontaneity and life of the project and indicate the learning taking place than the final pieces. Select work from a wide age and ability range, choosing work about which you can be positive and enthusiastic – it should be lively and engaging rather than slick, but it will not do any harm to include one or two examples from your most technically gifted pupils.

Your portfolio should ooze professionalism. If your old folder is covered in art-school graffiti or just looks old, invest in a new one. In the unlikely event of you not being invited to show your portfolio when on interview, politely suggest that you do so, but remember that most interviews do not last much more than 40 minutes, and could be less.

The next section offers suggestions about how to start forming your portfolio. It is important to point out from the beginning that it must reflect a personal approach; your personality and experience, what ever that may be, must come through. It would be pointless and counter-productive for us to suggest a formula for a successful portfolio, but we offer tried and tested ideas for you to develop.

GENERAL LAYOUT GUIDELINES

There are no strict rules for layout; in fact it depends entirely upon what you are presenting. However it helps if you can theme the layout to the content of the sheets. You may consider allowing the theme to suggest a complementary/contrasting background colour and materials. Using such ideas as fabric, textured paper or card, newspaper, drawings or prints for the background all of which can produce interesting effects. It is important for the sheets to have a focal point ensuring the viewers' eyes are led around the sheet (this may even be continued from adjoining sheets or throughout the whole portfolio). The focal point could be a title or header, while considering the usage of borders, headings, symbols, positioning, etc. as links, helping the story to unfold and flow. With regard to typeface we suggest something which is supportive and sympathetic to the content. Avoid hand-produced script unless you are very skilled in the craft; it looks unprofessional and there are many fonts to explore in your ICT software packages, too many in fact. Avoid mixing font types and constantly changing the colours; again it may seem wacky, but all too often looks tacky. Keep it uniform, consistent, simple and straightforward. (Consider Plates 5, 6 and 7 in relation to the above.)

It is advisable to plan out all ideas before sticking down the final versions. Work in small thumbnail sketches to devise a grid. Experiment with the position and size of the title and text, while considering how to present the images. You can then enlarge your preferred designs to one quarter of the finished size, i.e. A4, and position onto an A3 sheet. Use the space around the edges to note intended colours, effects, typefaces, etc. In the spirit of Matisse and using the notions of collage it is at this point that you may want to cut out elements and move them around, repositioning, changing, trying out different possibilities, positions and sizes for greater impact and emphasis. Remember constantly to stand back to get some distance between you and the sheets. Admire and ponder over time, rearranging their elements until the sheet is both pleasing and presents the image and text in a coherent way. We often advise students not to stick down the work until they have completed several sheets. It tends to take this long to actually get a clear feel for what you want or intend to do.

IDEAS FOR LAYING OUT TEXT

As mentioned above start by devising a grid. Simply divide the sheet into columns which will give you a structure to work from. This does not mean however that you cannot break out from this; it will provide a format that you can consciously choose to abandon for aesthetic and communicational reasons.

Use the grid to help you experiment with positioning; it should offer many combinations for you. Avoid filling the whole page. Try to keep it reasonably simple and uncluttered, and this should also prevent it being confusing and overwhelming. Use space to create impact, clarity and dynamism. It is often difficult to judge accurately by eye alone, so use set squares and rulers.

The size of typeface and images should be considered carefully; always keep in mind your theme and message. Keep it simple, clear and legible.

Begin by devising a grid. This is simply dividing the page into columns; it will give you a structure in which to work, and can be broken, but needs to be done with consideration.

Figure 13.2.1 Idea 1

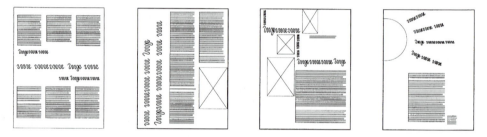

Positioning of type on the grid offers many variations. Avoid filling the whole page; it can be confusing and overwhelming. Use space to create impact and clarity.

Figure 13.2.2 Idea 2

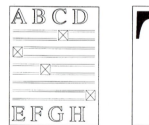  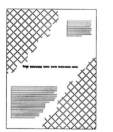 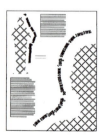

Size of type and pictures should be considered carefully, but keep in mind your message. It must be clear, simple and legible.

Figure 13.2.3 Idea 3

IDEAS FOR LAYING OUT IMAGES

If all the images are rectangular but varying in size, it is a good idea to create an outside border as a frame to hold the work in and together. Squared-up pictures can look a little predictable, so use the grid carefully to consider pattern and balance. An alternative to the uniformity and monotony of boxes and rectangles is to break them up by introducing contrasting soft shapes into the sheet.

Squared up pictures can look boring, but when the grid is used carefully, and a pattern or some balance is created, it can look very effective.

Figure 13.2.4 Idea 4

The shapes of images need not be square; they can be round, triangular and diamond.

Do not forget what we stated earlier about collage; cut-outs can be very interesting, either neatly cut or torn out for a more informal effect. Play around with positioning, or alternatively they may cover the whole background.

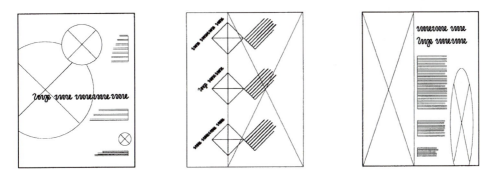

Shape: images need not be square; they can be round, triangular, diamond, in fact any shape at all.

Figure 13.2.5 Idea 5

Basically treat your sheets as you would a composition for a painting, juggling and manipulating shapes and colours until they come together comfortably. Just like painting the background plays a integral part in the overall composition, draw on it, paint on it, apply collage and write over it, until the backdrop and content of the sheet work together.

Cut outs can give interest, and be effectively used with squared up pictures, or torn out for a more informal approach. Alternatively they may cover the whole page.

Figure 13.2.6 Idea 6

Task 13.2.4 Portfolio: design and making

Design and plan a portfolio sheet with the rationale that it should be a visual exemplar of pupils' learning in Art & Design. Remember this is not to be just a record of pupils' outcomes. Try to ensure:

- clarity of objectives;
- a clear sense of audience;
- that a range of ideas, materials and approaches is explored;
- that it is National Curriculum/examination board specific;
- that you show evidence of how the Standards for NQTs have been considered;
- that attention has been paid to a professional standard of presentation and layout.

THE INTERVIEW

It is not unknown, in the dim and very distant past, for a job to be offered to someone who turns up late, in jeans and with a decidedly casual manner. Notwithstanding this, we can confidently advise you not to turn up late, in jeans and with a decidedly casual manner.

You must prepare yourself by anticipating questions. This does not mean that you should regurgitate pre-formed answers, but you should show that you know something about the school and you are familiar with the structure and aims of your ITE course. It is likely that you will be asked general questions such as 'What are your interests? or (sometimes more difficult) 'Tell us about yourself'. You could be asked to describe your most successful lesson or your ideal school.

As well as wanting to form an impression of you as a person and a future teacher the interviewers will want to be confident in your knowledge of the National Curriculum. Be prepared to answer questions on aspects of the National Curriculum for Art such as 'How do you integrate making, knowledge, investigating and understanding?' They will be looking for you to provide a view, have some knowledge of current thinking and, importantly, describe practical solutions with which you have had success. Remember to be constructively critical of yourself and your experience; be perceptive, balanced and produce evidence. Avoid relying on excessive praise or blanket criticism.

Task 13.2.5 Interview questions: subject-specific

Try answering the following typical subject-specific questions which might be asked of you at interview:

- Who is your favourite artist/craftsperson/designer? (Why?)
- Which examination syllabuses are you familiar with? What do you think are their strengths and weaknesses?
- Why do you think Art & Design is an important part of the school curriculum? How would you defend its position as an important part of a pupil's development?
- Do you think Art & Design should be taught in mixed-ability groups?
- How would you explain to a year 9 pupil the value of Art & Design?
- How could you motivate pupils who feel they are 'no good' at art?
- What are pupils' most common misconceptions and mistakes in Art & Design?

Remember that subject teaching is only one aspect of being a school teacher. You will be expected to contribute to the general life of the school. You should therefore expect questions about pastoral issues and professional/career development.

Task 13.2.6 Interview questions: general

Try formulating responses to the following questions:

- What do you feel to be the most important aspect of being a form tutor?
- What experience have you had of personal, social and health education?
- How would you seek to promote the spiritual, moral, social and cultural development of pupils?

It is expected that teachers show a willingness to develop professionally and so you could be asked questions such as:

- What do you see yourself doing in five years' time?
- How would you like your career to develop? (Your CEP will help you focus on these kind of questions; see the CEP section.)
- What INSET courses would you want to go on? Why?

You should note that, in general, there are no right or wrong answers; the way you reply is often just as important as what you say. Try to be confident (even though you may be shaking inside). Speak clearly and naturally. You will be asked if you have any questions. Do not ask questions just for the sake of it. You could, however, have a genuine question such as 'What is the school's policy on supporting in-service training?' or 'Are there any opportunities for cross-curricular work between Art & Design and Technology?' It is not a good idea at this point to ask about your pay (we advise you to investigate this before the interview at an appropriate stage). You will probably be asked 'If we offered you the job, would you accept it?' Unless you have serious reservations, you should reply with an emphatic 'Yes!'. If you have slight reservations, you could make a more cautious response, but this could go against you. If the school genuinely feels that you are the right person they may allow you time to consider, but often you are expected to decide on the day. A verbal contract is binding. You must not take another job if you have already agreed to accept one.

The myth of the romantic, rebellious, eccentric, somewhat disreputable artist continues to be perpetuated in films and novels. As indicated above, some headteachers and governors even expect their Art & Design teachers to be a little 'different'. If in doubt, as with letter writing, err on the side of the formal and conservative. Adapt yourself to the situation.

It is possible that you will be asked to teach a lesson. This can be quite unnerving, but bear in mind that not everything hangs on your performance. It is likely that your interviewers will be looking more at the way in which you interact with the pupils rather than the innovative and imaginative nature of your lesson. As always, the key is preparation. Choose a subject with which you are familiar; it is surprising how many people suddenly choose to take on something about which they know next to nothing, just for a change perhaps, or to add extra nervous energy. Find out as much as you can beforehand about the class you will teach, what they already know, what they have done in Art & Design recently and what their ability levels are. Find out about the space, materials, equipment and resources. If you are not given this information, at least you can assume that your erstwhile competitors do not have it either. You will need to remember all you have learned during your ITE with regard to lesson planning, principally, to be very clear about what you expect pupils to learn, and what you and they are going to do to ensure they learn it.

FURTHER READING

Capel, S., Leask, M. and Turner, T. (1999) *Learning to Teach in the Secondary School – A Companion to School Experience*, 2nd edn, London: Routledge.

Hickman, R. (1996) *Finding Your First Job – A Guide for Prospective Art & Design Teachers*, Corsham: NSEAD.

National Union of Teachers (annually) *Your First Teaching Post*, London: NUT.

Times Educational Supplement (annually, around the middle of January), 'First Appointments Supplement', TES.

Teaching as a Career Unit (TASC) (annually) *Getting a Teaching Job*, London: University of Greenwich.

13.3 CAREER ENTRY PROFILE AND INDUCTION FOR NQT STATUS

INTRODUCTION

From June 1998 all providers of initial teacher education have been required to provide newly qualified teachers with a Teacher Training Agency Career Entry Profile (CEP). The CEP is a summary of information collected during ITE and documents:

> . . . the new teachers' strengths and priorities for their further professional development, in relation to the standards for the award of QTS, from initial teacher training to their first teaching post. The TTA CEP is not intended to be a replacement or a substitute either for references from providers to prospective employers, or for schools' systems for monitoring and supporting NQTs.
>
> (TTA 1997: 2)

OBJECTIVES

By the end of this unit you should:

- be aware of the relationship between your CEP and FPD;
- understand the nature of the induction process for NQTs;
- be able to plan for your own appraisal as a NQT.

CAREER ENTRY PROFILE (CEP)

It is important to think of the CEP as a bridging document between your ITE and your first job. It is designed to make the transition from training to teaching smoother and less problematic. The CEP and induction period (see next section) are intrinsically linked and will be used to ensure NQTs get off to the right start in their new schools.

It is also important to remember that first and foremost the CEP is for you! You most take ownership of it and the responsibility for its production. Use it to identify strengths and areas for development and to pinpoint your interests and achievements. The TTA want it as an element of QTS and see it as a baseline assessment at the professional point of entry. Schools, on the other hand, will use it to identify starting points for induction, to target and stage your QTS development.

The profile has four sections:

1   Section A – a summary of the initial training experience, including any distinctive features.
2   Section B – an analysis of your strengths and priorities for further professional development agreed with by the provider and you.
3   Section C – your own targets for your first teaching post.
4   Section D – an action plan, negotiated between you and the school, which aims to set targets and ensure that there is a planned programme of monitoring and support in the induction period.

The first two sections are completed towards the end of ITE. The third section should be completed once training has been concluded, and the final section once you are in post.

It cannot be stressed enough the opportunity this presents for individualising the content and having the chance to 'look in the mirror' at yourself. Section A asks for 'distinctive features of training . . .', things which make your experience different from that of others. This is an opportunity for the Art & Design teachers to shine, celebrating achievements they have in common like the organising of gallery and museum visits, the range of disciplines which they can offer i.e. Design Technology, Craft, ICT and multimedia, the setting up of community projects and exhibitions, collaborating with outside agencies and artists, the running of after school/lunch-time art clubs, all of which will have made your experience different from the next NQT.

The task below will help you to sift through the information and documentation involved in Sections B and C. By the end of the task you should have prioritised your strengths and areas for development.

> **Task 13.3.1 Preparation for the CEP**
>
> In pairs, go to the Standards and select either section B, C or D on which to focus. (B, or C would probably serve as a more Art & Design specific focus.)
>
> Go through all the individual statements and place yourself along a continuum for each element using 1-10 as the scale, 10 being the position you intend to be at the end of the ITE year.
>
> Reflect upon and use this information to help you to identify areas of strength and areas for development. (Remember to use interim, final reports and teaching observation/assessments for additional guidance.)

INDUCTION FOR NEWLY QUALIFIED TEACHERS (NQTs)

By carefully completing Section D in the CEP, you will arrive at your first teaching post in a strong bargaining position to negotiate training needs with your mentor.

Schools will be legally bound to provide structured support and training to consolidate and extend the skills you have learned in ITE. The induction period aims to: ' . . . help newly qualified teachers develop into confident and competent professionals, able to make a full and distinctive contribution to raising standards in their schools' (DfEE 1998: 2).

With your completed CEP in hand and a national framework (see *Professional Standards for the Induction Period* from the DfEE) to guide you, start to plan career goals. All NQTs will be required to complete successfully an induction period to remain eligible for employment in a maintained school or non-maintained special school. There will be no such requirement for those seeking employment in independent schools.

It is envisaged that the following elements should be included in every new teacher's induction year programme:

- Opportunities to work with the best schools in the area – use the Internet and examine OFSTED reports of all your local schools' Art & Design departments. Check national examination results, league tables and general information which is known about Art & Design departments in your area. Attend seminars organised by those schools, or ITE providers who liaise with them, to establish links and communication avenues. There may be opportunities to collaborate in projects, off-site visits or organise joint exhibitions.
- Access to mentoring by, and observation of and by experienced teachers – this may be your head of department or other colleagues in Art & Design. There also may be the opportunity to visit schools that have been contacted as a part of the collaborative exercise above and experience neighbouring schools' approaches.
- Regular observations of teaching and feedback – to support the assessment and development of your practice in relation to the QTS and induction standards, and inform review discussions.
- Regular discussions – with those most closely involved in the induction programme to review professional practice, assess progress towards targets and set new ones, identify further professional development activities, and update the action plan in the CEP.
- Opportunities to attend training courses – organised by the LEA, higher education or others (such as the NSEAD or museums and galleries), seen as relevant to any specific development needs identified in the CEP or in review discussions. These could range from skills/process-based courses to ones exploring the role of critical and contextual studies in Art & Design education.
- Opportunities to visit special schools and work with the school SENCO and see and learn from effective practice in identifying and meeting the needs of pupils with special educational needs in mainstream schools.
- Opportunities to have their work observed and commented on by the headteacher and access to the headteacher to discuss any additional training needs or difficulties they may be experiencing.

- Opportunities to join and contribute to working groups – at your school, at higher education institutions or others, to join any networks for NQT established by groups of schools or by the LEA, and to meet and share experiences with other teachers.

You will be pleased to note that it is not expected that this will be done while teaching a full timetable! It has been proposed that a lighter timetable and workload be given to the NQT in the induction period, possibly 90 per cent of the average hours normally taught by other full-time teachers.

You will be assessed on the satisfactory completion of the induction period. NQTs must clearly demonstrate:

> . . . mastery of the Qualified Teacher Status standards on a consistent and sustained basis, in the very different situation of employment in a school – with the direct and personal responsibility and accountability for pupil performance that brings with it. They must also have shown that they have built on and progressed beyond the QTS standards such as managing pupil behaviour, contributing to pupils' learning and to the planning and achievement of their school's performance targets.
>
> (DfEE 1998: 9)

If you fail to meet these standards you will not be eligible to be employed as a teacher. Therefore it is crucial for you to take responsibility for collecting and documenting evidence of progress to support your regular meetings and reports/records.

Task 13.3.2 Standards for the induction period

Look at your copy of the 'Standards for the Induction Period'. Examine the different sections and while using your CEP identify your areas of strength and weakness.

Start to design a system for monitoring your progress, whether this is in the form of a notebook or an A3/A4 tracking sheet similar to Section D in the CEP: it is up to you. It is important that you devise a system which you can use effectively to record and document evidence on a regular basis which is parallel to the standards, formative and supportive. Avoid something which is time consuming or too complicated.

Along with your mentor you can start to focus on collecting and targeting of evidence on a daily/weekly basis to ensure success in the final assessment.

The successful completion of the induction period should not be seen as the end of your FPD. You will have a continuing need for opportunities to update subject knowledge and classroom skills and to continue to exchange good practice.

Our new induction proposals are intended to give new teachers the best possible start to their professional life but that should only be the beginning of a continuing process of career-long learning and development to enable the teacher to support and extend their pupils' learning and to play their part in raising standards across all our schools (DfEE 1998: 18).

Bear in mind that your professional development is much more than simply fulfilling your statutory obligations as a teacher, and complying with changing government directives. The teaching profession draws its strengths from the diversity of its workforce − individual teachers, whose personal beliefs, philosophies and approaches to teaching and learning in Art & Design are fundamental to ensuring pupils' success.

> . . . observation of several mid-career Art & Design teachers revealed that their teaching was at its best when there were strong connections between what they were teaching in the classroom and their wider belief systems. In those situations their motivation was springing from deep personal conviction. They cared about what they were teaching, and taught it well because they believed in it.
>
> (C. Robinson 1995: 133)

FURTHER READING

Capel, S., Leask, M. and Turner, T. (1999) *Learning to Teach in the Secondary School − A Companion to School Experience*, 2nd edn, London: Routledge.

DfEE (1998) *Consultation Document on the New Induction Arrangements*, London: HMSO.

DfEE (1997) *Teaching: High Status, High Standards*, London: HMSO.

Robinson, C. (1995) 'National Curriculum for Art: Translating it into Practice', in R. Prentice (ed.) *Teaching Art & Design. Addressing Issues and Identifying Directions*, London: Cassell.

TTA (1997) *Career Entry Profile − For newly qualified teachers*, London: University of Greenwich.

14 Critical Pedagogy in Art & Design Education

Nicholas Addison and Lesley Burgess

14.1 VISUAL AND AESTHETIC LITERACY AND THE CRITICAL CURRICULUM

> **OBJECTIVES**
>
> By the end of this unit you should be able to:
>
> - define the possible role of visual and aesthetic literacy;
> - critique exclusive definitions of the aesthetic;
> - consider arguments for an expanded definition of aesthetics in Art & Design education to include 'grounded' and critical domains.

A critical faculty is a terrible thing. When I was eleven there were no bad films, just films I wanted to see, there was no bad food, just Brussels sprouts and cabbage, and there were no bad books – everything I read was great. Then suddenly, I woke up in the morning and all that had changed. How could my sister not hear that David Cassidy was not the same as Black Sabbath? Why on earth would my English teacher think that the *History of Mr Polly* was better than *Ten Little Indians* by Agatha Christie? And from that moment on, enjoyment has been a much more elusive quality.

(Hornby 1992: 29)

VISUAL LITERACY

One of the aims of Art & Design education, reinforced in the National Curriculum Art Order (NC) (DFE 1995) is to enable pupils to become visually literate. The right to verbal literacy is never contested; it is seen as a prerequisite for participation in democratic society. Put simply, literacy is the ability to read and write. To be literate enables a person to participate fully in verbal culture. In usurping the terminology of language to proclaim a central aim of Art & Design education you might feel that educators are admitting defeat and bowing down to the primacy of language. But, as literacy is perceived as fundamental, it is worth considering what visual literacy, as a metaphor, implies. Literacy encompasses both reception and production, reading and writing. What is the visual equivalent of these activities? Learning to look, see and make? The concept of visual literacy clearly questions the privileged position of making in the Art & Design curriculum.

Boughton (1986) and Raney (1997) have redefined the concept of visual literacy by tracing its development from reaction to reciprocity. Initially, art educators felt threatened by the success of communication studies, its embrace of mass media and popular culture and its quest to demystify the sacrosanct practices of high art through new analytical methods. Art & Design educators felt the need to respond to this encroachment on their territory and quickly defended their subject by reinforcing its critical and historical elements and by emphasising its cognitive potential.

In recent years the concept of visual literacy has come under attack, particularly in the USA, because in adapting a linguistic model it has been seen to 'narrow visual meaning' (Freedman 1997: 7). However, we believe that this singular concern with the 'visual' is in itself a narrowing of the potential of Art & Design. If the term is extended to include the 'aesthetic', acknowledging the multi-sensory, multimedia nature of Art & Design, then 'visual and aesthetic literacy' more tellingly describes what you should be teaching.

AESTHETICS

The debate about the role and value of aesthetics in secondary Art & Design education has always been a contentious issue, one that, by and large, teachers have avoided or taken for granted. Rather than avoid the concept and its implications for teaching and learning we believe it is worth revisiting in order that it can be reconceptualised and used to inform work in the classroom (see Unit 3.4).

Task 14.1.1 The role of aesthetics

What role did aesthetics play in your own art education?
Was the term referred to during your degree course, if so how and in what context?
Record your answers and compare them with other members of your tutor group.
Try and identify any common experiences.

Steiner (1989) insists that rather than concentrate on critical theory you would do better to fall silent before the awesome presence of an art work. Abbs (1987) supports this belief arguing that the aesthetic field is antithetical to critical practice:

> The elevation of criticism brings about a further distancing from the aesthetic realm, for the question becomes not one of prolonged *aesthetic* engagement but one of *conceptual* meaning and the task not one of creation, or performance or appreciation, but one of ideological placement, or, more frequently, displacement.
>
> (pp. 30–31)

In Unit 3.4 we reiterated the definition of the aesthetic field as one which: 'refers to a particular form of sensuous understanding, a mode of apprehending through the senses the patterned import of human experience' (Abbs 1989: xi). Elsewhere Abbs suggests that for someone to be aesthetically engaged they must be involved in 'making, presenting, responding and evaluating' (Abbs 1987: 55). He continues: 'the parts are not self-contained but gain their meaning through connection with all the other parts' (*ibid.:* 56). The last of these processes, evaluation, is: '. . . an attempt to organize the complex elements of our aesthetic response – to state intellectually our relationship to a work of art, to formulate the aesthetic response (as near as we can get to it) conceptually . . . It is a critical act' (*ibid.:* 61). In this model Abbs reveals that critical engagement is a prerequisite in developing a deep understanding. However, his earlier statement warns that a disengaged approach to criticism, one that fails to return to the act of making, is liable to serve the interests of political education not aesthetic understanding.

If you believe that human perceptions and responses are shaped by social and cultural contexts then Abb's belief in a separate, apolitical realm is called into question. The impact of critical theory on Art & Design education has resulted in a sharper awareness of the historically determined and socially constructed nature of individual achievement and personal expression. In addition, its sphere of influence is not restricted to practices deemed historically significant, with the artefacts of high aesthetic culture, but with the expanded and expanding field of visual and material culture.

GROUNDED AESTHETICS

Willis (1990b) coined the term 'grounded aesthetics' in an attempt to retrieve the term 'aesthetic' from its traditional Western fine art preoccupations with 'abstract or sublimated qualities of beauty'. He takes pleasure in pointing out that 'other societies, earlier civilizations, "less developed" cultures have also revelled in a rumbustious, robust and profane, vested and testing of aesthetics in the everyday. So does ours if we would only look' (p. 14). He recalls how a particular pop song can suddenly evoke and come to represent an intense personal experience and how a dramatic episode on a television 'soap' can parallel and illuminate personal problems and dilemmas. Such moments can make personal experiences more understandable, more controllable.

Willis attempts to lift aesthetics off its pedestal. He insists that young people need to engage with aesthetics not only because of its intrinsic values but because of its capacity to produce social meanings. Grounded aesthetics ensures that learning takes place within an expanded field of visual and material culture. In the past, the hierarchical and canonic view of 'Art' has marginalised if not dismissed the home culture of most pupils. It is important that you question this separation by investigating issues of taste.

TASTE

What is taste? The dictionary definition (Oxford Shorter English Dictionary 1993) links it closely with judgement and aesthetics: 'mental perceptions of quality, judgement, discriminative faculty . . . aesthetic discernment'. Pateman (1991) offers three other possibilities:

1   Taste is a faculty of mind, innate and equal in virtually everyone, which is capable of education, and by means of which we distinguish (or are struck by) the difference between the beautiful and the ugly, the tasteful and the tasteless.

2   Taste is the faculty of the mind, innate and very much unequal, which is capable of education, etc.

3   Taste is a set of socially inculcated dispositions or culturally transmitted abilities to respond to and distinguish objects, events, etc., as beautiful or ugly, tasteful or tasteless. What counts as beautiful is itself socially and culturally constructed as part of the construction of taste, and may vary without any evident restriction. Taste is always differentially distributed: some people get it, some don't. It thus can act as a social discriminator or distinguisher, and is used by individuals to define and distinguish themselves socially.

(p. 175)

Task 14.1.2 Investigating taste

Which of these definitions do you find most sympathetic?

Devise a lesson which encourages pupils to look critically at a range of different objects and artefacts in order to identify their tastes, likes and dislikes.

How might you encourage pupils to consider the origins, personal and cultural, of their tastes and whether these preferences are 'innate' or the result of learning?

How is it possible to enable pupils to understand and value the tastes of others? Devise questions to support these aims.

CRITICAL AESTHETICS

Critical aesthetics refuses to accept that aesthetic experience is beyond articulation. It calls for reflection and deliberation proposing that you give reasons for your aesthetic judgements. Crowther (1996) suggests that the search for objectivity in 'critical aesthetics' can facilitate the deepening of aesthetic experiences. Lack of familiarity with the artistic conventions of a culture may make it difficult to fully appreciate and understand its material and visual artefacts; judgement may be rooted in personal taste based on different and conflicting criteria. This is not to suggest that the codes and conventions of producers are the only criteria for aesthetic judgement, rather that judgements of taste are contingent upon personal experience and exposure (acculturation), knowledge of familiar discourses. A richer understanding is afforded by knowledge of motivations, intentions and art as social production.

The claim that taste is universal is no longer tenable, not only in relation to determinants such as social class, race and ethnicity but also in relation to gender and sexuality. Abbs (1987: 31) defines feminism as a method of literary and artistic criticism, a movement away from the primary aesthetic engagement and encounter. But it is important to recognise that a cultural object addresses a gender-specific audience and at the same time contributes to the 'gendering' of that audience (McRobbie 1994).

It is all too easy to claim that the subjective nature of our responses to art makes it difficult, if not impossible, to teach aesthetic appreciation. Davey (1994) insists that it is important for aesthetics to have a place within art education, but that first teachers have to address some of the 'strangulating preconceptions about philosophical aesthetics' and understand:

> when art works are either historically or culturally juxtaposed . . . questions of aesthetics no longer involve the study of pure appearances but an understanding of how art facilitates the coming into appearance of meaning, an epiphany, which simultaneously reveals and binds us to the cultural horizons we are within.
>
> (p. 74)

Davey's article provides a compelling argument for the inclusion of aesthetics in the Art & Design curriculum, one that replaces the Kantian notion of 'pure perception' with a phenomenological account of 'significant perception'. He suggests that aesthetic experiences are highly relational or associative and only evoke a response when they are personally significant.

As you can see aesthetics is a difficult concept to define. A fixed, traditional definition confines it to the history books. Rather than resort to this, you need to consider its place in relation to contemporary practice by reflecting on possibilities and contradictions.

FURTHER READING

Pollock, G. (1988) *Vision and Difference*, London: Routledge.

Battersley, C. (1991) *Thinking Art: Beyond Traditional Aesthetics*, London: ICA.

Kaelin, E. F. (1989) *An Aesthetics for Art Educators*, New York: Teachers College Press.

Parsons, M. J. and Blocker, G. (1993) *Aesthetics & Education*, Champaign, IL: University of Illinois Press.

14.2 CONTEMPORARY PRACTICE: DISCONTINUITIES AND ALLEGIANCES

OBJECTIVES

By the end of this unit you should be able to:

● enter into and inform discussion about contemporary practice.

There has been much debate about whether or not a fundamental philosophical shift took place in the second half of the twentieth century, a debate that, in the arts, revolved around the terms, 'modern' and 'postmodern' (Foster 1983).

Task 14.2.1 Defining terms

In tutor groups discuss your understanding of the terms: modern and postmodern. How does your practice as maker and educator correspond to your definitions?

At the turn of the century many Art & Design educators are attempting to reconceptualise the curriculum by rejecting a modernist approach in favour of postmodern strategies (Efland 1996 *et al.*; Clark 1997; Hughes 1998a; Swift and Steers 1999). This binary opposition has led to oversimplifications and misrepresentations indicating fixed historical periods rather than a fluid dialogue between different approaches: for example the aesthetic with the political, Abstraction and Dada, the Lisson sculptors and the Guerrilla Girls. Table 14.2.1 indicates some of the characteristics of these interrelated tendencies in twentieth century art. The avant-garde ('historic' and 'neo') or 'the counter project' (Foster 1996) designates those artists and groups who have continuously questioned the universalising, hegemonic agendas of an official modernism. The term postmodernism, with its implication of an end to modernism, is not used so as to avoid associations with periodisation: this table recognises developments as well as discontinuities, allegiances as well as differences. It also suggests that the characteristics of the avant-garde may now be seen as the characteristics of

Table 14.2.1 Allegiance and difference

Counter project	Official modern
a critical inquiry: redefinition	creative action
b aesthetic play: testing categories	aesthetic refinement: ensuring excellence
c active viewer: construction of meaning; visual literacy	passive viewer: immanent and transcendental meaning; aesthetic literacy
d interest	quality
e of its time	universal
f temporal	spatial
g performative: theatre, spectacle	present: 'sublime instantaneousness'
h doubt	conviction
i conditional, strategic	essential
j reproduction, repetition	auratic, original, unique
k inclusive	exclusive
l textual or site-specific	autonomous
m allegorical, contingent, fallen	symbolic, holistic, utopian
n ephemeral, fugitive	eternal, immutable

an 'official' contemporary practice, that is, practice designated significant by the institutions of art.

Task 14.2.2 Allegiances in school practice

Discuss in your tutor group the relationship between these terms and practice in Art & Design education.

To what extent is school practice:
- firmly located in one set of characteristics;
- a balance between the two;
- not related to either?

To what extent do the terms for the counter project correspond to your understanding of postmodernism?

During the twentieth century the interrelationship between groups of artists, craftspeople and designers was occasionally confrontational. Often, reaction was couched in an ongoing critical project that prevented art from standing still or stagnating. Totalitarian and nationalist attempts to halt change were undermined by the international profile of the project, although, until recently, histories tended to exclude contributions outside the West. Additionally, art's dialectical relationship to developing technologies and emerging philosophies conditioned its constant shifts.

This is not to suggest some notion of art as evolving progressively through critical refinements, only that it constantly changed, and is still changing, through a series of critical relationships, revisions and revolutions as it simultaneously responds to and produces the needs of any one time or context. There are those who imply that contemporary art and craft, if not design, are an irrelevance (Sewell 1995) or at best an elitist enrichment within the structures of late capitalism dominating the turn of the century. We hope that this book indicates something of the multi-functional potential of the art, craft, design continuum and its role in education. We therefore propose that it is these critics' exclusive notion of what constitutes art that is the irrelevance, not the practice of artists. The extent to which practice in schools engages with this expanding field will determine the significance and relevance of Art & Design as a curriculum subject in the twenty-first century.

FURTHER READING

Foster, H. (1996) *The Return of the Real*, Cambridge, MA: MIT Press.

14.3 CRITICAL PEDAGOGY

OBJECTIVES

By the end of this unit you should be able to:

- consider models of critical pedagogy;
- explore strategies for critical pedagogy.

Grossberg (1994: 16–21) identifies four models of progressive pedagogy which he describes under the headings: hierarchical, dialogic, praxical, articulation and risk. The 'hierarchical' addresses problematic bodies of knowledge. The teacher retains a traditional role in a position of authority determining the frames of reference by which an issue can be discussed. Grossberg suggests that such an approach can assist 'emancipationary struggles' but that it is the teacher who defines what is right. The 'dialogic' shifts the balance of power, inviting pupils and students a participatory voice. Because this approach acknowledges differing positions it questions the authority of teachers and thus asks them to reflect on their own position. However, unless pupils and students are empowered with critical and discursive tools it is ineffectual. The 'praxical', while demanding dialogue, is not content to leave it there. It requires that the educational community effects institutional change by challenging the structures of power. It can only do so by forming alliances with the wider emancipatory community and joining in their political struggle. The difficulty with this model is that once the tools which secure emancipation have been established they can become the hierarchical tools of a new orthodoxy. The 'pedagogy of articulation and risk' in Art

& Design marks a subtle shift, bringing together the political idealism of the dialogic and the praxical while acknowledging that the expanding field of cultural practice requires people to make connections between what was once considered separate and ineffable and what is now theorised as social and personal. Grossberg defines this synthesis as an:

> affective pedagogy a pedagogy of possibility (but every possibility has to risk failure) and of agency. It refuses to assume that even theory or politics, theoretical or political correctness, can be known in advance. It is a pedagogy which aims not to predefine its outcome (even in terms of some imagined value of emancipation or democracy) but to empower its students to begin to reconstruct their world in new ways and to rearticulate their future in unimagined and perhaps even unimaginable ways.
>
> (Grossberg 1994: 18)

The implication here, of a wholesale abandonment of authorial control by the teacher has led critics to question the validity and practicability of this form of critical pedagogy. Ellsworth (1989) suggests it is all well and good in theory but it 'has developed along a highly, abstract and utopian line which does not necessarily sustain the daily workings of the education its supporters advocate' (p. 297). Todd refutes this: 'One gets the sense [from Ellsworth] that critical pedagogical discourse has remained virtually unchanged and the assumptions never questioned thereby encouraging unself-reflexive teaching practices' (Todd 1997: 73).

As Hall suggests, 'a theory is only a detour on the way to something more important' (in Giroux and McLaren 1994: 17). This notion of an 'affective pedagogy' corresponds to hooks's 'engaged pedagogy' (1994) which is one of continuous flexibility and flux. She insists that 'engaged pedagogy' does not offer a blueprint rather it 'recognises each classroom as different, that strategies must constantly be changed, invented, reconceptualised to address each new teaching experience' (hooks 1994a: 10).

How, as Art & Design teachers, can you make sense of this rhetoric so that it has some impact in the classroom? Throughout this book we have recommended methods and strategies to enable experiential learning and interactivity (Unit 3.3), discursive and critical inquiry (Chapter 10), the use of inclusive resources (Chapter 5) and a culture of reciprocity (Chapters 3, 5, 12). You may have observed and used one or more of these practices on your course, but it is unlikely that you have consistently used them all. If you find ways to connect them so that they converge, albeit it at unexpected nodal points, this collective and collaborative pedagogy can radically alter the culture of the classroom.

STRATEGIES FOR A CRITICAL PEDAGOGY

Interventions

Pollock (1988) asserts:

> The structural sexism of most academic disciplines contributes actively to
> the production and perpetuation of a gender hierarchy. What we learn about
> the world and its peoples is ideologically patterned in conformity with the
> social order within which it is produced.
>
> (p. 1)

In Chapter 5 Burgess identifies ways in which the 'repressed question of sexual
difference' can be raised so that 'the differentiation which is so manifest and so
symbolic [is seen to be] produced' (Pollock in Addison and Allen 1997). This awareness impacts on other areas of difference, whether those of class, race, disability, age,
sexuality, by using strategies of visibility and intervention. In addition to external
interventions which may be institutional (galleries and museums) or human (artists,
craftspeople and designers, local communities) or publications (teaching-packs, slides,
CD-ROMS), you must remember that the contribution of pupils, particularly in
bringing in their home cultures, can act as an intervention into the official curriculum, what Shohat and Stam (1998) refer to as 'affective investments' (p. 14).

The work of Adrian Piper can be cited as an example of multiple intervention. Her
video art, a medium that is in itself a technical intervention within the traditional Art
& Design curriculum, confronts viewers with issues that raise repressed questions of
identity: racial, sexual, class. We do not suggest that these issues are unproblematic. On
the contrary, they may make the classroom a place of challenging interaction. To
manage this it is essential that teachers construct a safe environment within which to
orchestrate the classroom dynamic using systems such as debating structures, small-
group and whole-class discussion and presentations (Unit 3.6). We do not advise you
to attempt this type of interventionary strategy until you know pupils and their
contexts well and have a confident knowledge of the work and issues under discussion. To begin, you could try these strategies with classroom teachers and experienced
artists/educators. They are the sorts of strategy that could usefully be deployed in a
curriculum development project towards the end of your course.

Mediations

The preceding section makes claims for the position of contemporary practice in Art
& Design education. This emphasis on looking across practice at any one given time
encourages an interdisciplinary and dialogic approach. However, it is also important
to consider histories, the means by which the past is preserved, presented and perpetuated. Material and visual culture is increasingly used as a primary form of evidence in the construction of lineages, histories, trajectories and notions of essential
identities. Paley (1995) recommends the use of 'mediations' recognising that the past

can be a powerful base for developing thinking about the contemporary, but at the same time, insisting that you need to:

> Simultaneously honour and reformulate experiences associated with the historical without being trapped in either it or its objectification . . . don't position history as an absolute to be memorised and consumed in the manner of formal educational exercises, but as a point of 'mediation' that requires continuous sustained rereading, reworking and reconstruction.
>
> (p. 178)

Chapter 12 asks you to question essentialist constructions used to preserve the power of vested interests which have become naturalised as 'truths'. Gretton (forthcoming), amongst others, is wary of this tendency recognising that it is 'certainly both a material power and an institutional one, damping change, reproducing traditions'. Giroux (1992) reminds you:

> The radical educator deals with tradition like anything else. It must be engaged and not simply received. Traditions are important. They contain great insights, both for understanding what we want to be and what we don't want to be. The question is: in what context do we want to judge tradition? Around what sense of purpose? We need a referent to do that. If we don't have a referent then we have no context to make sense of tradition.
>
> (pp. 17–18)

The difficulty in Art & Design is to manage the wealth of traditions afforded by museum culture. Material and visual culture not only 'represents' the past; it is a part of the visual field of the present, as much available for configuring the future as it is for reconfiguring the past. We have asked you to be aware that appropriation can all too easily decontextualise difference, past and present, because it is one of the primary means for reproducing stereotypes (see Chapters 5 and 11). As a strategy, we suggest that in extending your historical knowledge you focus on points of cultural interaction. This way, you discover that plurality is not entirely new: cultures are rarely discrete and autonomous, rather they are fluid and dynamic, mediated through continuous interaction, whether confrontational or collaborative.

Transformations

The turn of the century marks a symbolic point of transition, a time for you to choose between the centring and essentialist strategies of traditional pedagogies or the uncertainties of critical transformations, what Giroux (1992) describes as 'border pedagogy':

> border pedagogy necessitates combining the modernist emphasis on the capacity of individuals to use critical reason to address the issue of public life, with the postmodern concern with how we might experience agency in a

world constituted in differences unsupported by transcendental phenomena or metaphysical guarantees.

(p. 29)

Border pedagogy decentres as it remaps. The terrain of learning is inextricably linked to shifting parameters of place, power, identity and history. This terrain can only be traversed by crossing borders, can only be understood by questioning the forces by which its borders have been generated, can only be transformed by creating new borderlands. To be effective pupil and teacher must become 'border crossers'; they must link arms together and skip along the 'yellow brick road'. However, this is not Hollywood. Do not always anticipate being in step, nor singing in unison or conventional harmony: the travellers may each be singing their own tune, common goals different needs. Like Dorothy in the *Wizard of Oz* (1939) who only comes to self-understanding by helping others, engaging the pupil in border pedagogy requires that you reflect on your own position and practices, and reconsider them in relation to the narratives of others (including pupils). Border pedagogy can thus be considered a process, not of transmission, but of imaginative transformation: ultimately the borders are not out there but they are in you. As Giroux (1992) asserts: 'What border pedagogy makes undeniable is the relational, constructed situated nature of one's own politics and personal investments' (p. 35).

FURTHER READING

Giroux, H. and McLaren, P. (eds) (1994) *Between Borders*, London: Routledge.

hooks, b. (1994a) *Teaching to Transgress*, London and New York: Routledge.

Paley, N. (1995) *Finding Art's Place: Experiments between Contemporary Art and Education*, New York and London: Routledge.

Bibliography

Abbs, P. (ed.) (1987) *Living Powers: the Arts in Education*, London: Falmer Press.
—— (1989) *A is for Aesthetic*, London: Falmer Press.
Adams, E. (1989) 'Learning to see', *Journal of Art and Design Education* 8(2): 183–196.
—— (1997) 'Public art and art & design education in schools', *Journal of Art and Design Education* (16): 3.
Adams, E. and Ward, C. (1982) *Art and the Built Environment*, London: Longman.
Adams, J., Meecham, P. and Orbach, C. (1998) *Working with Modern British Art: A Guide for Teachers*, Liverpool: Tate Gallery and John Moores University.
Addison, N. (1997) *In Defence of Decoration*, London: Institute of Education.
—— (1999) 'Who's afraid of signs and significations', *Directions: Journal of Art and Design* (18) 1.
Addison, N. and Allen, E. (1997) 'Art histories in action', Engage.
Ades, D. (1976) *Photomontage*, London: Thames and Hudson.
Ades, D. and Bradley, F. (1998) *Salvador Dali: A Mythology*, London: Tate Gallery Publishing.
Adorno, T. (1941) 'On popular music', in A. Easthope and K. McGowan (eds) (1992) *A Critical and Cultural Theory Reader*, Buckingham: Open University Press.
Allen, D. (1996) 'As well as painting', in L. Dawtrey *et al.* (eds) *Critical Studies and Modern Art*, Milton Keynes: Open University Press.
Allison, B. (1974) 'Professional art education', *Journal of the NSAE* 1 (1): 3–9.
—— (1986) 'Some aspects of assessment in art and design', in M. Ross (ed.) *Assessment in Arts Education – A Necessary Discipline or a Loss of Happiness?*, Oxford: Pergamon Press.
—— (1997) 'Professional attitudes to research in art education', *Journal of Art and Design Education* 16 (30): 211–215.
Anderson, D. (1997) *A Common Wealth: Museums and Learning in the UK*, London: Department of National Heritage.
Anderson, T. (1992) 'Drawing upon the eye, the brain and the heart', *Art Education* 45 (5): 45–50.
Appignanesi, R. (1997) *Postmodernism for Beginners*, Cambridge: Icon, Audio.
Appignanesi, R. and Garratt, C. (1995) *Postmodernism for Beginners*, Cambridge: Icon, Books.

Araaen, R. (1989) *The Other Story*, London: Hayward Gallery.

Association of Advisers and Inspectors in Art and Design (AAIAD: Midland Group) (1994) *Sketchbooks*, Bromsgrove: AAIAD.

Association of Advisers and Inspectors in Art and Design (AAIAD) (1996) *Art: a review of inspection findings 1995–96*, AAIAD.

Attfield, J. and Kirkham, P. (eds) (1995) *A View from the Interior: Women and Design*, London: Women's Press.

Aubrey, C. (ed.) (1994) *The Role of Subject Knowledge in the Early Years of Schooling*, London: Falmer Press.

Audette, A. (1993) *The Blank Canvas: Inviting the Muse*, Boston and London: Shambala.

Avery, H. (1994) 'Feminist issues in built environment education', *Journal of Art and Design Education* 13 (1): 65–69.

Baddeley, O. (ed.) (1994) *New Art from Latin America*, London: Art and Design.

Ball, L. (1998) *Crafts 2000: A Future in the Making*, London: Crafts Council.

Bancroft, A. (1995) 'What do dragons think about their dark lonely caves?', *Journal of Art and Design Education* 14 (1).

Barrett, M. (1979) *Art Education: a Strategy for Course Design*, London: Heinemann Educational Books.

—— (1982) *Art Education: A Strategy for Course Design*, London: Heinemann.

—— (1990) 'Guidelines for assessment in art and design education: 5–18 years', *Journal of Art and Design Education* 9 (3).

Barthes, R. (1957) *Mythologies*, trans. A. Lavers and C. Smith (1990 edition), London: Jonathan Cape and New York: Hill and Wang.

—— (1965) *Elements of Semiology*, trans. A. Lavers and C. Smith (1975 edition), London: Jonathan Cape and New York: Hill and Wang.

—— (1991) *The Responsibility of Forms*, trans. R. Howard, Berkeley and Los Angeles: University of California Press. (Original work published 1982.)

Bastick, T. (1982) *Intuition: How We Think and Act?*, Chichester: Wiley.

Battersley, C. (1991) *Thinking Art: Beyond Traditional Aesthetics*, London: ICA.

Baynes, K. (1982) 'Beyond design education', *Journal of Art and Design Education*, 1 (1): 105–114.

—— (1990) 'Defining the design dimension of the curriculum', in D. Thistlewood (ed.) *Issues in Design Education*, London: Longman.

BECTa for DfEE (1998) *Choosing and Using CD-ROMs for Art & Design*, London: HMSO.

Berkeley, A. (1987) Academic Inflation, London: Crafts Council.

Bernadac, M-L. (1996) Louise Bourgeois, Paris, New York: Flammarion.

Bernal, M. (1987) *Black Athena: The Afroasiatic Roots of Classical Civilization*, New Brunswick: Rutgers University Press.

Bhabha, H. (1994) *The Location of Culture*, London and New York: Routledge.

Binch, N. (1997) Foreword in D. Allen and A. Rogers *What's happening to Photography?*, London: Arts Council of England.

Binch, N. and Robertson, E. (1994) *Resourcing and Assessing Art, Craft and Design – Critical Studies in Art at Key Stage 4*, Corsham: NSEAD.

Bird, J. *et al.* (1996) 'The block reader', in N. Mirzoett (ed.) *Visual Culture*, London: Routledge.

Birkert, S. (1994) *The Gutenberg Elegies: The Fate of Reading in an Electronic Age*, London: Faber and Faber.

Blackburn, J. (1998a) 'It's not all looms and sad crochet', *Observer* 25 January, 27.

—— (1998b) 'Arts and crafty', *Metro*, 4 July 1998.

Blair, T. (1998) Foreword in the Green Paper *Teachers Meeting the Challenge of Change*, London: DfEE, HMSO.

Boime, A. (1990) *The Art of Exclusion, Representing Blacks in the Nineteenth Century*, London: Thames and Hudson.

de Boisbaudran, H. L. (1862) *The Training of Memory in Art*, trans. L. D. Luard (1914 edition), London: Macmillan.

Boughton, D. (1986) 'Visual literacy: implications for cultural understanding through art education', Journal of Art and Design Education 5 (1 and 2).

Boughton, D., Eisner, E. and Ligtvoet, J. (eds) (1996) *Evaluating and Assessing the Visual Arts*, New York: Teachers College Press.

Brandt, G. (1991) Paper given at Commonwealth Institute Conference on Cultural Diversity and NC Art.

Brett, G. (1986) *Through our Eyes*, London: Heretic Books.

Britton, A. (1991) 'The manipulation of skill and their outer limits of function', in *Beyond the Dovetail: Craft, Skill and Imagination*, London: Crafts Council.

Brophy, J. (ed.) (1991) *Introduction to Vol. 2 Advances in Research on Teaching*, Greenwich: CTJAI Press.

Broude, N. and Garrard, M. D. (1994) *The Power of Feminist Art*, New York: Harry N. Abrams Inc.

Bruner, J. (1960) (1977 edition) *The Process of Education*, Cambridge, MA and London: Harvard University Press.

Bryson, N. (1991) 'Semiology and visual interpretation', in N. Bryson *et al.* (eds) *Visual Theory*, Cambridge: Polity Press.

Buchanan, M. (1990) 'Design and technology: issues for implementation and the role of art and design', in *Design & Technology Teaching*, Stoke-on-Trent: Trentham Books.

—— (1993) MA Lecture, London: Institute of Education.

—— (1995) 'Making art and critical literacy: a reciprocal relationship', in R. Prentice (ed.) *Teaching Art and Design: Addressing Issues and Identifying Directions*, London: Cassell.

Buck, L. (1993) *On the Edge: Art meets Craft*, London: Crafts Council.

Buckingham, D. (ed.) (1998) *Teaching Popular Culture*, London: University College Press.

Buckingham, D. and Sefton-Green, J. (1994) *Cultural Studies goes to School*, London: Taylor and Francis.

Budd, M. (1995) *Values in Art*, London: Penguin.

Burgess, L. (1995) 'Human resources: artists, craftspersons, designers', in R. Prentice (ed.) *Teaching Art and Design: Addressing Issues and Identifying Directions*, London: Cassell.

Burgess, L. and Holman, V. (1993) 'Live art', *Times Educational Supplement*, 2 June.

Burgess, L. and Reay, D. (1999) 'Postmodernism and postfeminism: problematic paradigms', (forthcoming).

Burgess, L. and Schofield, K. (1998) 'Shorting the circuit', in *Ideas in the Making*, London: Crafts Council.

Button, V. (1998) *The Turner Prize*, London: Tate Gallery, London.

Buzan, T. (1982) *Use Your Head*, London: Ariel Books, BBC.

Camp, J. (1981) *Draw: How to Master the Art (foreword by D. Hockney)*, London: Dorling Kindersley.

Capel, S., Leask, M. and Turner, T. (1999) *Learning to Teach in the Secondary School – A Companion to School Experience*, 2nd edn, London: Routledge.

Caruana, W. (1993) *Aboriginal Art*, London: Thames and Hudson.

Causey, A. (1998) *Sculpture since 1945*, Oxford: Oxford University Press.

Cézanne, P. (1985) *A Cézanne Sketchbook*, London: Dover Publications.

Chadwick, W. (1990) *Women, Art and Society*, London: Thames and Hudson.

—— (1991) *Women Artists and the Surrealist Movement*, London: Thames and Hudson.

Chalmers, G. (1996) *Celebrating Pluralism Art, Education, and Cultural Diversity*, Los Angeles: Paul Getty Trust.

Cheng, F. (1994) *Empty and Full: the Language of Chinese Painting*, London: Shambahla.

Chicago, J. (1996) *The Dinner Party*, New York and London: Penguin.

Chipp, H. (1968) *Theories of Modern Art*, Berkeley, Los Angeles and London: University of California Press.

Clark, K. (1985) *Ruskin Today*, Harmondsworth: Penguin.

Clark, R. (1997) 'Purposeful uncertainty', in R. Irwin and K. Graver (eds) *Readings in Canadian Art Teacher Education*, Ontario: Aylmer Press.

Clark, T. J. (1999) *Farewell to an Idea: Episodes from a History of Modernism*, New Haven and London: Yale University Press.

Clark, G. and Zimmerman, E. (1984) *Educating Artistically Talented Students*, Syracuse, NY: Syracuse University Press.

Clement, R. (1986) *The Art Teacher's Handbook*, London: Hutchinson Education.

—— (1993) (second edition) *The Art Teacher's Handbook*, Cheltenham: Stanley Thornes.

Clifford, J. and Marcus, G. E. (1986) (eds) *Writing Culture: the Poetics and Politics of Ethnography*, Berkeley and London: University of California Press.

Clive, S. and Binch, N. (1994) *Close Collaborations: Art in Schools and the Wider Environment*, London: ACE.

Clive, S. and Geggie, P. (1998) *Unpacking Teachers' Packs*, London: Engage.

Cohen, P. (1998) 'Tricks of the trade: on teaching arts and "race" in the classroom', in D. Buckingham (ed.) *Teaching Popular Culture*, London: University College Press.

Cole, A. (1992) *Perspective*, London: Dorling Kindersley and National Gallery.

Concise Oxford Dictionary (1985) Oxford: Oxford University Press.

Conran, T. (1993) in *Sunday Express Magazine*, 3 October.

Coomaraswamy, A. K. (1934) (1956 edition) *The Transformation of Nature in Art*, New York: Dover.

Coombes, A. (1991) 'Ethnography and the formation of national and cultural identities', in S. Hiller (ed.) *The Myth of Primitivism*, London: Routledge.

—— (1994) *Reinventing Africa*, New Haven and London: Yale University Press.

—— (1998) in D. Preziosi (ed.) *The Art of Art History*, Oxford: Oxford University Press.

Courtney, C. (1995) 'Private Views and Other Containers', artists' books reviewed by Cathy Courtney for *Art Monthly 1983–1995*, London: Estamp Editions.

Crafts Council (1982) *Crafts Conference for Teachers*, London: Crafts Council.

—— (1996) (exhibition catalogue) *Recycling: Focus for the Next Century – Austerity for Posterity*, London: Crafts Council.

—— (1998) *Definitions of Craft as provided by the Crafts Council*.

Crafts Council Survey (1995) *Pupils As Makers, Part 1*, London: Crafts Council.

—— (1998) *Pupils As Makers, Part 2*, London: Crafts Council.

Crafts: The Decorative and Applied Arts Magazine, published bi-monthly.

Creber, P. (1990) *Thinking Through English*, Milton Keynes: Open University Press.

Crow, T. (1996) *Modern Art in Common Culture*, New York, London: Yale University Press.

Crowther, P. (1996) *Critical Aesthetics and Post Modernism*, London: Clarendon Press.

Cunliffe, L. (1996) 'Art and world view: escaping the formalist labyrinth', *Journal of Art and Design Education* 15 (3).

—— (1999) 'Learning how to learn: art education and the "background"', *Directions: Journal of Art and Design Education* 18 (1): 115–121.

Dabydeen, D. (1987) *Hogarth's Blacks: Images of Blacks in Eighteenth Century English Art*, Manchester: Manchester University Press.

Dahl, D. (1990) *Residencies in Education*, Sunderland: AN Publications.

Dalton, P. (1987) 'Housewives, leisure crafts and ideology: deskilling in consumer craft', in G. Elinor *et al. Women and Craft*, London: Virago.

—— (1999) *Dismantling Art Education: Modernism, Art Education and Critical Feminism*, Oxford: Oxford University Press.

Dash, P. (1999) Thoughts on a relevant art curriculum, *JADE* 18 (1).

Davey, N. (1994) 'Aesthetics as the foundation of human experience', *Journal of Art and Design Education* 13 (1).

Davison, J. and Dowson, J. (eds) (1998) *Learning to Teach English in the Secondary School*, London: Routledge.

Dawtrey, L., Jackson, T., Masterson, M., Meecham, P. and Wood, P. (eds) (1996) *Critical Studies and Modern Art*, Milton Keynes: Open University Press.

Dearing Report (1997) *Higher Education in the Learning Society*, London: HMSO.

Deepwell, K. (ed.) (1995) *New Feminist Art Criticism*, Manchester University Press.

de Monchaux, C. (1997) interviewed by L. Buck in *Tate: The Art Magazine* Summer, in M. Parkin (1998) *The Turner Prize 1998*, London: Tate Gallery.

Denvir, B. (1986) *The Late Victorians: Art, Design and Society 1852–1910*, London: Longman.

DES (1985) 'Education for all: the report of the committee of inquiry into the education of children from ethnic minority groups', in *The Report*, London: HMSO.

—— (1987) *Task Group on Assessment and Testing (TGAT)*, London: HMSO.

—— (1988) *National Curriculum Design & Technology Working Group, Interim Report*, London: HMSO.

—— (1991a) *Art for Ages 5 to 14*, London: HMSO.

—— (1991b) *NC Art Working Group, Interim Report*, London: HMSO.

—— (1992) *Art in the National Curriculum*, London: HMSO.

—— (1995) *NC Design and Technology Order*, London: HMSO.

Design Council, (1994) *Design Focus in Schools, Consultation Document*, London: Design Council. *Design Review* 13 (4).

Dewdney, A. (1996) 'Art photography: continuing historical narratives', *Journal of Art and Design Education* 15 (1).

Dewey, J. (1934) (1979 edn) *Art as Experience*, New York: Paragon Books.

DFE (1994) *Education Act 1993: Sex Education in Schools*, circular number 5/94, London: HMSO.

DFE (1995) *Art in the National Curriculum*, London: HMSO.

DfEE (1995) *A Guide to Safe Practice in Art and Design*, London: HMSO.

—— (1998) *Consultation Document on the New Induction Arrangements*, London: HMSO.

—— (4/98) *Circular 4/98 Teaching: High Status, High Standards*, London: HMSO.

Department of National Heritage (DNH) (1996) *Setting the Scene: the Arts and Young People*, London: HMSO.

Dickson, M. (1995) *Art with People*, Sunderland: AN Publications.

Dinham, J. (1989) 'Drawing: what is it and why has it traditionally held a special place in the art programme?', *Journal of Art and Design Education* 8 (3): 317–329.

Dissanayake, E. (1992) *Homo Aestheticus: Where Art Comes from and Why*, New York: Macmillan.

Donald, J. and Rattansi, A. (1992) *'Race', Culture and Difference*, London: Sage Publications.

Dormer, P. (1982) 'Is the crisis in British crafts the next boom in art criticism?', *Journal of Art and Design Education* 2 (1): 37.

—— (1991a) 'Beyond the dovetail: crafts, skills and imagination', in C. Frayling (ed.) *Beyond the Dovetail: Crafts, Skills and Imagination*, London: Crafts Council.

—— (1991b) *The Meanings of Modern Design*, London: Thames and Hudson.

—— (1994) *The Art of the Maker*, London: Thames and Hudson.

—— (1997) *The Culture of Craft*, London: Crafts Council.

Duffin, D. (1995) 'Exhibiting strategies', in K. Deepwell (ed.) *New Feminist Art Criticism*, Manchester: Manchester University Press.

Duncan, C. (1995) *Civilising Rituals: Inside Public Art Museums*, London: Routledge.

Duncum, P. (1984) 'How 35 children between 1724 and 1900 learned to draw', *Studies in Art Education* 26 (2): 93–102.

Dyson, A. (1989) 'Art history in schools: a comprehensive strategy', in D. Thistlewood (ed.) *Critical and Contextual Studies in Art and Design Education*, Harlow: Longman.

Easthope, A. and McGowan, K. (eds) (1992) *A Critical and Cultural Theory Reader*, Buckingham: Open University Press.

EDEXCEL (1995) *BTEC GNVQ Unit Booklets*, London: EDEXCEL.

—— (1998) *Foundation Press Release*, 5 January, London: EDEXCEL.

Education Reform Act (ERA) (1988) London: HMSO.

Edwards, A. and Furlong, V. (1978) *The Language of Teaching: Meaning in Classroom Interaction*, London: Heinemann Education.

Edwards, B. (1986) *Drawing on the Artist Within: A Guide to Innovation, Invention, Imagination and Creativity*, New York: Simon and Schuster.

Efland, A. (1970) 'Conceptualising the curriculum problem', *National Society for Education in Art and Design Bulletin*, April.

—— (1976) 'School art's style of functional analysis', *Studies in Art Education* 17 (2).

—— (1979) 'Conceptions of teaching in art education', *Art Education* 32 (4): 21–33.

Efland, A., Freedman, K. and Stuhr, P. (eds) (1996) *Postmodern Art Education: An Approach to Curriculum*, Reston VA: The National Art Education Association.

Eggleston, J. (1996) (2nd edition) *Teaching Design & Technology*, Milton Keynes: Open University Press.

—— (ed.) (1998) *Learning through Making: A National Enquiry into the Value of Creative Practical Education in Britain*, London: Crafts Council.

Ehrenzweig, A. (1967) (1993 edition) *The Hidden Order of Art*, London: Weidenfeld.

Eisner, E. (1972) *Educating Artistic Vision*, New York: Macmillan.

—— (1982) *Cognition and Curriculum: A Basis for Deciding what to Teach*, Harlow: Longman.

—— (1998) 'Does experience in the arts boost academic achievement?', *Journal of Art and Design Education* 17 (1).

Elinor, G., Richardson, S. *et al.* (eds) (1987) *Women and Craft*, London: Virago Press.

Elliot, J. (1991) 'A model of professionalism and its implications for teacher education', in *British Educational Research Journal* 17 (4): 309–318.

Ellsworth, E. (1989) 'Why doesn't this feel empowering? Working through repressive myths of critical pedagogy', *Harvard Educational Review* 59 (3): 297–324.

Feldman, E. (1987) *Varieties of Visual Experience*, New York: Harry N. Abrams.

Ferguson, B. (1995) 'Race, gender and a touch of class', in R. Prentice (ed.) *Teaching Art and Design: Addressing Issues and Identifying Directions*, London: Cassell.

Fernie, E. (1995) *Art History and Its Methods*, London: Phaidon.

Field, R. (1970) *Change in Art Education*, New York: Routledge and Kegan Paul.

Fisher, M. (1994) 'Labour plans to get architecture taught in schools', *Building Design Magazine*, 1202.

Fiske, J. (1982) *Introduction to Communication Studies*, London: Methuen.

—— (1989) *Understanding Popular Culture*, London: Routledge.

Foster, H. (1983) *The Anti-Aesthetic: Essays on Postmodern Culture*, Seattle, WA: Bay Press.

—— (1992) 'The primitive unconscious', in F. Frascina and J. Harris (eds) (1992) *Art in Modern Culture*, London: Phaidon.

—— (1996) *The Return of the Real*, London and Cambridge, MA: MIT Press.

Frascina, F. and Harris, J. (eds) (1992) *Art in Modern Culture*, London: Phaidon.

Frascina, F. and Harrison, C. (eds) (1982) *Modern Art and Modernism: A Critical Anthology*, London: Open University Press.

Frayling, C. (1982) *Crafts Conference For Teachers*, Crafts Council.

—— (1990) 'Some perspectives on the crafts revival in the twentieth century', in D. Thistlewood (ed.) *Issues in Design Education*, Harlow: Longman and NSEAD.

Freedman, K. (1994) 'Interpreting gender and visual culture in art classrooms', *Studies in Art Education* 35 (3).

—— (1997) 'Visual art/virtual art: teaching technology for meaning', in *Art Education* 50 (4); Freedman refers to N. Brown, (1989) 'The myth of visual literacy', *Australian Art Education* 13 (2): 28–32.

Freire, P. (1985) *The Politics of Education: Culture, Power and Liberation*, Westport, CT: Bergin and Garvey.

Fuller, P. (1983) 'The necessity of art education', in *The Naked Artist*, London: Writers and Readers.

—— (1993) *Modern Painters; Reflections on British Art*, London: Methuen.

Gablik, S. (1991) *The Reenchantment of Art*, London: Thames and Hudson.

—— (1995) *Conversations Before the End of Time*, New York: Thames and Hudson.

Gardner, H. (1983) *Frames of Mind: The Theory of Multiple Intelligences*, New York: Basic Books.

—— (1993) *The Unschooled Mind*, London: Fontana.

Garner, S. W. (1990) 'Drawing and designing: the case for reappraisal', *Journal of Art and Design Education* 9 (1): 39–55.

GCSE CDT Guide for Teachers, Milton Keynes: The Open University in collaboration with SEC.

Gentle, K. (1988) *Children and Art Teaching*, London: Routledge.

—— (1990) 'The significance of art making for individuals', *Journal of Art and Design Education* 9 (3).

Gillborn, D. (1995) *Racism and Anti-Racism in Real Schools: Theory, Policy, Practice*, Buckingham: Open University Press.

Gipps, C. (1994) *Beyond Testing: Towards a Theory of Educational Assessment*, London: Falmer Press.

—— (1997) *Principles of Assessment*, keynote lecture, London: Institute of Education.

Gipps, C. and Stobart, G. (1993) *Assessment: A Teacher's Guide to the Issues*, Hodder and Stoughton.

Giroux, H. (1992) *Border Crossings*, New York, London: Routledge.

—— (1994) *Disturbing Pleasures*, New York: Routledge.

Giroux, H. and McLaren, P. (eds) (1994), *Between Borders*, London: Routledge.

Glenister, S. (1968) *The Technology of Craft Teaching*, London: Harrap.

Glimcher, A. and M. (eds) (1986) *Je Suis le Cahier: The Sketchbooks of Picasso*, New York: The Pace Gallery.

Golby, J. M. (1986) *Culture and Society in Britain 1850–1890*, Oxford: Oxford University Press in association with The Open University.

Goldwater, R. (1986) *Primitivism in Modern Art*, Cambridge, MA: Belknap Press.

Golomb, C. (1989) 'Sculpture: the development of representational concepts in three-dimensional medium', in D. H. Hargreaves (ed.) *Children of the Arts*, Milton Keynes: Open University Press.

Gombrich, E. (1992) 'Absolute Standards', *Crafts*, April.

—— (1950) *The Story of Art*, London: Phaidon.

Green, P. (1982) *Design Education: Problem Solving and Visual Experience*, London: Batsford.

Greenberg, C. (1965) 'Modernist painting', in F. Frascina and C. Harrison, (eds) (1982) *Modern Art and Modernism*, London: The Open University Press.

Greenhalgh, M. (1997) 'The History of Craft', in P. Dormer (ed.) *The Culture of Craft*, Manchester: Manchester University Press.

Greenhalgh, M. and Megaw, V. (1978) *Art in Society*, London: Duckworth.

Gregory, R. (1977) *Eye and Brain: The Psychology of Seeing*, London: Weidenfeld and Nicolson.

Gretton, T. (1986) 'New lamps for old', in A. L. Rees, and F. Borzello (eds) *The New Art History*, London: Camden Press.

—— (forthcoming) 'Loaded canons', in N. Addison and E. Allen (eds) *Art Histories in Action*, London: Institute of Education.

Grossberg, L. (1965) 'Introduction: bringin' it all back home', in H. Giroux and P. MacLaren (eds) *Between Borders*, London: Routledge.

Grossman, P. L., Shulman, L. and Wilson, S. (1989) 'Teachers of substance: subject matter knowledge for teaching', in M. Reynolds *The Knowledge Base for the Beginning Teacher*, Oxford: Pergamon Press.

Grosz, E. (1995) *Space, Time and Perversion*, London: Routledge.

Grumet, M. (1990) 'On daffodils that come before the swallow dares', in E. Eisner and A. Peshkin (eds) *Qualitative Inquiry in Education: The Continuing Debate*, New York: Teachers College Press.

Guardian (1998) 'Schools education syllabuses', 1 September.

Guerilla Girls (1995) *Confessions of Guerilla Girls*, London: Pandora.

Hargreaves, D. H. (1983) 'The teaching of art and the art of teaching: towards an alternative view of aesthetic learning', in M. Hammersley and A. Hargreaves (eds) *Curriculum Practice: Some Sociological Case Studies*, London: Falmer.

—— (1989) *Children and the Arts*, Milton Keynes: Open University Press.

Harland, J., Haynes, J., Kinder, K. and Schagen, I. (1998) *The Effects and Effectiveness of Arts Education in Schools, Interim Report on Research Organised by the National Foundation for Educational Research (NFER) and the Royal Society for the Encouragement of the Arts, Manufacturers and Commerce (RSA)*, Slough: NFER.

Harris, J. (1997) 'Art education and cyber-ideology; beyond individualism and technological determinism', in P. Meecham (ed.) *Fall Art Journal* (Oct 1998), Liverpool: John Moores University.

Harrison, C. and Wood, P. (eds) (1992) *Art in Theory: 1900–1990*, Oxford: Blackwell.

Harrod, T. (ed.) (1997) *Obscure Objects of Desire*, London: Crafts Council.

Hauser, A. (1951) *The Social History of Art*, J. Harris (ed.) (1999 edition), London: Routledge.

Hebdidge, D. (1988) *Hiding in the Light*, London: Routledge.

Heidegger, M. (1954) *What is called Thinking?*, trans. J. Gray (1968 edition), London, New York: Harper and Row.

Hickman, R. (1996) *Finding Your First Job – A Guide for Prospective Art and Design Teachers*, Corsham: NSEAD.

—— (ed.) (forthcoming) *Meanings, Purpose and Directions*, London: Cassell.

Hiett, C. and Orbach, S. (1996) *Venus Re-Defined*, Resource Pack, Liverpool: Tate, Liverpool.

Hill, E. (1966) *The Language of Drawing* (revised edition), Englewood Cliffs, NJ: Prentice-Hall.

Hiller, S. (ed.) (1991) *The Myth of Primitivism*, London: Routledge.

—— (1996) *Thinking About Art: Conversations with Susan Hiller*, Manchester: Manchester University Press.

Hodgson, A. and Spours, K. (eds) (1997) *Dearing and Beyond: 14–19 Qualifications, Frameworks and Systems*, London: Kogan Page.

Holt, J. (1996) 'Art for earth's sake', in Dawtrey *et al.* (eds) *Critical Studies and Modern Art*, Milton Keynes: Open University Press.

Honour, H. and Fleming, J. (1982) *A World History of Art*, London: Macmillan.

hooks, B. (1992) 'Representing whiteness on the black imagination', in M. L. Grossberg *et al.* (eds) *Cultural Studies*, New York, London: Penguin.

—— (1994a) *Teaching to Transgress*, New York, London: Routledge.

—— (1994b) *Outlaw Culture: Resisting Representations*, New York: Routledge.

Hoopes, J. (ed.) (1991) *Peirce on Signs; Writings on Semiotics*, Chapel Hill and London: University of North Carolina Press.

Hornby, N. (1992) Fever Pitch, London: Victor Gollancz.

Hughes, A. (1989) 'The copy, the parody and the pastiche: observations on practical approaches to critical studies', in D. Thistlewood (ed.) *Critical Studies in Art and Design Education*, Harlow: Longman.

—— (1998a) 'A further reconceptualisation of the art curriculum', in *Broadside 1*, UCE: ARTicle Press.

—— (1998b) 'Reconceptualising the art curriculum', *Journal of Art and Design Education* 17 (1).

—— (1999) 'Art and Intention in Schools: Towards a New Paradigm', *Journal of Art and Design Education* 18 (1).

Hughes, R. (1993) *Culture of Complaint*, New York and Oxford: New York Public Library and Oxford University Press.

Hurwitz, A., Wilson, B. and Wilson, M. (1987) *Teaching Drawing from Art*, Worcester, MA: Davis Publications, Inc.

Hyde, S. (1997) *Exhibiting Gender*, Manchester: Manchester University Press.

Ilies, C. and Roberts, R. (eds) (1997) *In Visible Light: Photography and Classification in Art, Science and The Everyday*, Oxford: MoMA.

IWM (1999) *Education Service Leaflet*, London: IWM.

Jefferies, J. (1995) 'Weaving across borderlines', in K. Deepwell (ed.) *New Feminist Art Criticism*, Manchester: Manchester University Press.

Journal of Art and Design Education (JADE) three editions yearly, is the journal of the NSEAD (National Society for Education in Art and Design).

Journal of Art and Design Education, (1983) (1).

—— (1995) 'Naming of parts', Crafts 34: 42–45.

—— (1997a) *Crafts Council Conference*, London: Crafts Council.

—— (1997b) 'Crafts', *Crafts Magazine* 146, June.

—— (1998) 'Material culture', *Blueprint* 147, February: 24–26.

Johnson, P. (ed.) (1998) *Ideas in the Making*, London: Crafts Council.

Jones, A. (1996) *Sexual Politics*, Los Angeles: University of California Press.

Jung, C. (1964) *Man and His Symbols*, London: Aldus Books Ltd.

Kaelin, E. F. (1989) *An Aesthetics for Art Educators*, New York: Teachers College Press.

Kelly, G. (1955) *The Psychology of Personal Constructs*, New York: Norton.

Kennedy, H. (1997) *Learning Works: Widening Participation in FE Education*, FEFC.

Kennedy, M. (1995a) 'Issue-based work at KS4: Crofton School – a case study', *Journal of Art and Design Education* 14 (1).

—— (1995b) 'Approaching assessment', in R. Prentice (ed.) *Teaching Art and Design: Addressing Issues and Identifying Directions*, London: Cassell.

Ker, M. (1997) 'Grounds for design: improvement of school and college grounds', *Journal of Art and Design Education* 16 (1).

Kimbell, R., Stable, K. and Green, R. (1996) *Understanding Practice in Design and Technology*, Milton Keynes: Open University Press.

Kress, G. and Leeuwen, T. (1996) *Reading Images: the Grammar of Visual Design*, London: Routledge.

Lacy, S. (1995) *Mapping the Terrain: New Genre Public Art*, Seattle: Bay Press.

La Trobe Bateman, R. (1997) 'Frontispiece to introduction', in P. Dormer (ed.) *The Culture of Craft*, Manchester: Manchester University Press.

Lavie, S. and Swedenburg, T. (1996) *Displacement, Diaspora and Geographies of Identity*, Durham, NC and London: Duke University Press.

Levi-Strauss, C. (1963) 'Do dual organisations exist?', in *Structural Anthropology*, Harmondsworth: Penguin.

Lippard, L. (1995) 'Looking around: where we are, where we could be', in S. Lacy (ed.) *Mapping the Terrain*, Seattle: Bay Press.

Lloyd, F. (1999) *Contemporary Arab Women Artists: Dialogues of the Present*, London: WAL.

Locke, E., Latham, G., Saari. L. and Shaw, K. (1981) 'Goal setting and task performance', *Psychological Bulletin* 90 (1): 125–152.

Loeb, H. (1984) Changing Traditions, Birmingham: UCE.

Lowenfeld, V. and Brittain, W. (1966 and 1987) *Creative and Mental Growth*, London: Macmillan.

Lowry, P. (1997) *The Essential Guide to Art and Design*, London: Hodder and Stoughton.

Lyons, J. (ed.) (1985) *Artists' Books: A Critical Anthology and Sourcebook*, New York: Visual Studies Workshop Press.

Mac an Ghaill, M. (1995) *The Making of Men: Masculinities, Sexualities and Schooling*, Buckingham: Open University Press.

MacDonald, N. and McCullogh, K. 'Power from the past', *Design Review* 13 (4): 39–41.

MacDonald, S. (1970) *The History and Philosophy of Art Education*, London: London University Press.

MacLeod, B. (1986) *Cross-cultural Art Booklets*, Nottingham: Nottingham Educational Supplies.

MacPherson, W. (1999) *The Stephen Lawrence Inquiry: Report of an Inquiry*, London: HMSO.

McRobbie, A. (1994) *Postmodernism and Popular Culture*, London: Routledge.

—— (1998) 'But is it art?', *Marxism Today* November/December: 55.

Maharaj, S. (forthcoming) 'Interculturalism', in N. Addison and E. Allen (eds) *Art Histories in Action*, London: Institute of Education.

Make is the quarterly journal of the Women's Art Library London.

Malik, R. (1995) *Portrait: Education Pack*, London: InIVA.

Manser, S. and Wilmot, H. (1995) *Artists in Residence: a Teachers' Handbook*, London: London Arts Board.

Mason, R. (1995) *Art Education and Multiculturalism*, NSEAD, London: Croom Helm.

Mayer, M. (1998) 'Can philosophical change take hold in the American art museum?', *Art Education: The Journal of the National Art Education Association*, Reston, VA, USA.

Meecham, P. (1996) 'What's in a national curriculum?', in Dawtrey *et al.* (eds) *Critical Studies in Modern Art*, Milton Keynes: Open University Press.

—— (1999) 'Of webs and nets and lily ponds', *Journal of Art and Design Education* 18 (1).

Miles, M. (1998) 'Strategies for the convivial city', *Journal of Art and Design Education* 17 (1).

Millar, S. (1998) 'Artists' holes start mother of all rattles in krazy golf war', in the *Guardian*, 1 August, Home News.

Mintz, S. and Price, R. (1976) *The Birth of African American Culture*, Boston; MA: Beacon Press.

Moon, B. and Mayes, A. S. (eds) (1994) *Teaching and Learning in the Secondary School*, London: Open University and Routledge.

Moore, H. (1952) 'Notes on sculpture from M. Evans, "The Painters Object"', in *The Creative Process: A Symposium*, London: Mentor Books, The New English Library Ltd.

—— (1972) *Henry Moore's Sheep Sketchbook*, London: Thames and Hudson.

Morgan, S. and Morris, F. (eds) (1995) *Rites of Passage: Art for the End of the Century*, London: Tate Gallery.

Morley, D. (1992) *Television Audiences and Cultural Studies*, London: Routledge.

Morrison, T. (1990) *Playing in the Dark: Whiteness and the Literary Imagination*, London and Basingstoke: Picador.

Myerson, J. (1993) *Design Renaissance*, London: Open Eye.

National Union of Teachers (annually) *Your First Teaching Post*, London: NUT.

NC (1992) *Art Non-Statutory Guidance Booklet*, York: National Curriculum Council.

NCET (1998) *Fusion: Art and IT in Practice Supporting National Curriculum Art at Key Stage 3 and Beyond*, Coventry: NCET (now BECTa).

Neperud, R. (ed.) (1995) *Context, Content and Community in Art Education: Beyond Postmodernism*, New York: Teachers College Press.

Nesbitt, J. (ed.) (1993) *Robert Gober*, Liverpool and London: Serpentine Gallery and Tate Liverpool.

New Shorter Oxford Dictionary, (1993) Oxford: Clarendon Press.

Nochlin, L. (ed.) (1991) *Women, Art and Power and Other Essays*, London: Thames and Hudson.

Nochlin, L. (1999) Representing Women, London: Thames and Hudson.

Objects of Desire (exhibition catalogue) (1997) London: Hayward Gallery.

Oddie, D. and Allen, G. (1998) *Artists in Schools: A Review*, London: OFSTED House.

OFSTED (1995a) *The Framework for the Inspection of Schools*, London: HMSO.

—— (1995b) *Art: A Review of Inspection Findings 1993–94*, London: HMSO.

—— (1998) *The Arts Inspected*, Oxford: Heinemann.

Oguibe, O. (1993) 'In the "Heart of Darkness"', in E. Fernie (ed.) (1995) *Art History and its Methods*, London: Phaidon.

Osbourn, R. (1991) 'Speaking, listening and critical studies in art', *Journal of Art and Design Education* (10) 1.

Paley, N. (1995) *Finding Art's Place: Experiments between Contemporary Art and Education*, New York, London: Routledge.

Palmer, F. (1989) *Visual Elements of Art and Design*, Harlow: Longman.

Panting, L. (1999) 'Debating gender', *MAKE* 33.

Papanek, V. (1971) *Design for the Real World*, London: Thames and Hudson.

—— (1995) *The Green Imperative*, London: Thames and Hudson.

Pariser, D. (1984) 'Two methods of teaching drawing skills', in R. MacGregor (ed.) *Readings in Canadian Art Education*, (143–158), Vancouver: WEDGE, University of British Columbia.

Parker, R. (1984) *The Subversive Stitch: Embroidery and the Making of the Feminine*, London: Women's Press.

Parker, R. and Pollock, G. (1981) *Old Mistresses: Women, Art and Ideology*, London: Pandora.

Parsons, M. (1987) *How We Understand Art: A Cognitive Developmental Account of Aesthetic Experience*, Cambridge: Cambridge University Press.

Parsons, M. J. and Blocker, G. (1993) *Aesthetics and Education*, Champaign IL: University of Illinois Press.

Pascall, D. (November 1992) *Speech at the Royal Society of Arts*, London.

Pateman, T. (1991) *Key Concepts: A Guide to Aesthetics, Criticism and the Arts in Education*, London: Falmer Press.

Pearson, A. and Aloysius, C. (1994) *The Big Foot: Museums and Children with Learning Difficulties*, London: British Museum.

Perry, L. (1998) Conversation with the authors at the Institute of Education.

Petersen, K. and Wilson, J. (1976) *Women Artists: Recognition and Reappraisal from the Early Middle Ages to the 20th Century*, London: The Women's Press.

Phillips, T. (1980) *A Humument*, London: Thames and Hudson.

Picasso, P. (1974) 'La Tête d'obsidienne', A. Malraux, in E. Cowling and J. Golding (eds) (1994) Picasso Sculptor/Painter, London: Tate Gallery.

Piaget, J. (1962) *Judgement and Reasoning in the Child*, London: Routledge and Kegan Paul.

Pietersie, J. (1992) *White on Black: Images of Africa and Blacks in Western Popular Culture*, London: Yale University Press.

Piper, K. (1997) *Relocating the Remains*, A Monograph and CD-ROM, London: InIVA.

Plato (c. 380 BCa) *The Republic*, trans. H. D. P. Lee (1955 edition), Harmondsworth: Penguin.

Plato (c. 380 BCb) 'The Republic', in S. Buchanan (ed.) *The Portable Plato*, Harmondsworth: Penguin.

Pliny the Elder, (1st century AD) *Natural History: A Selection*, trans. J. F. Healy (1991 edition), Harmondsworth: Penguin.

Pocket Oxford Dictionary of Current English, (1992) Oxford: Clarendon Press.

Pointon, M. (1986) *History of Art: A Student Handbook*, London: Unwin Hyman.

—— (1990) *Naked Authority*, Cambridge: Cambridge University Press.

Polanyi, M. (1964) *Personal Knowledge*, New York: Harper and Row.

Pollock, G. (1982) 'Vision, voice and power', in *BLOCK* 6.

—— (1988) *Vision and Difference: Femininity, Feminism and the Histories of Art*, London and New York: Routledge.

—— (1996a) 'Art, art school, culture', in J. Bird *et al. The BLOCK reader in Visual Culture*, London: Routledge.

—— (1996b) (ed.) *Generations and Geographies in the Visual Arts: Feminist Readings*, London: Routledge.

—— (1996c) 'Art, artists and women', in Dawtrey *et al.* (eds) *Critical Studies and Modern Art*, Milton Keynes: Open University Press.

Poupeye, V. (1998) *Caribbean Art*, London: Thames and Hudson.

Powell, R. (1997) *Black Art and Culture in the 20th Century*, London: Thames and Hudson.

Prentice, R. (1995) 'Learning to teach: a conversational exchange', in R. Prentice (ed.) *Teaching Art and Design: Addressing Issues and Identifying Directions*, London: Cassell.

Press, M. (1996) 'Recycling', in *Crafts Council Catalogue*, Birmingham: Craftspace Touring.

—— (1997) 'A new vision in the making', in *Crafts* 147, August: 42–45.

Preziosi, D. (1998) *The Art of Art History: A Critical Anthology*, Oxford: Oxford University Press.

Price, M. (1989) 'Art history and critical studies in schools: an inclusive approach', in D. Thistlewood (ed.) *Critical Studies in Art and Design Education*, Harlow: Longman.

Proctor, N. (1996) 'Is women's art homeless?', *Make*, September, 71.

Pye, D. (1969) *The Nature of Design*, London: Studio Vista.

Pye, J. (1988) *Invisible Children: Who Are The Real Losers at School?*, Oxford: Oxford University Press.

QCA (1997) *A Survey of Teachers' Reference to the work of Artists, Craftspeople and Designers in Teaching Art*, London: QCA.

—— (1998a) 'A survey of artists, craftspeople and designers used in teaching art research', in *Analysis of Educational Resources in 1997/98*, research coordinated by R. Clements (1996–97) CENSAPE: University of Plymouth.

—— (1998b) *Education for Citizenship and the Teaching of Democracy in Schools*, London: QCA on behalf of the DfEE.

—— (1999) *QCA's Work in Progress to Develop the School Curriculum: Pack B*, London: QCA.

Ramaprasad, A. (1983) 'On the definition of feedback', in *Behavioural Science* 24: 4–13.

Rampton Report (1981) 'West Indian children in our schools: interim report of the Committee of Inquiry into the Education of Children of Ethnic Minority Groups', London: DES, HMSO.

Raney, K. (1997) *Visual Literacy: Issues and Debates*, London: Middlesex University.

Read, H. (1943 and 1958) *Education through Art*, London: Faber and Faber.

Recycling: Focus for the Next Century – Austerity for Posterity, (1996) Crafts Council Exhibition Catalogue.

Reeve, J. (1995) 'Museums and galleries', in R. Prentice (ed.) *Teaching Art and Design: Addressing Issues and Identifying Directions*, London: Cassell.

Reid, L. A. (1969) *Meaning in the Arts*, London: Allen and Unwin.

—— (1986) *Ways of Understanding and Education*, London: Heinemann.

Rhodes, C. (1996) *Primitivism and Modern Art*, London: Thames and Hudson.

Robinson, C. (1992) Unpublished paper given at the NSEAD Conference in Brighton.

—— (1995) 'National curriculum for art: translating it into practice', in R. Prentice (ed.) *Teaching Art and Design: Addressing Issues and Identifying Directions*, London: Cassell.

Robinson, G. (1993) 'Tuition or intuition? Making and using sketchbooks with a group of ten-year-old children', in *Journal of Art and Design Education* 12 (1): 73–84.

—— (1995) *Sketchbooks: Explore and Store*, London: Hodder and Stoughton.

Robinson, K. (ed.) (1982) *The Arts in Schools: Principles, Practices and Provision*, London: Calouste Gulbenkian Foundation.

Rogers, C. (1969) (1983 edition) *Freedom to Learn*, New York: Macmillan.

Rogoff, I. (1998) 'Studying visual culture', in N. Mirzoeff (ed.) *Visual Culture Reader*, London: Routledge.

Rose, C. (1991) *Design After Dark the Story of Dancefloor Style*, London: Thames and Hudson.

Rosenberg, H. (1952) 'The American action painters', in D. Shapiro (ed.) (1990) *Abstract Expressionism: A Critical Record*, Cambridge: Cambridge University Press.

—— (1967) 'Art – where to begin', *Encounter*, March.

Rosenthal, N. *et al.* (1997) *Sensation: Young British Artists from the Saatchi Collection*, London: Royal Academy of Arts.

Ross, M. (1993) *Assessing Achievement in the Arts*, Milton Keynes: Open University Press.

Rothenstein, J. (ed.) (1986) *Michael Rothenstein: Drawings and Paintings Aged 4–9, 1912–1917*, London: Redstone Press.

RSA (1998) *The Effects and Effectiveness of Arts Education in Schools*, Slough: NFER.

Rowley, A. (1996) 'On viewing three paintings by Jenny Saville: rethinking the feminist practice of painting', in G. Pollock (ed.) *Generations and Geographies in the Visual Arts: Feminist Readings*, London: Routledge.

Rubin, W. (1969) 'Some reflections prompted by the recent work of Louise Bourgeois', *Art International*, 13 (4): 20 April.

Rubin, W. (ed.) (1984) *Primitivism in Twentieth Century Art*, New York: MOMA.

Rycroft, C. (1968) *A Critical Dictionary of Psychoanalysis*, Harmondsworth: Penguin.

Said, E. (1980) *Orientalism*, Harmondsworth: Penguin.

—— (1993) *Culture and Imperialism*, London: Chatto and Windus.

Said, E. in R. Hughes (1993) *The Culture of Complaint*, New York and Oxford: Oxford University Press.

Salmon, P. (1988) *Psychology for Teachers: An Alternative View*, London: Hutchinson.

Salmon, P. (1995) 'Experiential learning', in R. Prentice (ed.) *Teaching Art and Design: Addressing Issues and Identifying Directions*, London: Cassell.

Saussure, F. (1974) *Course in General Linguistics*, trans. W. Baskin, New York: Fontana, Collins.

SCAA (1996) *Consistency in Teachers' Assessment – Exemplification of Standards KS3*, London: SCAA.

—— (1997a) *Art and The Use of Language*, London: SCAA.

—— (1997b) *Survey and Analysis of Published Resources for Art (5–19)*, London: SCAA.

Schneider, N. (1990) *The Art of Still Life*, Köln: Taschen.

Schodt, F. (1993) *GManga, Manga, the World of Japanese Comics*, New York: Kodansha International.

Schofield, K. (1995) 'Objects of desire', in R. Prentice (ed.) *Teaching Art and Design: Addressing Issues and Identifying Directions*, London: Cassell.

Schon, D. (1987) *Educating the Reflective Practitioner*, San Francisco: Jossey Bass.

Searle, A. (ed.) (1993) *Talking Art 1*, London: ICA.

Secondary Examinations Council (1986) *CDT GCSE Guide for Teachers*, London: HMSO.

Sefton-Green, J. (ed.) (1998) *Digital Diversions: Youth Culture in the Age of Multimedia*, London: University College Press.

Selwood, S., Clive, S. and Irving, D. (1994) *Cabinets of Curiosity*, London: Art and Society for ACE.

Sensation (exhibition catalogue) (1997) London: Royal Academy of Arts.

Sewell, B. (1995) *An Alphabet of Villains*, London: Bloomsbury Press.

Shahn, B. (1967) *The Shape of Content*, Cambridge, MA: Harvard University Press.

Sharp, C. and Dust, K. (1998) *Artists in Schools*, Slough: NFER.

Shohat, E. and Stam, R. (1995) 'The politics of multiculturalism in the postmodern age', in *Art and Cultural Difference: Hybrids and Clusters*, Art & Design Magazine, 43.

—— (1998) 'Narrativizing visual culture: towards a polycentric aesthetics', in N. Mirzoeff, *The Visual Culture Reader*, London: Routledge.

Shreeve A. (1998) 'Tacit knowledge in textile crafts', in P. Johnson (ed.) *Ideas in the Making*, London: Crafts Council.

Shulman, L. (1986) 'Those who understand: knowledge growth in teaching', *Educational Researcher*, February, 4–14.

Simpson, A. (1987) 'What is art education doing?', *Journal of Art and Design Education* 6 (3).

Smith, M. (1995) 'Joseph Beuys: life as drawing', in D. Thistlewood (ed.) *Joseph Beuys: Diverging Critiques*, Liverpool: Liverpool University Press and Tate Gallery, Liverpool.

Sontag, S. (1977) *On Photography*, London: Penguin.

Sori, M. (1994) 'The body has reasons: tacit knowledge in thinking and making' *Journal of Aesthetic Education*, 28 (2).

Sparke, P. (1995) *As Long as it's Pink: The Sexual Politics of Taste*, London: Pandora.

Sparrow, F. (ed.) (1996) *Bill Viola: The Messenger*, Durham: The Chaplaincy to the Arts and Recreation in North East England.

Spencer, R. (1989) *Whistler: A Retrospective*, New York: Hugh Lanter Levin Associates.

Spender, D. (1989) *Invisible Women: The Schooling Scandal*, London: Women's Press.

—— (1995) *Nattering on the Net: Women, Power and Cyberspace*, Melbourne: Spinifox.

Stanley, N. (1986) 'Anthropology and the visual arts', *Journal of Art and Design Education* 5 (1 and 2).

—— (1996) Photography and the politics of engagement, *Journal of Art and Design Education* 15 (1).

—— (1998) *Being Ourselves for You: The Global Display of Cultures*, London: Middlesex University Press.

Stanworth, M. (1987) 'Girls on the margins: a study of gender divisions in the classroom', in M. Arnot and G. Weiner (eds) *Gender and the Politics of Schooling*, London: Hutchinson.

Steers, J. (1994a) 'Art and design: assessment and public examinations', *Journal of Art and Design Education* 13 (3).

—— (1994b) 'Technology in the national curriculum: the 1994 draft order', paper given at NSEAD Conference, Westminster, 2 July.

—— (1996) 'GNVQs: the content, the rhetoric and the reality', *Journal of Art and Design Education* 15 (2): 201–213.

—— (1998) Lecture given to PGCE students at the Institute of Education, University of London.

—— (February 1999) 'NSEAD Conference Address', Somerset.

Steers, J. and Swift, J. (1999) 'The manifesto for art', *Directions: Journal of Art and Design Education* 18: (1).

Steiner, G. (1989) 'Real presences', in P. Abbs, (ed.) *The Symbolic Order*, London: Falmer Press.

Stevenson, M. (1983) 'Problems of assessment and examinations in art education', *Journal of Art and Design Education* 2 (3).

Stibbs, A. (1998) 'Language in art and art in language', *Journal of Art and Design Education* 17 (1).

Stuhr, P., Petrovich-Mwaniki, L. and Wasson, R. (1992) 'Curriculum guidelines for the multicultural classroom', *Art Education* 45 (1): 16–24.

Stungo, N. (1998) 'Smart dressing', in *Blueprint*, February.

Sturrock, J. (1993) *Structuralism*, London: Fontana Press.

Swann, M., *The Swann Report* (1985) *Education for All: The Report of the Committee of Inquiry into the Education of Children from Ethnic Minority Groups*, London: HMSO.

Swift, J. (1992) 'Marion Richardson's contribution to art teaching', in D. Thistlewood (ed.) *Histories of Art and Design Education*, Harlow: Longman.

Swift, J. and Steers, J. (1999) 'A manifesto for art in schools', *Directions: Journal of Art and Design Education* 18 (1).

Talboys, M. (March 1999) 'QCA Seminar Address', London.

Tate, N. (1996) Introductory address to Information Technology, communications and the future, Curriculum Conference SCAA/NCET.

—— (1997) 'Keynote address', in *The Arts in the Curriculum*, London: SCAA.

Taylor, B. (1989) 'Art history in the classroom: a plea for caution', in D. Thistlewood (ed.) *Critical Studies in Art and Design Education*, Harlow: Longman.

—— (1995) *The Art of Today*, London: Everyman Art Library, Weidenfeld and Nicolson.

Taylor, R. (1986) *Educating for Art*, London: Longman.

—— (1989) 'Critical studies in art and design education', in D. Thistlewood (ed.) *Critical Studies in Art and Design Education*, Harlow: Longman.

—— (1991) *Artists in Wigan Schools*, London: Calouste Gulbenkian Foundation.

—— (1992) *Visual Arts in Education*, London: Falmer Press.

Taylor, R. and Taylor, D. (1990) *Approaching Art and Design: A Guide for Students*, London: Longman.

Teaching as a Career Unit (TASC) (annually), *Getting a Teaching Job*, London: University of Greenwich.

Third Text is a quarterly journal presenting perspectives on contemporary art from the developing world.

Thistlewood, D. (ed.) (1990) 'Editorial: the essential disciplines of design education', *Journal of Art and Design Education* 9 (1): 6.

—— (ed.) (1992) Histories of Art and Design Education, Harlow: Longman.

—— (1996) 'Critical development in critical studies', in L. Dawtrey *et al.* (eds) *Critical Studies and Modern Art*, Milton Keynes: Open University Press.

Thurber, F. (1997) 'A site to behold: creating curricula about local urban environmental art', *Art Education*, November.

Times Educational Supplement (annually, around the middle of January) First Appointments Supplement, TES.

Todd, S. (1997) 'Psychoanalytical questions', in H. Giroux (ed.) with P. Shannon *Education and Cultural Studies*, New York: Routledge.

Tomlinson Report (1996) *Inclusive Learning*, Report of the Learning Difficulties and/or Disabilities Committee, London: HMSO.

Tomlinson, S. (1990) *Multicultural Education in White Schools*, London: Batsford.

Troyna, B. and Hatcher, R. (1993) *Racism in Children's Lives*, London: Routledge.

TTA (1997) *Career Entry Profile – For Newly Qualified Teachers*, London: University of Greenwich.

Tucker, W. (1974) 'What sculpture is?' *Studio International* in A. Gouk, 'Carbon, diamond', in *Have you seen Sculpture from the Body?*, London: Tate Gallery.

Turner, J. M. W. (1987) *The 'Ideas of Folkestone' sketchbook, 1845*, London: The Tate Gallery.

Turner, S., Tyson, I. and Courtney, C. (eds.) (1993) *Facing the Page: British Artists' Books, A Survey, 1983–1993*, London: Estamp Editions.

Usher, R. and Edwards, R. (1994) *Postmodernism and Education*, London: Routledge.

Usherwood, P. (1998) 'Park life', *Art Monthly* 219, September.

Van Der Volk, J. (1987) *The Seven Sketchbooks of Vincent Van Gogh*, London: Thames and Hudson.

van Manen, M. (1991) *The Tact of Teaching: The Meaning of Pedagogical Thoughtfulness*, Albany, NY: SUNY.

Victoria and Albert Museum, London (1985) *John Constable: Sketchbooks 1813–14*, London: Victoria and Albert Museum.

Viola, W. (1936) *Child Art and Franz Cizek*, Vienna: Austrian Junior Red Cross.

Vygotsky, L. (1986) *Thought and Language*, trans A. Kozulin (revised edition), Cambridge, MA: The MIT Press.

Walker, J. (1983) *Art in the Age of Mass Media*, London: Pluto Press.

—— (1992) *A History of Design and Design History*, London: Pluto Press.

—— (1998) 'Marcus Harvey's sick disgusting painting of Myra Hindley: a semiotic analysis', *The Art Magazine* 14: Spring.

Warnock, M. (1978) *Special Educational Needs: the Report of the Committeee of Enquiry into the Education of Handicapped Children and Young People*, London: DES, HMSO.

Wells, L. (ed.) (1997) *Photography: A Critical Introduction*, London: Routledge.

White, M. (1998) 'The pink's run out', in N. Yelland (ed.) *Gender and Early Childhood*, London: Routledge.

Whiteley, N. (1993) *Design for Society*, London: Reaktion Books.

Willats, J. (1997) *Art and Representation*, Princeton, NJ: Princeton University Press.

Williams, G. (1994) 'Inaugural lecture, Royal College', in G. Norman, 'Work Fuel Debate on Meaning of "Sculpture"', *The Independent*, 1 May.

Williams, R. (1971) *The Long Revolution*, Harmondsworth: Pelican.

—— (1976) *Keywords*, London: Fontana.

—— (1979) *Television: Technology and Cultural Form*, Glasgow: Fontana.

Williams, S. (1969) *Voodoo and the Art of Haiti*, Nottingham: Morland Lee Ltd.

Williamson, J. (1982) 'How does girl number 20 understand ideology?', *Screen Education* 40: 80–87.

Willis, P. (1990a) *Moving Culture*, London: Calouste Gulbenkian Foundation.

—— (1990b) *Common Culture*, Milton Keynes: Open University Press.

Wilson, B. and Wilson, M. (1977) 'An iconoclastic view of the imagery sources in the drawings of young people', *Art Education* 30 (1): 5–12.

Wilson, S. M., Shulman, L. and Richert, A. (1987) '150 ways of knowing: representation of knowledge in teaching', in J. Calderhead (ed.) *Exploring Teacher Thinking*, London: Cassell.

Witkin, R. (1974) *The Intelligence of Feeling*, London: Heinemann.

Wolheim, R. (1983) *Art and its Objects*, Cambridge: Cambridge University Press.

—— (1987) *Painting as an Art*, London: Thames and Hudson.

Wolff, J. (1990) *Feminine Sentences: Essays on Women and Culture*, Cambridge: Polity Press. Reprint 1995 in association with London: Blackwell Press.

Woods, P. (1996) *Contemporary Issues in Teaching and Learning*, London: Routledge.

Yeomans, R. (1993) 'Islam: the abstract expressionism of spiritual values', in D. Starkings (ed.) *Religion and the Arts*, Sevenoaks: Hodder and Stoughton.

Zimmerman, E. (1990) 'Questions about multiculture and art education or I'll never forget the day M'Blawi stumbled on the work of the post-impressionists', in *Art Education* 43 (6): 8–24.

RESOURCE CENTRES

The Artists' Book Fair is held in London every November.

The Hardware Gallery, 162, Archway Road, Highgate, London N6 5BB, specialises in artists' books.

Name and author index

Subject index

Note: Page numbers in **bold** type refer to **figures**. Page numbers in *italic* type refer to *tables*. Page numbers followed by 'T' refer to Tasks, e.g. 258T.